# A Study on Legal Regulation of Food Safety and Standards in India

By

**SUMITHRA R.**

## Book Name: ***A Study on Legal Regulation of Food Safety and Standards in India***

# TABLE OF CONTENTS

## LIST OF ABBREVIATIONS

- **A.P** : Andhra Pradesh
- **AIR** : All India Reporter
- **All** : Allahabad
- **Art.** : Article
- **Bom** : Bombay
- **C.J** : Chief Justice
- **CAC** : Codex Alimentarius Commission
- **Cal.** : Calcutta
- **Co** : Company
- **Corp.** : Corporation
- **CPA** : Consumer protection Act 1986
- **Cr.L.J** : Criminal Law Journal
- **edn.** : Edition
- **F C** : Federal Court
- **FAO** : Food and Agricultural Organisation
- **FCI** : Food Corporation of India
- **FSSA** : Food Safety and Standards Act, 2006
- **FSSAI** : Food Safety and Standards Authority of India
- **Govt** : Government
- **Guj** : Gujrat
- **HC** : High Court
- **IARSI** : Indian Agricultural Research Statistics Institute
- **Ibid** : Ibidem (in the same place)
- **ICAR** : Indian Council for Agricultural Research
- **ILR** : Indian Law Report
- **Jour.** : Journal
- **Kar** : Karnataka
- **KLJ** : Karnataka Law Journal

- **Ltd** : Limited
- **M.P** : Madhya Pradesh
- **Mad.** : Madras
- **NFSA** : National Food Security Act, 2013
- **NSSO** : National Sample Survey Organization
- **Oris** : Orissa
- **Ors** : Others
- **P & H** : Punjab and Haryana
- **P C** : Privy Council
- **p** : Page
- **Para** : Paragraph
- **Pat** : Patna
- **PFA Act** : Prevention of Food Adulteration Act, 1954
- **Punj** : Punjab
- **SC** : Supreme Court
- **SCC** : Supreme Court Cases
- **SCJ** : Supreme Court Journal
- **SCR** : Supreme Court Reporter
- **Sec** : Section
- **SFC** : State Food Commissioner
- **St.** : State
- **T N** : Tamil Nadu
- **U P** : Uttar Pradesh
- **U.N.O** : United Nations Orgainization
- **UAS** : University of Agricultural Science
- **UDHR** : Universal Declaration of Human Rights
- **UK** : United Kingdom
- **UN** : United Nations
- **UNDP** : United Nations Development Programme
- **UOI** : Union of India

- **US** : United States
- **v.** : Versus
- **Vol** : Volume
- **WTO** : World Trade Organisation

## TABLE OF CASES

- *A.P.G & S. Merchants Association* v. *Union of India,* (1971) 1 SCJ 518
- *Ahmedabad Municipal Corporation* v. *Nawab Khan Gulab Khan & Ors,* AIR 1997 SC 152
- *Air India Statutory Corporation* v *United labour Union*, AIR1997SC645
- *Akhil Bharathiya Grahak Panchayat* v *Life Insurance Corporation of India,* (1991)1CPR 112 (Maha CDRC)
- *Amrinder Chopra* v. *State of Punjab*, 2014(1) FAC 444 at p 445 (P&H)
- *Anil Kumar Srivastava* v. *State of Bihar*, 2013(2) FAC 173 at p 174(Pat)
- Asiad Games case *PUDR* v. *Union of India*, (1982)2SCC 161
- *BalKrishan* v. *State of* Rajasthan,1997 RLW (2) 1082
- *Bandhua Mukthi Morcha* v *Union of India*, (1984)3SCC 161
- Beggar's Case, (1987)3 SCC 430
- *Bhagawan Das Jagdish Chandra* v. *Delhi administration*, AIR 1975 SC 1309
- *Board of Trustee of the Port of Bombay* v. *Dilip Kumar Raghavendranath Nadkarni,* (1983) 1 SCC 124
- *Brahma Das* v. *State of H.P*, 1988 SC 1789.
- *Budha Ram* v. *State of Punjab*, 1988 E.F.R 558
- *C.E.S.C Ltd.* v. *Subhash Chandra Bose*, (1992) 1 SCC 441
- *Cargill India Private Ltd* v. *State of Uttarakhand*, 2013 Cr.L.J(NOC)140(Uttar)
- *Centre for Public Interest Litigation* v. *Union of India*, 2013(2)FAC 135 at p. 139 (SC)
- *Centre for public interest litigation v. Union of India,* (2013) 16 SCC 279
- *Chairman Railway Board v. Chandrima Das*, 2000(2)SCC 465,
- *Chameli Singh v State of UP,* 1996(2) SCC 549
- *Chand v. State of Punjab,* 1987 E.F.R. 61
- *Cotterill v Pern, (1936) 1K.B.* 53 (T.A.C).
- *Crown Alluminium Work v. Their Workmen,* AIR 1958 SC 30.
- *Danisco (India)Private Ltd, v. Union of India* AIR, 2015 (NOC)219(Del)
- *Dasaudha Singh and others* v. *State of Haryana*, (1973) 2 SCC 393

- *Delhi Transport Corporation* v. *DTC Mazdoor Congress*, AIR 1991 SC 101: Supp(1) SCC 600
- *Dharampal Satyapal Ltd* v. *State of Kerala*, AIR 2014 Ker 51 at p.53
- *Dhariwal Industries* v. *State of Maharashtra*, 2013(1)FAC 26 at p.43(Bom)
- *Dinesh Bhagat* v *Balaji Auto Limited*, (1992) IIICPJ 272 (Delhi CDRC)
- *Dinesh Chandra* v. *State of Gujarat*, AIR SC 1011
- *Dr. Shivarao Shantaram Wagle and Others* v. *Union of India and Others*, AIR 1988 SC 952.
- *Election Commission of India* v. *St.Mary's School*, (2008)2 SCC 390
- *Express Newspaper Private Ltd* v. *Union of India*, AIR 1958 SC 578.
- *F.I* v. *Narayanan*, 1980 K L.J. 454
- *Farook Haji Ismail* v *Gavabhai Bhesania*, (1991) II CPJ452 (Guj.CDRC)
- *Food Inspector* v. *State of Kerala*, 1978 KLJ. 830.
- *Fowler* v. *Padjet*, 1978 JTR 509,511
- *Francis Coralie Mullin* v. *The Union Territory of Delhi*, (1981) 4 SCC 494
- *Ganesh Pandurang Jadhao* v. *State of Maharashtra*, 2016 Crl.L.J 2401 at p.2407(Bom)
- *Ganeshmal Jeshraj* v. *Government of Gujarat*, (1980) 1 SCC, 363
- *Gauranga Aich* v. *State of Assam*, 1990 (2) FAC41.
- *Glass Studio* v *Collector of Central Excise*, (1991)II CPJ 585 (Orissa CDRC)
- *Gopal* v. *State of MP*, 1992 Cr. L.J. 2135(2141
- *Gujarat Agricultural University* v. *Rathod Labhu Bechar*, AIR 2001 (SCW) 351,
- *Hydro (Engineers) Pvt. Ltd* v. *Workman*, AIR 1969 SC182
- *Sodan Singh* v *New Delhi Municipal Committee*, AIR1989 SC1988
- *Indian Council of Legal Aid and Advice & others* v. *State of Orissa*, (2008) (1) SCC Supreme 421, Writ Petition (Civil) 43 of 1997.
- *International Spirits and Wine Association of India* v. *Union of India*, AIR 2013 Bom 178 p 183
- *Jaydip Paper Industries* v. *Workmen*, (1974) 4 SCC 316
- *Joshy K V* v. *State of Kerala*, 2013 Cri L.J 2789 at 2795(Ker)
- *Kacheroomal's Case* 1976 SC 394

- *Kailash Chandra* v. *State of U.P*, 2001 All LJ 2753
- *Kapila Hingorani* v. *State of Bihar*,1999 Cri LJ 1082 (Del).
- *Kat* v. *Dimenz Lord Goodard*, (1952) 1K.B.34.
- *Keshavananda Bharati* v. *State of Kerala*, (1973) 4 SCC 225 (paragraph 1700).
- *Kishan Lai* v. *State of Rajasthan*, 2001 Cri LJ 3617 (Raj).
- *Kishen Pattanaik* v. *State of Orissa*, 1989 Supp(1) SCC 258
- *Life Insurance Corporation of India* v *Uma Devi*, (1991) 3 Comp LJ 171 (NCDRC)
- *Lochamesh B hugr* v. *Union of India*, 2013(1) FAC 239 at p.262 (Kant)
- *M. Eswaraiah* v. *State of A.P*,1999 (1) ALJ 682
- *M/S Chromachemie Laboratory Private Limited, Bangalore* v. *Authorized Officer*, Chennai Seaport and Airport, Food Safety and Standards authority of India, Ministry of Health and Family Welfare, Chennai, AIR 2016 (NOC) 238 (Mad)
- *M/S Garima Milk and food Products Limited* v. *State of Rajasthan*, 2017(1)FAC 65 p66(Raj)
- *M/S Nestle India Ltd* v. *Food safety and Standards Authority of India, Mumbai*, AIR 2016 (N.O.C) 225 (Bom)
- *M/S Omkar Agency* v. *Food Safety and Standards Authority of India*, AIR 2016 Pat.160 at p173
- *M/s Raj Traders* v. *State of H.P*, 1995 Cri LJ 1990 (HP).
- *Maneka Gandhi* v. *Union of India*, AIR 1978 SC 597
- *Maya Prakash* v. *State of U.P.* and Another,1998 All LJ 116
- *Morgan Stanley Mutual Fund* v *Kartick Das & others*, (1994) 11 CPJ 7 (SC)
- *Motumal* v. *State of M.P*,1999 Cri LJ 4038.
- *Municipal Corpn of the City of Ahmedabad* v. *Jan Mohd. Usmanbhai*, (1986) 3 SCC 20, 31.
- *Municipal Corporation Delhi* v. *Ved Prakash*, (1974) Cr LJ 189 (Delhi) (DB)
- *Municipal Corporation of Delhi* v. *Shanti Prakash*, (1974) Cr LJ 1086 (Delhi) (DB)
- *Municipal corporation* v. *Jan Mohammad*, AIR 1986 SC 1205
- *Munn* v. *Illinois* 94 U.S, 113(1877)
- *Nakara's case*, AIR 1983 SC 130

- *Narayan Malviyas* v. *State of M P*, 2013(2)FAC 463 at p 464 (MP)
- *Narayani Flavours & Chemicals Pvt Ltd* v. *The Auth. Officer, Food Safety Standard Authority of India,* 2014(1) FAC 440at pp 441,442(Cal)
- *Neeraja Choudhury* v. *State of Madhya Pradesh*, (1984) 3 SCC 243
- *Northan Mal* v. *State of Rajasthan*, 1995 CriLJ 2661.
- *Olga Tellis* v. *Bombay Municipal Corporation*, AIR 1986 SC 180
- *P. Unnikrishan* v. *Food Inspector*, Palghat Municipality, Palghat, 1995 Cri LJ 3638.
- *P.G. Gupta* v. *State of Gujrat*, (1995) Supplementary 2 SCC 182
- *Pavement Dweller's case, Olga Tellis* v. *Bombay Municipal Corporation*, (1985)3SCC 545
- *Pearks, Gunsten Tee Ltd* v *Ward*, (1902)2 K.B.I(Kansas Bureau of Investigation)
- *Peerless General Finance and Investment Co Ltd* v. *Reserve Bank of India*, AIR 1992 SC 1033
- *Per Majority, Syno Textiles Pvt Ltd* v. *Greaves Cotton & Company Ltd*, (1991)1 CPR 615 (NCDRC)
- *PK Tejani* v. *MR Dange*, AIR 1974 SC 228
- *Pourushottam Khemuka* v. *The State of Madhya Pradesh*, 2014(1)FAC 433 at p 434 (MP)
- *Prabhakaran Nair's case*, AIR 1987 SC 2117
- *Pradeep Kumar Gupta* v. *State of UP*, 2016 Crl.L.J 122 at p128 (All)
- *Priti Sinha* v. *Bihar State Housing Board*, (1996) 1 CPR 369 (Bihar CDRC)
- *PUCL* v. *Union of India*, 2000 (5) SC ALE30
- *Pyarali K. Tejani* v. *Mahadeo Ramchandra Dange and Others*, AIR 1974 S.C. 228
- *R* v. *Allday*,(1837) 8 C&P 136
- *Ram Dayal's* case A 1924 All.214
- *Ram Labhaya* v. *Municipal Corp of Delhi &others*,AIR 1974 SC 789
- *Ramakant Gupta* v. *State of Chhattisgarh*, 2016Crl.L.J 3386 at p.3390(Chhatt)
- *Ranvir Saran* v. *State (Delhi Administration)*, 1999 Cri LJ 1082 (Del).
- *Rickshaw pullers case Azad Rickshaw Pullers Union* v. *State*, (1980) SSC Supp 601

- *Samatha* v. *State of Andhra Pradesh*, AIR 1997 SC 3297,3330
- *Sawaran Singh* v. *State of Haryana*,1998 CriLJ 204 (P&H).
- *Shantistar Builders* v. *Narayan Khamalal Totame*, (1990) 1 SCC 520.
- *Sherras* v. *De Rutzen*, (1895) 1 Q. B. 91,
- *Sivani* v. *State of Maharashtra*, AIR 1995 SC 1770
- *Smt Mani Bai* v. *State of Maharashtra*, (1974) 1 SCJ 712
- *Sodan Singh* v *New Delhi Municipal Committee*, AIR1989 SC1988
- *Sri Lakshmi Narayana Rice Mills* v. *The Food Corporation of India*, (1995) IIICPR 630 (NCDRC)
- *State of 'Maharashtra* v. *Chandrabhan*, (1983) 3 SCC 387
- *State of Maharashtra & Anr.* v. *Sayyed Hassan Sayyed Subhan & Ors*, Criminal Appeal No.1195 of 2018 arising out of Special Leave Petition (Criminal) No.4475 of 2016
- *State of Punjab* v. *Kerwal Krishnan*, 1992 Cr. L.1. 743
- *State of Rajasthan* v. *Hat Singh*, (2003) 2 SCC 152
- *State of Rajasthan* v. *Ladu Ram*, 2002 Cri LJ 426 (Raj).
- *State of Tamilnadu* v. *Abhi*, 1984 SC 326
- *State of Uttar Pradesh* v. *Uptron Employees Union CMD*, (2006) 5 SCC 319.
- *State* v. *Puram Lal*, A1985 SC
- *Subhash Chander* v. *State of Punjab*, 1997 RLW (2) 1082
- *Sudarshan Chits(India) Ltd* v. *Official Liquidator*, (1992) 1 Comp LJ 34 (Mad.HC)
- *Swami Achyuthanand Tirth* v. *Union of India*, AIR 2016 S.C.3626 at 3634
- *Tek Chand Bhatia's case Delhi Municipality* v. *Tek Chand Bhatia*, A 1980 SC 360
- *U. Unichoyi and others* v. *State of Kerala*, AIR 1962 SC 12
- *Uma exports Ltd* v. *Union of India*, 2013(1) FAC 422 at p 424 (Cal)
- *United Distribution Incorporation* v. *Union of India*, AIR 2015 Delhi 31 at p36
- *V K Srinivasan* v. *Food Inspector*, (1965)1 M L J 58
- *Vincent* v *Union of India*, AIR1987 SC 990
- *Vishnu Pouch packaging Pvt Ltd* v. *State of Maharashtra*, 2013(1) FAC 172 at p. 177 (Bom)

- *Vital Neutraceuticals Pvt Ltd* v. *Union of India*, 2014(1) FAC 1 at p.24 (Bom)
- *Warner* v. *Metropolitan Police Commissioner*, (1968) 2 All. E R 356
- *Yamuna Sah* v. *State of Bihar*, 1990 (2) FAC 16 (Pat).
- *Yash* v. *State of Punjab*, 2014(1) FAC 445 at p.446

# CHAPTER-I
# INTRODUCTION

## 1.1 Introduction

Food[1] is the fundamental need of all living creatures. '*One cannot think well, love well, sleep well, if one has not dined well'*[2]. Admittedly, a man cannot live by food alone but yet he cannot do without that even. Beginning from the hunting, gathering era till date, the story of civilization is man's pursuit to feed himself and his family. Access to a minimum amount of nutritious and safe food enjoys sanction by all communities and nations. All faiths the world over believe that feeding the needy and hungry as the greatest virtue and quite a few charities offer food on daily basis like in Amritsar's Golden temple which is called langar and which means free food service, operates round the clock throughout the year without a break and in most of the well known pilgrimage centers in South India, like Dharmasthala and Tirupati, pilgrims are provided with free food through out the year. Lot of sanctity and divinity is attached to the act of feeding the hungry in India.

A hungry population is an economic burden. Food is inevitable for the existence of life on earth. Food has been defined as "any substance, whether processed, semi processed or raw, which is intending for human consumption and includes drinks, chewing gums, and any substance which has been used in the manufacture, preparation or treatment of food"[3]. Hence, the term food implies that it is meant for human consumption and it ought to be safe and standard. It is only off late that we have begun to use the term '*safe food*' owing to the wide prevalence of unsafe food in the market in the form of adulterated food, misbranded food, food packed without adequate information on the label about their nutritive value and ingredients that have gone into its preparation, manufacturing date expiry/best before date. Hence, when any judgement, legislation, or provision of law mentions the term food, it implies safe food and food that meets the dietary and nutritive requirements of a person.

---

1 Sec 3 (j) Food Safety Standards Act, 2006 'Any substance, whether processed, partially processed or unprocessed, which is intended for human consumption and includes primary food, genetically modified or engineered food or food containing such ingredients, infant food, packaged drinking water, alcoholic drink, chewing gum and any substance including water used in food during its manufacture, preparation or treatment but does not include any animal feed, live animals unless they are prepared or processed for placing on the market for human consumption, plants prior to harvesting, drugs and medicinal products, cosmetics, narcotic or psychotropic substance'.

2 Virginia Woolf was a British writer, considered one of the most important modernist 20th-century authors and a pioneer in the use of stream of consciousness as a narrative device.

3 *Webster's Millenium College Dictionary*, Dreamstech India, New Delhi (4th edn, 2004).

It is important to note that community health is national health. A prospective society must have healthy people and adequate good food is inevitable for achieving good health. So the food laws are enacted to protect the consumers against unsafe articles of food and adulteration as well as to protect the honest producers and traders of food. These food laws which fix the standards for quality and safety for different articles of food on par with global standards facilitate the movement of food within and between countries. Food laws have generally been considered to be both for public good and responsibility to the public. The consumers in India are generally most unorganised and helpless victims of the society coupled with a neutral and soft legal system which is not able to curb the exploitation by the manipulations and machinations of vested interests. Hence, our Indian Constitution has cast a duty on the state to raise the level of nutrition of the people, standard of life of the people and to improve the public health as among its primary duties and in particular, the state shall undertake to bring about prohibition of consumption of intoxicating drinks except for medicinal purposes.

It is a universal truth that even if a person is starving, he does not consume anything knowing fully well that it is not safe, jeopardizing his very existence. In the olden times, when barter system was in vogue, people in general and traders in particular were god fearing and law abiding, all commodities in the market and specifically the food and the articles of food were always safe. After the advent of money, people in general became greedy and wanted to maximize their profits and hoard money. Traders were no exception to this, the social evil of adulteration slowly began to raise its head, spreading its tentacles to all sectors not sparing the food even, ultimately engulfing the entire food sector. Articles of food are more vulnerable for the menace of adulteration as it is the basic need and a consumable, will be in demand all the time as everybody will have to buy it, across all sections of the society irrespective of the barriers like rich, poor, upper caste and lower caste. Due to the weak economic condition of the people in India where major section of the people were economically challenged, were naturally lured by a comparatively lesser price of adulterated food due to its low quality, but they were oblivious of the fact of adulteration.

Over a period of time, market was flooded with spurious and adulterated articles of food thus affecting the lives of common men. Adulteration of food, reached such alarming heights that the then British government realized the grave need for a legislation for curbing the same and took upon itself the responsibility of passing suitable legislations to prevent and prohibit adulteration of food, stressing on the need to make only the safe food available for human consumption. After India became independent, Prevention of Food Adulteration Act, 1954 was passed. This Act only spoke about the adulteration of food and did not deal with the safety aspect of it. Prevention of Food Adulteration Act only examined if the available food is safe for consumption but did not have provisions to tackle the menace at the initial stage of production or manufacture which resulted in wastage of food. The Researcher strongly believes that when the food is scarce, we cannot afford to waste it.

To plug this lacuna, legislature passed the Food Safety and Standards Act, 2006 not only to curb adulteration and prohibit the availability of unsafe food but also stressing on the need to make the articles of food more safe, the Act is self-explanatory. But what is worrying the Researcher is the kind of substance used to adulterate the articles of food. The traders of food in a scurry to maximize their profits have ceased to be humane towards the society and have overlooked their responsibility to the human kind. Even various international instruments have dealt with the aspect of right to food as a basic human right. When they refer to food, it is implied that it is meant for human consumption and it ought to be safe food only. Any unsafe food for that matter, becomes fodder and not food. But in the present days even fodder will have to be safe and standard as the owners of the farm animals are interested in maximizing the produce, due to which the intake of the cattle is carefully chosen so that output can be bettered. When so much care is accorded to the animals, human beings deserve a better treatment.

## 1.2. Meaning of Food

Food is inevitable for the existence of life on the earth. Food, a basic necessity of life, derives its importance from the fact that it stimulates the appetite and supplies a variety of ingredients that give energy (carbohydrates, fat, dietary fibre); replace worn out tissues, thus promoting growth (protein); and help in preventing and curing diseases (vitamins and minerals). Food is a building block for the growth of both mental and physical body. When we use the term 'food', it is always implied that it is something edible by human beings or is meant for human consumption because of which it has to be safe, nutritious and standard as per the accepted norms by the law. Hence, the word 'food' always connotes that it is 'safe' though it is not evident in its terminology due to which food always means 'safe food' only. The concept of healthy eating for healthy living and longevity is not new. Apart from serving a biological need, food has become an economic and political weapon.

### 1.2.1 Meaning of Food Security

*"It is the condition in which all people at all times have physical, social and economic access to sufficient, safe and nutritious food that meets their dietary needs and food preferences for an active and healthy life"*[4]. *"Food security means that: all people at all times have both physical and economic access to enough food for an active, healthy life; the ways in which food is produced and distributed are respectful of the natural processes of the earth and thus sustainable; both the consumption and production of food are governed by social values that are just and equitable as well as moral and ethical; the ability to acquire food is ensured; the food itself is nutritionally adequate and personally and culturally acceptable; and the food is obtained in a manner that upholds human dignity"*[5]. Food security has been defined by the World Bank as 'access by all people at all times to enough food for an active and healthy life'. Food security can be categorised into Community food security and Household food security. "Community food security exists when all citizens obtain a safe, personally acceptable, nutritious diet through a sustainable food system that maximizes healthy choices, community self reliance and equal access for everyone."[6].

A household is said to be food secure when all members of the household or a family have an access to the food which is adequate in terms of quality, quantity, and safety and culturally acceptable, needed for a healthy life and when the household is not at undue risk of losing such access to food. 'The economics of food security will, therefore, not only have to be viewed in terms of budgetary outflow by way of food subsidy and other costs, but also in terms of strengthening the food entitlement of the people, particularly those belonging to the vulnerable sections of society'[7]. There cannot be food security without self sufficiency and control of local resources.

Human right to food, which in essence means food security, however remains a hollow paper postulate if it rests on economic, political and social conditions which make its realisation impossible. Food security and food sovereignty cannot be

---

[4] United Nations Committee on World Food Security
[5] World Health Organisation
[6] Public Health Association of British Columbia
[7] Partha Pratim Mitra, "Economics of Food Security: The Indian Context", Social Action[January-March, 1996] Vol 46 No1

guaranteed by either huge private concerns or centrally controlled states. The Researcher thinks that all of us need to realise that food security cannot be expected from those who are politically responsible but we as consumers can create it in cooperation with farmers. '*In Japan, such clubs, called Seikatsu Clubs, have been primarily founded by house wives. .......The women's market in West Africa, Ghana. The entire food supply is in the hands of these market women. They are very powerful. When they close the market, everything grinds to a halt. Most importantly, although they supply the population well with foods. They do not do so according to the capitalistic principles of profit maximising. According to a statement of Aba Amissah Quainoo from Ghana, the market women cannot be moved to sell just any anonymous product on the market just because it was brought from somewhere and is cheap. The market women have their arrangements with their producers, and they keep their word. A good relationship to the farmers-often women-is more important to them than easy money. Although there is a market here, it is not capitalistic not aimed at accumulation. Because of these market women and their principles of "a moral economy", food security for the people of Ghana even in times of crisis has been achieved'*[8].

Issues of food security and poverty have been recognized as necessary conditions for the creation of a stable socio-political environment for sustainable economic development. 'Food for all', the ultimate mission of all food strategies from micro to macro level seek to make sure the sustainable availability and affordability of food for everyone on earth. It is, therefore, not surprising that eradication of extreme poverty and hunger was one of the eight millennium development goals set to be achieved by 2015. But, ironically even today over a billion people in the world remain unfed and malnourished. 'In fact, the world food production at 3.9 billion tonnes is adequate to feed its 7.7 billion population.

Yet globally, around a billion are reported to suffer from hunger. Ditto with India-with 277.49 million tonnes of food production as on February 2018, a 0.9% growth over previous year, millions still go to bed hungry'[9]. The Researcher wonders if it can be realistic to achieve the mandate of 'Zero Hunger World by 2030'set by the Food and Agriculture Organisation of UN when we have been witnessing the crushing effects of hunger and malnutrition on the lives of the poorest and the most vulnerable

[8] Maria Mies, "An Ecofeminist Analysis of the World Food Summit", Mainstream [November 30 1996] Vol XXXIV No 52, p 31-34

[9] '*Food for all: saving sharing is caring'*, Deccan Herald dated 24/10/2018, Bengaluru Edition

people. The advisory of FAO this year reads as 'Don't waste food, produce more with less. Adopt a more healthy and sustainable diet'.

### 1.2.2 Dimensions of Food Security

Food security has got three dimensions namely food availability, access to food and utilisation of the food. The food availability deals with supply side of the food security, makes sure that sufficient quantity of quality food either from import or domestic market is available for a common man. This is only to ensure that the food available in certain territory/place/country is adequate in quantity to feed the people taking into consideration local agriculture production, stock level of food grains and net import and export.

The second dimension deals with access to food, which means that just because the food is available in adequate quantity in a particular territory/country/region does not ensure that they have enough food to eat. This dimension of food security considers income, expenditure and buying capacity of individuals and households. It is necessary to ensure that people have enough purchasing power to acquire the food that they need. Another dimension of food security is food utilization, which considers not how much food, the people eat but what and how they eat. For healthy life, food available should be of desirable quality as well as quantity to meet nutritional requirements. In essence it encompasses preparation of food, distribution among different households, sanitation and health care practices. In a nutshell food security involves production, availability, access and utilisation.

### 1.2.3 Meaning of Food Insecurity

Food insecurity is a situation when people are deprived of access to food at the outset or lack access to food which meets their dietary requirements: food which is not culturally acceptable for that particular section of the society: food which does not meet the nutritional requirement. "it has become common practice to estimate the number of food insecure households by comparing their calorie intake with required norms. The government has been implementing a wide range of nutrition intervention programmes for achieving food security at the household and individual levels"[10]. Dearden and Cassidy, the authors, who have authored 'Food Security: an ODA View', state that "food insecurity is the most fundamental manifestation of absolute poverty".

---

[10] Radhakrishna, "Food and Nutrition Security of the Poor, Emerging Perspectives and Policy Issues", Economic and Political Weekly, [April 30-May 6, 2006] Vol XL No18, p 1817-1821.

Food Insecurity can be categorised into Chronic food security and Transitory food insecurity. "Lack of minimum requirement of food to the people for a sustained period of time due to extended periods of poverty, lack of assets and inadequate access to productive or financial resources can be called as Chronic Food Insecurity"[11]. "Sudden lack of food or reduction in the ability to produce or access minimum requirement of food due to short-term shocks and fluctuations in food availability and food access, including year-to-year variations in domestic food production, food prices and household incomes can be defined as Acute or Transitory Food Insecurity and requires emergency measures such as rushing of food supplies"[12]. In the context of the present research, the Researcher is dealing with Chronic food insecurity which involves long term strategies for accomplishing sustainable food security encompassing increasing agriculture production, employment generation, infrastructure improvement and reduction in regional imbalances/disparities. The Researcher is convinced that 'sustainability of the country's food security depends on the sustainability.

### 1.2.4 Meaning of Food Safety

'Food safety' means assurance that food is acceptable for human consumption according to its intended use[13], which means that it is a guarantee that the food meant for human consumption is safe for consumption by human beings and the same is devoid of any adulteration or any substance which may be detrimental for health of human beings.

### 1.2.5 The Right to Food

The historical and political background of the Right to Food is much more than the history and politics of malnutrition[14]. It concerns the development of the notion of access to Food as a Right. As a Right it sets obligations on the State, which have been established as enforceable through centuries of social struggle for a democratic state in the service of the people. Traditionally, people had no remedy other than revolt against a king or state that failed to meet its obligations. Right to Food is one of the basic

---

[11] http://www.foodandenvironment.com/2013/01/basic-concept-of-food-security.htm(accessed 7/8/2018 at 12 noon).
[12] *Ibid*
[13] Food Safety and Standards Act,2006 Sec 3(q)
[14] http://www.legalserviceindia.com/article/l76-Right-to-Food-and-Development-in-India.html (accessed on 22/12/2016 at 9.30 am).

Human Rights[15]. The idea of the Human Right to Food is to establish procedural and legal means for seeking remedies against authorities when they fail to guarantee access to food[16]. Right to food has been playing a pioneering role in the renaissance of Economic and Social Rights during the past three decades. This was the first of the Economic Social and Cultural Rights to be studied by the United Nations Human Rights System. In 1987 a report titled 'The Right to Food as a Human Right became the starting point for a series of investigations into the Rights contained in the ICESCR[17].

The Right to food is a basic Human Right derived from the ICESCR 1966[18], recognizing the "Right to an adequate standard of living, including adequate food," as well as the "fundamental right to be free from hunger". States have the obligation to "respect, protect and fulfil" that is, first, the state must not itself deprive anyone of access to adequate food; second, it must protect everyone from being deprived of such access in any other way and third, when anyone is in fact without adequate food, the state must proactively create an enabling environment where people become self-reliant for food or, where people are unable to do so, must ensure that it is provided. Every individual is a Rights-holder, fully entitled to demand that the state perform these duties. In the absence of means of enforcement, there can be no right at all. The right to food creates legal obligations and affords the holder to demand the redressal of the infringement of the right.

In the modern period the Right to food is treated as a basic Human Right and is backed by various International conventions and Constitutional principles. Universal Declaration of Human Rights 1948 enshrines "Everyone has a Right to a standard of living adequate for the health and wellbeing of himself and his family, including food". International Covenant on Economic, Social and Cultural Rights 1966 urged 'States parties to recognise the fundamental right of everyone to be free from hunger'. World Food Conference 1974, declared that 'Every man, woman and child has the inalienable Right to be free from hunger and mal-nutrition'. World Food Security Convention 1985, stressed that 'The hungry cannot wait'. International Conference on nutrition 1992 gave a call saying that 'We pledge to act in solidarity to ensure that freedom from hunger becomes a reality'. These conventions have ensured that all member nations should

---

[15] Human rights are those minimal rights which every individual must have against the state or other public authority by virtue of his being a member of the human family irrespective of any other consideration. DD Basu, *Human Rights in Constitutional Law*, Lexisnexis (3rd edn 2008), P.5.

[16] Supra note 7

[17] International Covenant on Economic Social and Cultural Rights 1966.

[18] *ibid*

adopt and practice food safety norms and people get enough food to eat. These conventions tried to make sure that the people are supplied with enough food and articles of food which are safe.

India has been quite successful in ensuring ample availability of food in the country as the same is produced within the country to meet the consumption demand, with some surplus being exported. But making food available is only one aspect of food security, though an important one, hence food availability and food security will have to be differentiated. The others are economic access to food and its absorption by people for better nourishment. "India is glorified as Annapurna but 'poorna meal' has become an unfulfilled desire for poor little Indian" [19].It is here that India has faced its biggest challenge and paradox.

Despite buoyant economic growth in recent years, around one-third of India's population, i.e. 400 million people, still live below the poverty line[20], India ranks at 75 among 109 countries (2011), indicating the extent of deprivation in terms of living standards, health, and education. 20 per cent of Indian children under five years old were acutely malnourished and 48 per cent were stunted (chronically malnourished)[21]. One of the surveys conducted across 112 rural districts of India in 2011 showed that 42 percent of children below the age of five years are underweight and 59 percent are stunted[22]. All these estimates point to the existence of food insecurity at the micro-level in terms of either lack of economic access to food or lack of absorption of food for a healthy life.

The current population of India is 1,354,213,231 as on July 7th, 2018, based on the latest United Nations estimates[23]and accounts for a third of the world's poor as remarked by the World Bank[24]. The proportion of people living below poverty line[25] in India has come down from 37.2 per cent in 2004-05 to 21.9 per cent in 2011-12 - a decline of 15.3 percent as reported by the Planning Commission[26], thanks and kudos to the various social and poverty alleviation measures initiated by the government. 'The

---

[19] Ravulapti Madhavi, "Is Food Safety Lurking in The Food Safety and Standards Act 2006?", Supreme Court Journal, [8th May to 12th June, 2008] Vol 4, p17-20.

[20] 2010, World Bank's definition of US Dollars 1.25/day.

[21] National Family Health Survey (NFHS-3) conducted in 2005-06.

[22] The HUNGaMA (Fighting Hunger and Malnutrition) HUNGaMA Survey Report by Naandi Foundation available at https://hungamaforchange.org/HungamaBKDec11LR.pdf (accessed on 1/4/2015 at 3.50pm).

[23] http://www.worldometers.info/world-population/india-population/ (accessed on 7/7/2018 at 7.50 pm).

[24] planningcommission.nic.in/reports/genrep/rep_hasim1701.pdf(accessed on 7/7/2018 at 8.00am).

[25] An income of less than $1.25 per day per head of purchasing power.

[26] planningcommission.nic.in/reports/genrep/rep_hasim1701.pdf (accessed on 21/12/20146 at 8.00am).

poor are not poor simply because they are less human or because they are physiologically or mentally inferior to others whose conditions are better off. On the contrary, their poverty is often a direct or indirect consequence of society's failure to establish equity and fairness as the basis of its social and economic relations'[27].

'Acute starvation is a problem of catastrophic proportions in India. It has led some to steal out of weekly markets and passing trucks, otherwise have gone to the extent of selling their children so as to be able to afford food. Some have eaten even inedibles, hunger is a disturbing problem'[28]. It is hard for vast number of people in India to meet their daily requirements of food grains due to which their diet is deficient in terms of nutrition. This is not the problem with just the poor but also the problem faced by those who are the victims of natural calamities as well as manmade disasters.

The problem of starvation deaths is increasing though the granaries are overflowing. 'The paradox of plenty. There is plenty and more of it and yet there are those who have none of it. It is there and yet it is not where it is most required. 'It' here could represent that base of survival and existence - Food lures and evades the hungry. Some live to eat; more than that, eat to live and most don't eat and die'[29]. The right to food is like any other economic and social right has been recognised only in the recent past. The debate regarding the specific nature of the right, moral or legal, individual or collective, absolute or conditional right is still going on. Many development schemes and poverty alleviation programmes are enforced by the executive. But bureaucratic apathy and red-tapism by the government machinery has denied the people of these benefits. The Researcher has observed that the stark irony is that, the producers of food grains are denied food and means of livelihood, kept hungry and left to die due to the apathy of government policies. 'Mahatma Gandhi during the freedom struggle movement introduced 'constructive programme in India' which included healthy eating habits and he denounced drinking alcohol'[30].

When India became independent, the Constitution declared it to be socialist, secular, democratic republic. The fundamental right under the Constitution sets down

---

[27] Dr Justice AR. Lakshmanan, *Voice of Justice*, Universal Law Publishing Co. Pvt. Ltd., Delhi (2006) p 121.

[28] Abhinav Chandrachud, "Malnutrition and the Writ of Continuing Mandamus: The Remedy Befitting the Right", All India Reporter [2006] Rajasthan Section Vol 93, p19-28.

[29] Atul Vishwanathan and Ketan Makhija, "Arrest Hunger The Right to Food", Lawyers collective, [February 2004] Vol 19 No 2, p13-17.

[30] K Sreeranjani Subba Rao & Chenna Gowri Shankar, "Gandhian Food Habits and Environment", Resurgent India, [February 2014] Vol 12 No 10, p 14-15.

that every citizen has a right to life. This has been interpreted by the highest court of the land as every citizen's right to a life in dignity, good health and free speech in fraternity of communal harmony and national integrity. These rights are possible only if people are not starving in the first place. Unless poverty is eradicated our socialist credo will remain just a pretence. Conceptually, parens patriae theory is the obligation of the State to protect and take into custody the rights and privileges of its citizens for discharging its obligation. The Constitution makes it imperative for the State to secure to its citizens the rights guaranteed by the Constitution and where the citizens are not in a position to assert and secure their rights, the State must come into picture and protect and fight for the rights of the citizens. Therefore, the State can be activated and approached to effectively come on the scene[31].

It concerns the development of the notion of access to Food as a Right. As a Right it sets obligations on the State, which have been established as enforceable through centuries of social struggle for a democratic state in the service of the people. Right to Food is one of the basic Human Rights[32]. In order to give effect to this basic human right, National Food Security Act, 2013 was passed to ensure the availability of food for the targeted public group, living below poverty line.

Erstwhile, the Indian food regulations comprised of various food laws that were enacted at different points of time under the ambit of various ministries of Government of India. Historically they were introduced to complement and supplement each other in achieving total food safety and quality. The result was that the food sector in India was governed by a number of different statutes rather than a single comprehensive enactment. Each ministry prescribed its set of rules and standards under relevant Acts and Orders, often creating confusions and sometimes contradictory environment for the industry. In general, this regulatory system resulted in lack of comprehensive, integrated food law under single regulatory authority that ensures public health, safety and also failed to specify quality norms for meeting the globally recognized standards.

[31] Law Commission Of India, 'Need for Ameliorating the lot of the Have-nots' – Supreme Court's Judgments Report No. 223 April 2009 available at lawcommissionofindia.nic.in>report223(accessed on 1/3/2017 at 5.00pm).

[32] Human rights are those minimal rights which every individual must have against the state or other public authority by virtue of his being a member of the human family irrespective of any other consideration. DD Basu, Human Rights in Constitutional Law, Lexisnexis 3rd edn 2008, P.5.

The Food Safety and Standards Act was promulgated in 2006, which integrated eight different food laws[33], and is a major transformation that brought paradigm shift in the food regulatory scenario of the country. The food processing industry is one of the largest sectors in India in terms of production, growth, consumption, and export. Food processing industry is widely recognized as the 'sunrise industry' in India and is of enormous significance for India's development because of the vital linkages that it promotes between the two pillars of the economy, that is, industry and agriculture. Initiating a new era in food safety, the Food Safety and Standards Act, 2006 (FSSA) came into force across the country making it at par with international standards. The Act ensures improved quality of food for the consumers by following the general principles of Food Safety[34] and censures misleading claims and advertisement by those who are in food business[35].

In our country, to ensure Right to food for the people living below poverty line[36] National Food Security Act, 2013 has been enacted and to ensure that its citizenry consumes safe food which meets the standards set by the law, has enacted FSSA. The cumulative effect of these two legislations is that they have ensured food security[37] and safe and standard food for a common man which culminates in the realization of Right to Safe Food[38]. These two are the major laws pertaining to food in India.

## 1.3 Statement of the Research Problem

Though there are numerous international instruments guaranteeing the Right to food, yet the same has eluded major section of the society especially in the developing countries specifically India. When the states are queried in this regard, they have been

[33] Prevention of Food Adulteration Act of 1954, Fruit Products Order of 1955, Meat Food Products Order of 1973, Vegetable Oil Products (Control) Order of 1947, Edible Oils Packaging (Regulation) Order of 1988, Solvent Extracted Oil, De- Oiled Meal and Edible Flour (Control) Order of 1967, Milk and Milk Products Order of 1992 and also any order issued under the Essential Commodities Act, 1955 relating to food.

[34] FSSA Sec 18

[35] *ibid* Sec 24

[36] A minimum income level used as an official standard for determining the proportion of a population living in poverty.

[37] Food security as defined in the 1974 World Food Summit underlines this: "availability at all times of adequate world food supplies of basic foodstuffs to sustain a steady expansion of food consumption and to offset fluctuations in production and prices". United Nations. 1975. Report of the World Food Conference, Rome 5-16 November 1974. New York. as quoted in Trade Reforms and Food Security, FAO, 2003 as stated by Ashok Gulati, jyothi Gujral, T Nandakumar in National Food Security Bill-Challenges and Options in Discussion paper no 2 of Commission for Agricultural Costs and Price.

[38] A human right protecting the right for people to feed themselves in dignity, implying that sufficient food is available, that people have the means to access it, and that it adequately meets the individual's dietary needs. The right to food protects the right of all human beings to be free from hunger, food insecurity and malnutrition-Universal declaration of Human Rights.

evasive in their answers and have only initiated schemes and programmes that have resulted in superficial contentment of having achieved cent percent success. Though there is a guarantee of Right to food, yet safe food has evaded the people. When the state mentions right to food, it should have been imperative that it means safe food only. In spite of Food Safety and Standards Act 2006 been passed, yet safe food has become a far cry to the people in general in India, even today. We witness people being affected by the unsafe food that they consume. In this work, the Researcher has made an attempt to study the forces that compelled the promulgation of Food Safety and Standards Act, 2006 and its effective implementation and the suggestions for the full bloom implementation.

### 1.4 Research Questions

1. Whether food can be demanded as a right?
2. Is the Right to food an extension of Right to life?
3. Do food and safe food mean one and the same?
4. Whether the consumer of food can proceed under the Consumer Protection Act 1986 against the erring food business operators?
5. Whether the right to food enjoys the support at the international level?
6. Whether the Food Safety and Standards Act 2006 can cure the short-comings of Prevention of Food Adulteration Act 1954?
7. Whether Food Safety and Standards Act 2006 is able to curb the prevalence of large scale unsafe food in the country?

### 1.5 Objectives of the Study

- To examine different international instruments supporting the right to food
- To study the genesis and development of right to food in India
- To examine the provisions of the Indian Constitutional with respect to right to food.
- To discuss important verdicts of the courts relating to right to food in India.
- To trace the history of adulteration both at the international and national level
- To discuss the food laws in India and factors responsible for promulgation of the same
- To know the evolution of anti-food adulteration laws in India reckoning from pre-independence period

- To examine the issues before the Food Safety and Standards Act 2006 to be tackled
- To study the changes brought about by the Food Safety and Standards Act
- To discuss the arduous efforts of Food Safety and Standards Authority of India, to curb adulteration of food which is omnipresent and omnipotent in India.
- To discuss the challenges faced in the implementation of Food Safety and Standards Act 2006
- To devise means to overcome the inevitable implementation hiccups of Food Safety and Standards Act 2006 taking the heterogeneous nature of food business in India into consideration

## 1.6 Hypotheses

I. This Research lies on the premise that though, Right to qualitative and quantitative food in India is the upshot of innumerable international instruments and constant judicial interpretation of the Indian Constitution, yet, the human right to food can be effectively ensured only when a parliamentarian explicitly recognizes it under Part III of the Indian Constitution.

II. The Research is based on the assertion that though FSSA has been carefully crafted in such a way that it could subsume Prevention of Food adulteration Act, 1954 and eight other different minor legislations[39], due to the blatant lacunae that they all suffered from, yet, the provisions of the Act are inadequate to provide panacea in India, for the bane of food adulteration and unsafe food which is a curse for the humankind.

III. In the light of the hostile heterogeneous nature of the food business in India and the perennial challenges encountered in the implementation of the Food Safety and Standards Act, instances of people suffering from consumption of unsafe food question the commitment of Food Safety and Standards Authority of India in ensuring safe food to the people.

[39]Prevention of Food Adulteration Act of 1954, Fruit Products Order of 1955, Meat Food Products Order of 1973, Vegetable Oil Products (Control) Order of 1947, Edible Oils Packaging (Regulation) Order of 1988, Solvent Extracted Oil, De- Oiled Meal and Edible Flour (Control) Order of 1967, Milk and Milk Products Order of 1992 and also any order issued under the Essential Commodities Act, 1955 relating to food.

## 1.7 Significance of the Study

The Research problem deals with the basic need of a common man. A human being in order to live should not only eat but eat healthy food. But in India, where 65% of the population i.e. three fourth of rural population and half of urban population live below poverty line, the state claims to have made an earnest effort to supply clean, standardized food for this major chunk of population. This Research goes a long way in protecting basically the human right of Right to food i.e. not just food but safe and standard food to the major section of population and ensures that the Right of consumers who consume the food supplied by the state is also protected in case of any negligence or omission by the state. The Research may bring about general awareness among the people to demand for safe and standard food from the food business operators in general and the state in particular. In the light of the findings and the suggestions made in this study, this research helps the policy makers, law makers and judicial officers to plug the loopholes and to bring an amendment. It also helps the academicians and the research scholars who might take the study to a further new height in the coming days to keep the legislation abreast of the latest needs. The research may also help the NGO's, law students and the general public.

## 1.8 Scope and Limitation of the Study

The Research has dealt with detailed study of the provisions of FSSA, limited study of International instruments namely International Conventions, Declarations, Conference dealing with Right to food, The Researcher has confined her studies of International instruments only to the extent of its application to Food and Food articles only. Limited study of Constitution of India dealing with some of the Fundamental Rights, Directive Principles of State Policy, distribution of legislative powers between the centre and state as for as the matter of food is concerned. The Research deals with the partial study of National Food Security Act 2013 and Consumer Protection Act 1985.

## 1.9 Research Methodology

The present Research is mainly based on Doctrinal work in the light of available data like Judgments, Statutes, Rules and Regulations, Law Commission Reports, International Conventions, Declarations, Journals, Magazines, and Text Books analysing the issues relating to availability of food, safety of food, protection of the Right of consumers of food, Constitutional provisions providing for the same.

### 1.10 Sources of Data Collection

The Researcher has made use of both primary and secondary sources of information. Primary sources include authoritative Judgments, United Nations Charter, Law Commission Reports, official websites of the authorities, UN Digital library, International Conventions/Declarations/conference, Policies, Plans, Programmes and Indian Constitution. Secondary sources include Articles, News paper reports, Reports, Journals, Magazines, Text Books, and Internet Official websites.

### 1.11 Review of Literature

The Review of Literature is one of the prominent process of research, which requires the researcher to discuss about various studies and available information relating to research area. In the instant study, the researcher has examined and reviewed the observation and approach adopted by various eminent scholars relating to food safety legislations. Some of the books are discussed herein such as *'Food Safety and Standards Act 2006 along with Rules and Regulations 2011'* and *'Indian Constitutional Law'* by Prof. M. P. Jain (2014). The authors in these books have explained only the doctrinal aspect of Food Safety and Standards Act, 2006. They have dealt with important issues pertaining to the implementation of the food safety and standards and the Right to Safe Food. The authors have also dealt with many landmark rulings of the higher courts which are very much valuable for this research. The authors have meticulously dealt with the case laws strengthening the Right to Safe and Standard Food. The authors have also noted the different provisions under which a consumer can proceed against the food business operators, in the event of injury by consuming the unsafe food.

In spite of having the legislations for ensuring safe food, we still witness people being affected by unsafe and adulterated food. The Researcher has reviewed many authoritative books authored by the other eminent scholars. The Researcher has also undertaken the study of Law Commission Reports, Report by Special Rapporteur on Right to Food, Articles/write-ups penned by various authors across the country. But no author/scholar has dealt with the topic/subject that the Researcher has dealt with in this

Research work, hence this study. Till today no research work has been done with the approach that the researcher has adopted in this work.

After a thorough study of the above mentioned books and the available literature on the research topic it is clear that no single author has made a comprehensive work regarding the availability of the safe and standard food, implementational difficulties with respect to ensuring safe food and also the challenges encountered for the same. No author has made an attempt to recommend measures which will cut down the unsafe food in market and ensure the availability of only the safe food.

## 1.12 Scheme of the Study

This research work titled "***A Study on the Legal Regulation of Food Safety and Standards in India***" has been divided into the following Seven chapters including the Introduction and the Conclusion.

**Chapter One Titled '*Introduction*'**

The First Chapter deals with definitions and meaning of certain important terms which are relevant to this study. This chapter also contains scheme of study encompassing the introduction to the topic of Research, objectives of the study-enunciating the main purpose of the Research, Hypothesis which forms the framework around which the Research work revolves, Significance of the Research-shows as to the way in which this study would be useful to the society and to the people in general and why should there be research in this area of study, Scope and Limitation of the study embodies as to the extent of study made of different statutes, rules and regulations, different International instruments in the form of Conventions, Conferences, Declarations, Research methodology speaks about the methodology adopted. The instant Research work is mainly Doctrinal, where the Research depends on the available data and literature. Sources of Data collection is from primary sources like statutes, legislations, law commission reports, judgements and also from Secondary sources like books, articles, research papers, magazines, reports. This chapter also deals with the compartmentalisation of the entire Research into different chapters, their headings, sub-headings. This chapter also deals with review of literature.

**Chapter Two Titled '*Historical Evolution and Conceptual Development of Right Food in India: An Assessment*'**

In this chapter the Researcher has made an attempt to throw light on the importance of food and how it is crucial for basic sustenance. An attempt has been made to study the practices about food habits and how food was treated with utmost fervour by the people of ancient times and for this Vedas and Upanishads have been examined. Verses from Mahabharata have been discussed, in which Bhishma advices Yudhishtra to guarantee that every single person in his kingdom is fed and nobody goes to bed with empty stomach. Manusmriti and Yajnavalkya smriti have also been referred to see how they have contributed to make laws for the development of Right to food. In 2013, National Food Security Act (NFSA) was passed according fundamental Right status to Right to Food. The Researcher in this chapter has highlighted the salient features of this Act and the challenges that are encontered in the implementation of the Act.

Emergence of adulteration in India and the laws to curb the same is also discussed. Kautilya's Arthashatra has been elaborately discussed to show how adulteration was discouraged by levying heavy penalty in those days. A study of the same in medieval period is carried out. In the recent times Food Safety and Standards Act was passed in 2006 in order to ensure that safe and standard food is made available to the people at large. This Act was passed to consolidate the laws relating to food and to establish Food Safety and Standards Authority of India for laying down science based standards for articles of food and to monitor and regulate their manufacture, storage, distribution, sale and import. The main objective of this Act is to ensure availability of safe and wholesome food for the people for consumption.

**Chapter Three Titled '*Constitutional Perspective and Judicial Dictum on Right to Food in India- An Evaluation*'**

In this chapter the provisions of our Indian Constitution pertaining to right to food are discussed along with the cases decided by various courts in India. It is only the Constitution that plays a fundamental role in the realization of the Right to food as it is the supreme law of the land and the source of all political power within a nation. The constitutionality of every law and act of government is one of the most important political principles of democracies and universally accepted rule of law norms. In this chapter at the outset, an effort is made to study the values imbibed in the Preamble of the Indian Constitution. A thorough study of various Articles of Indian Constitution supporting the Right to food, may be expressly or impliedly along with Article 32 and 226 has been under taken. A detailed study of the interpretation of Article 21 by the judiciary has been included in this chapter.

Different kinds of Rights are examined to ascertain the nature of Right to food. Nature of the Right to food is discussed elaborately. The trail of judicial precedents that ultimately led to the fructification of the Right to food is meticulously followed. Limitation of the judicial approach to the Right to food is also being examined, where the Researcher has opined that the efforts of the judiciary alone is not enough but follow up action by the authorities as well as the stake holders is also needed. This chapter contains the case analysis of popularly called 'the Right to food cases', *Kishen Pattanaik* v. *State of Orissa*[40] and *PUCL* v. *Union of India*[41]. Some of the Law Commission Reports supporting (1) the protection of the interest of the consumers of food and (2) Right to food concentrating on providing early child hood development in terms of nutritional needs of the child and (3) Ameliorating the lot of under privileged people deriving the principle from 'parens patraie' which means that a duty is cast on the State under the Constitution to take up the cause of its citizens, when they are not in a position to do so by themselves. This chapter also contains the observations made by the Researcher in the form of Findings, which might be helpful in the proper implementation of the Right to food, if the loopholes as identified by the Researcher are plugged at the earliest.

**Chapter Four Titled '*Legal Regulation of Food Safety and Standards in India-An Overview*'**

This chapter contains the history of adulteration in India and the growth of legislations to curb the same. A limited study of Prevention of Food Adulteration Act, 1954 has been embarked upon to study the lacunae suffered by it thus creating a vacuum which demanded the promulgation of a better law to tackle and deal with adulterated and unsafe food. In the recent past there were many legislations governing food and food articles. They were the Prevention of Food Adulteration Act,1954, The Fruits Product Order, 1955, the Milk and Milk Products Order,1992, The Meat Products Order, 1973, The Vegetable Oil Products (Control) Order, 1947. The result was that the food sector in India was governed by a number of different statutes rather than a single comprehensive enactment. Each Ministry had prescribed its set of rules and standards under relevant Act and Orders which were most of the times confusing and contradictory to each other. Due to the mounting pressure from the industry and stakeholders for a single regulatory body and for an integrated modern food law, Food

[40] 1989 Supp(1) SCC 258.
[41] 2000(5) SC ALE30.

Safety and Standards Act, 2006 (FSSA) was enacted and the Food Safety and Standards Authority of India(FSSAI), a single regulatory body under the Act came to be setup.

In this chapter an elaborate study of the FSSA has been made to study the legal regulation of food safety and standards for food and articles of food in India. This Act came to replace eight different legislations mainly the Prevention of Food Adulteration Act,1954 (PFA Act). This Act borrowed the basic principles of food safety and standards from the erstwhile prevention of food adulteration Act, 1954 as it was a well thought of and an articulated Act and lot of initial ambiguities were addressed in this Act. Over a period of time, due to the development in science and technology and surge in the buying capacity of the people, adulteration became rampant in the greed of making more money by the traders. This did not spare the articles of food even.

Food, meant for human consumption and different articles of food became an easy target for the adulterators as food is something that is in demand on a daily basis, may not have a long shelf life as in the case of other articles, encouraged the manufacturers and traders of food to adulterate. The fact that in India, three fourth of the population live below poverty line is the main factor for adulteration to spearhead as the reality is 'poor will consume anything'. Food Safety and Standards Act was passed in order to ensure that safe and standard food is made available to the people at large, to consolidate the laws relating to food and to establish Food Safety and Standards Authority of India for laying down science based standards for articles of food and to monitor and regulate their manufacture, storage, distribution, sale and import. This Act's main objective is to ensure availability of safe and wholesome food for the people.

FSSA has 12 chapters containing 101 sections and two schedules. The Act, inter alia, incorporates the salient features of the Prevention of Food Adulteration Act 1954 which is now repealed and is based on international legislations, instrumentalities and Codex Alimentarius Commission. This Act with its three tier structure, an apex food safety and standards authority, a central advisory committee and various scientific panels and committees lays more emphasis on science based and participatory decisions while adopting the contemporary approach in both standard setting and implementation.

***An Overview of the Act***

The Act established a single reference point for all matters relating to food safety and standards, by moving from multi-level, multi-departmental control to a single line of command. The Act established the Food Safety and Standards Authority of India (FSSAI) as an apex regulatory authority consisting of a Chairperson and 22 members[42].

The Act prescribes general provisions for Food additives[43] and processing aids are to be added only in accordance with provisions / regulations under the Act[44]. Foods are not to contain residues of any insecticides or pesticides, veterinary drugs, antibiotic, solvent, pharmacological active substances and micro-biological contaminants in excess of limits prescribed under the regulation[45]. Regulations to be made for the manufacture, distribution or trade of any novel foods, genetically modified foods, irradiated foods, organic foods, foods for special dietary uses, functional foods, nutraceuticals, health supplements, proprietary foods[46], etc. The Act provides the general administrative principles to be followed by the Central Government, State Governments and FSSAI while implementing the provisions of this Act[47] and also prohibits advertisements which are misleading or deceiving or contravenes the provisions of this Act[48] and prohibits unfair trade practices[49].

No person shall import any unsafe or misbranded or sub-standard food or food containing extraneous matter[50]. Responsibility is imposed on the food business operator to ensure that the articles of food satisfy the requirements of this Act at all stages of production, processing, import, distribution and sale within the businesses under his control[51]. The Act also imposes certain liabilities on the manufacturers, packers, wholesalers, distributors and sellers if an article of food fails to meet the requirements of this Act[52]. The Act compels the establishment of food recall procedures[53] and also the licensing and registration of food business[54]. The FSSAI and the State Food Safety Authorities are responsible for the enforcement of this Act[55]. Small business operators

---

[42] FSSA Sec 4

[43] *Ibid* Sec 3 (k) 'Food additive' means any substance not normally consumed as a food by itself or used as atypical ingredient of the food, whether or not it has nutritive value, the intentional addition of which to food for a technological purpose in the manufacture, processing preparation, treatment, packing packaging……..but does not include contaminants or substances added to food for maintaining or improving nutritional qualities.

[44] *Ibid* Sec 19

[45] *Ibid* Sec 21

[46] *Ibid* Sec 22

[47] *Ibid* Sec 4

[48] *Ibid* Sec 24

[49] *Ibid*

[50] *Ibid* Sec 25

[51] *Ibid* Sec 26

[52] *Ibid* Sec 27

[53] *Ibid* Sec 28

[54] *Ibid* Sec 31

[55] *Ibid* Sec 29

and temporary stall holders are exempted from the license but need to get their businesses registered with the local municipality[56]. There are general provisions relating to offences and penalties for failure to comply with the requirements of this Act. The Act makes provision for graded penalties where offences like manufacturing, storing or selling misbranded or sub-standard food is punished with a fine, and more serious offences with imprisonment[57].

**Chapter Five Titled '*Food Safety and Standards and Consumer Protection Laws in India –A Study*'**

In this chapter the researcher has traced the evolution of consumer law dating back to ancient times, beginning from Dharma, sruthis smrtitis, manu smriti, and Kautilya's arthashastra. In the present day everybody wants to lead a comfortable and luxurious life which in turn has given birth to so many products which have rendered our lives easy and made these products inevitable. This development has not spared the consumers of food. There is a plethora of food and articles of food which have flooded the market. The consumer in order to emulate others has fallen prey to these articles without knowing the ill effects of them. Taking the advantage of the ignorance of the consumers, the unscrupulous manufacturers, distributors, and other food business operators have resorted to various undesirable and dangerous means of adulteration of food articles.

In a developing country like ours, the life of a common man has been rendered risky and worse by the adulteration of the basic need-food articles. In the present day, the need to supply the unadulterated food-stuffs to the citizens in this socialistic welfare state has naturally assumed great importance. It is an accepted truth that the consumers of all goods need to be protected in India but the protection of those consuming food-stuff is all the more of greater importance as Roscoe Pound has classified social interests under six heads and has placed public health at the top of it. According to him, public health is the most important social interest, as the food products are to make or mar the health of the citizens.

The guidelines of United Nations on consumer protection have been discussed. Consumer protection is deep rooted in the rich Indian civilisation dating back to 3200 BC. In this chapter, the researcher has defined various terms under the Consumer Protection Act 1986(CPA). Jurisdiction of various Consumer courts/forums has been

---

[56] *Ibid* Sec 31(2)
[57] *Ibid* Sec 48 to 67

discussed. The procedure before the consumer court and other courts has been discussed and a Comparison between Section 72 of FSSA and Section 3 of CPA is made. The Researcher has also made a Comparison of Food Safety and Standards Act with Consumer Protection Act. The Researcher has found out some of the loopholes in the working of consumer protection Act and has made some observations and recommendations for better working of Consumer Protection Act.

**Chapter Six Titled '*International Conventions and Declarations on Food Safety and Standards: An Analysis*'**

In this chapter, the Researcher has made an elaborate study on the various international instruments in the form of Conventions, Declarations, Summits and conferences dealing with the right to food. Though initial instruments did not use the term 'safe food', in effect they meant safe food only. After some time the countries found the necessity to use the term safe food in these international instruments and then began the use of the same. The Researcher has begun this chapter with the United Nations Charter, Universal Declaration of Human Rights which forms the basis of all human rights today and for generations to come and has dealt with the International Covenant on Economic, Social and Cultural Rights, 1966, the International Convention on Civil and Political Rights,1966, General Comment 12, Rights of the Child,1989, World Food Conference 1974, World Food Security Convention, 1985, International Conference on Nutrition 1992, UN Millennium Declarations: 2000, World Food Summit 2002, United Nations Sustainable Development Summit: Goal 2: 2015 and the latest one being High-level Side Event on Pathways to Zero Hunger: 2016.

The Researcher has discussed the nature of the human rights. Hungry person of any nationality in any part of the world is also a part of the world community and he has rights claims not only in relation to his own nation but also in relation to the world at large. The human right to adequate food will be hollow and narrow in its application if obligations to honour this right are restricted only to one's own state as the Researcher strongly believes that 'children born into poor countries are not born into a poor world'. The Researcher has discussed in detail the obligations of the states at different levels to ensure the right to food.

These Conventions have to ensure that all member nations should adopt and practice food safety norms and people get enough food to eat. But unfortunately inspite of there being number of attempts at the international level to combat the menace of hunger, there is no major breakthrough. Even today we find people dying of starvation.

**Chapter Seven Titled '*Conclusion and Suggestions*'**

In this chapter, the Researcher has dealt with the Conclusion for the entire study, proving of hypotheses and also the observations made by the Researcher in the form of findings while carrying out the entire study. The Recommendations which is an important part of the study can be divided into two parts, first to strengthen the Right to food and then to make the available food safe and fit for human consumption. As the Right to safe food entails in-limine availability of food, talking about safe food without adequate availability of safe food is like groping in the dark sans any fruits. The cumulative effect of these observations and recommendations can culminate in the Realisation of Right to safe food and the availability of only the safe food in the market, which meets the standard requirements as stipulated by the laws of the land i.e. FSSA.

## 1.13 Conclusion

The Right to safe food is essentially a second generation human right and is plagued with the same issues of enforceability as all other economic, social and cultural rights[58]. Hunger is caused by decisions made by human beings and can be ended by making different decisions. For the existence, growth and development of the society, the interest of the members should be protected. It is, therefore in the interest of the existence and growth of a society that consumer protection is extended as a right of the members of such society[59]. In case, it is not protected, the result may be chaos in the society which may result in disintegration and degeneration of the society. In order to strengthen the Right to safe food the countries should first of all encourage agriculture which is the source of safe food.

Only if there is enough supply of food, we can contemplate making the food safe. This may not be possible when we have scarcity. Quenching the immediate need of hunger is more important to the society than to see if the available food is safe. May be in India, we have a long way to go to attain that complacency of having only the safe and standard food in circulation. In order to increase agricultural production, India

---

[58]Ms. Uma Sud, "The Right to Food as an International Human Right", Social Action, [January-December 2004] Vol 54 No4, p 296-310.

[59]Dr Subhash C Sharma, "Consumer protection and the Prevention of Food Adulteration Act", Central India Law Quarterly, Annual Index [1995] Vol VIII, p 187-194

should, at the outset sensitise people about the need to grow more food as there cannot be anything parallel to self sufficiency. The preoccupation of all the states is to see that there is enough supply of food in general. Only if there is sufficient food in supply, we can think of safe food and not in a scenario where the very existence is a challenge. People should also be responsible as to not to waste the available food. Food safety should become a way of life. Healthy trade practices by the food business operators are crucial and need of the hour and the offenders should be brought to book mercilessly. The procedure in the courts must be more result oriented in terms of punishing the guilt and not prolonging the case endlessly. If the iron is not beaten when it is red hot, you have to heat it again and wait till it becomes heat, by which time lot of water would have flown down the river. Traders should be sensitized about their responsibility to the society. In the corporate world, it is not enough only if there is not enough if the social responsibility is cast on the corporate, each one of owe a duty to the society.

There should be overall development of all regions, which cuts down the migration of people from rural areas to urban areas in search of better pastures. If almost same facilities are available even in rural areas, why will people migrate to cities which encourages the people to take up/continue their original occupation. The states should strive to create independent economy than creating a dependent-on-government-for-food economy, which is not the answer to our problem. Creating an independent-self-sustaining one is. In India, food industry is of different sizes and the requisites of standards in each sector are different. The protocol for standardisation of food articles should keep in mind the actual users of these standards, environment, the culture and the present infrastructure of the country which seems to have been overlooked by the legislators.

It is very disheartening to find that adulteration- 'a demon' is still a big nuisance and menace to the public health despite there being stringent and harsh provisions in the FSSA and the strict interpretation of the same by the judiciary. Due to the scarcity of food articles, adulteration of food continues to be a serious problem in India. Poor purchasing power of the common man coupled with ignorance has made it difficult for the state to effectively administer trading activities of the private and public undertakings engaged in production, distribution and sale of food stuffs. Serving safe food is not an option, but an obligation. "Each and every member of the food industry, from farm to fork, must create a culture where food safety and nutrition is paramount"[60].

Law is a means to an end, which fails only sometimes to organise an orderly society. Though the executive and judiciary are playing prominent roles in curbing adulteration, the public participation in the same direction cannot be ignored. Unless there is a concerted effort by the people, by the state and by the authorities, legislations alone cannot bring about the desired effect. The Researcher is convinced that unless a common man becomes aware of the gravity of the quandary of adulteration of food and his rights and corresponding duties to the society in which he lives, 'the Dracula of adulteration pervades the society eternally and engulfs it's preys-the healthy consumers'. The Researcher believes that '*there can be no salvation without cooperation from all quarters and all stake holders*'.

[60]William Bill Marler, An accomplished attorney and national expert in food safety, he has become the most prominent foodborne illness lawyer in America and a major force in food policy in the US and around the world. He has represented thousands of individuals in claims against food companies whose contaminated products have caused life altering injury and even death.

# CHAPTER-II

# HISTORICAL EVOLUTION AND CONCEPTUAL DEVELOPMENT OF RIGHT TO FOOD IN INDIA: AN ASSESSMENT

## 2.1 Introduction

The historical and political background of the Right to Food is much more than the history and politics of malnutrition. It concerns the development of the notion of access to Food as a Right. As a Right, it sets obligations on the State, which have been established as enforceable through centuries of social struggle for a democratic state in the service of the people. Traditionally, people had no remedy other than to revolt against a king or state that failed to meet its obligations. Right to Food is one of the basic Human Rights[1]. The ancient Indian concept of Dharma laid extraordinary emphasis on individual and social action for growing and sharing food in abundance. Dharma played a cardinal role towards maintenance of social order, general well being, progress of mankind and social stability[2].

Dharma, which is a way of life never visualised disobedience to it. The State in ancient India laid only the positive sanctions ordaining the people to do something. As dictated by Dharma, the main endeavour of our ancient human values and virtues was only to improve the quality of one's life. In tune with this, people in ancient India were god fearing and law abiding. Because of which Dharma never spoke about punishments as the necessity did not arise at all. This is the reason most of the ancient scriptures use the term 'food' and not 'safe food' as in the present times. It was beyond the imagination and application of Dharma, that one day food which is meant for the consumption of human beings could be adulterated. Dharma could not foresee that such a heinous act could happen.

This is the reason we have no explicit reference to safe food in many ancient scriptures, due to which the Researcher urges that the food means safe food only. When these scripture were authored, the menace of adulterated or unsafe food did not exist at all. These scriptures obligated the king to provide his people with food, which was safe by itself for human consumption as adulteration did not exist at all. Taking the analogy

---

[1] Human rights are those minimal rights which every individual must have against the state or other public authority by virtue of his being a member of the human family irrespective of ant other consideration. DD Basu, *Human Rights in Constitutional Law*, (3rd edn 2008) , P.5

[2] P V Kane, '*History of Dharmashastra*', Bhandarkar Oriental Research institute, Poona(2nd edn,1973) p 749

from the obligation of the king, today the State is also obligated to make the food available and accessible on one hand and safe and standard on the other. The ancient scriptures believed in sharing of food, hence required the people to do charity before one could eats.

Dharma emphasised that this act of an individual earned religious gain(punya) for him. So the food had to be safe and did not affect the consumer adversely, lest the offerer or giver of food would incur the wrath(Paapa) of the almighty and the curse of the person/people who have consumed the food, believing it to be safe. So, a duty was cast by Dharma and the human values to see that the people who do charity should make sure that the food is safe for human consumption. Dharma also said that one should offer food to the needy not when he has excess of it nor he could give whatever he did not want, but share whatever he has and whatever he is wanting to consume. Indirectly it meant that, a person who gives the food in charity cannot give food which is of an inferior quality, putrid, rotten or anything that is not conducive to human health.

Whatever the offeror/giver consumes, the same food had to be shared, here indirectly the scriptures have stressed on the standard or quality of food. No person will eat anything whose quality is not good. Finally these scriptures have cast a duty on the State to see that the people are provided with safe food which is conducive for human health. At the same time, these scriptures have urged the people to share the food with needy in the form of charity. The main intention is to see that everybody has food to eat, which in turn results in social stability. On the converse, in the presence of lurking hungry bellies, the society will be breeding only the disgruntled and anti social elements, who can alarm and sound a threat to the peace of the society in general. In the modern times, similar obligations have been cast on the State in the form of Constitutional principles and Constitutional morality in their respective Constitutions. Different laws have been enacted which confers a Rights claim on the people to demand the Right to food from the State.

It was realized that 'life arises from food and world is sustained by food[3]. '*Do not turn away any one who comes seeking your hospitality which is the inviolable discipline of life*'[4]. "In the old testament, the Prophets declared that the sacrifice pleasing the god was to share your bread with the hungry(Isaiah 58.7), while the New Testament

---

[3] '*Annadbhavanti bhutani*' as stated in Bhagavadgita 3.4. G V Sharma Pandith's '*Dharma Shastra*' ,Divya Chandra Publishers, Bangalore (2011)

[4] *Ibid Tatriyopanishad* III 7 and III 10

amplified this message by posting that feeding the poor was the same as feeding the Christ(Mt 25:35). While justice requires more than people being fed, Food is necessary to enjoy the right to life which the church maintains is universal and inviolable"[5]. Later, the thinkers realised that the most effective way to feed the poor and hungry and the most fitting of their dignity was to create situations where people could provide for themselves and in ways appropriate to their cultural context.

## 2.2 Importance of Food

Food is the third most important thing for living beings, after air and water. We eat food because we need to survive. All living things need food. Even if a human being does not have a shelter to live or clothes to cover their body, we would still survive if we get wholesome food. This is the reason we have been motivated to search and seek food. Food is a fuel for the body of a living organism to survive and do its activities. Food is an indispensable part of everybody's life as it gives energy, body shape & form and nutrients essential for the body to grow, to be healthy both physical and mental health. The food can also act as medicines as the food that we consume has medicinal properties and cures many diseases which in turn helps the body to overcome diseases. Food is equally important to overcome stress. The body needs proteins, carbohydrates, fats, vitamins and minerals and the body sources all these from the food that we eat. The body has a beautiful capacity of repairing itself in case of any injury by sourcing the needed elements from the food that we eat. To stay healthy, balanced diet is very essential. When the body does not get enough food it results in under nourishment.

If we examine the human history, it is evident that food has been a catalyst for societal transformation, societal organisation, expansion and conflict in the society. Food is also an essential part of one's culture. It operates as an expression of one's cultural identity as each one of us may have different food cultures which may not be the same as others. Food culture means the practices, beliefs and attitudes and also the institutions and networks surrounding the production, distribution and consumption of food. In the Indian scenario eating can be a social activity where people meet up only to eat together. Of course, the key driver for eating is one's hunger, but in India and other under developed countries what we eat is also determined by economic factors like cost, income and availability. Eating is considered as one of the most important ritual in the

[5] Dr K V Ravikumar, "Right to food in India-Whether a protection under fundamental rights", Indian Bar review, [July-September, 2012] Vol XXXIX No3, p 39-48

Indian way of life. Food also formed the basis of the earliest forms of religions and their practices. In the early ancient times, Hindu gods were thought to be directly responsible for the production of food like, the earth, the rains, and the river.

On a careful examination, it is evident that food includes any substance meant for human consumption and also those substances which goes into the preparation, manufacturing or processing the food including water. When an article of food is meant for human consumption, we presume it to be safe and presumed to conform with the requirements of standards set by law in action. Initially when the menace of adulteration and unsafe food was unknown to the society there was no necessity to make it obvious by calling food as safe food because food was always considered safe.

But due to the onslaught of the evil of adulteration and unsafe food flooding the market, to curb the same, the thought process began to ensure that the food is made safe and hence the apart from provincial laws, the Prevention of Food Adulteration Act, 1954(PFA Act) was passed. But, due to the shortcomings in interpretation and implementation of PFA Act, a need was felt for a modified law, thus the law in action today, Food Safety and Standards Act 2006(FSSA) was passed. The word food includes the primary food i.e. the produce of agriculture or horticulture which has not undergone any artificial process. Due to this, it is beyond the control of human agencies to maintain the required standards as per law. If a sample of primary food is found to be sub-standard due to natural causes and factors which are beyond the human agency, it shall not be considered adulterated as long as the variation in its quality or purity does not render it injurious to health. But, if it is injurious to health, the sample will be treated as adulterated notwithstanding the fact that the variation is due to the natural causes.

Subsequent to this in 2013, in order to ensure the right to food, National Food Security Act, 2013(NFSA) was passed whose beneficiaries were targeted public group living below poverty line. Under this Act the beneficiaries are provided with free food grains which take care of their nutritional levels apart from filling their hungry bellies. These are the major food laws that are operating in India. But the Researcher thinks that there should have been a reversal in the chronology or promulgation of these laws. First the law ensuring the Right to food, i.e NFSA, should have been passed and then the FSSA could have been passed.

The Researcher believes that at the outset, the basic duty of the state is to feed its people and at the same time ensure that the articles of food it is distributing through the public distribution system and also available in the market are safe for a common man to consume. We have witnessed that 'poor will consume anything'. Hence it is the duty of the state to ensure right to safe food which meets one's dietary requirements. Here, the concern of the state is common man as he cannot afford to buy safe food at the exhorbitant cost as against the affluent class. The cumulative effect of these two legislations is that the Right to safe food which meets the dietary and nutritional requirements of people is ensured.

## 2.3 References of Reverence to Food in Indian Scriptures

In India, we have a reference to the adage *'Annadaatho Sukhinobhavanthu'* which means let the good things be bestowed on the people who feed others. In the modern period the same is treated as a basic Human Right and is backed by various constitutional principles and International conventions. Our ancient scriptures then itself had recognized and accepted this right, though may not have given a fancy expression as 'human rights'.

### 2.3.1 Dharma

"We are what we eat" is an old proverb. The concept of right to food can be traced to the ancient period wherein Dharma laid extra ordinary emphasis on individual and social action in growing and sharing food in abundance. Dharma is a way of life which deals with duties, rights, laws, conduct, virtues and *'right way of living'*. Great importance was attached to purity of food from very ancient times.Though it was not explicitly slated to be a right, yet lot of importance was accorded for ensuring that everyone is fed and nobody goes hungry. A duty was cast both on the society as well as the king to see that the people are provided with the food which is culturally acceptable by them.

### 2.3.2 Hindu Samskaras

It was realised that life arises from food and the world is sustained by food[6]. In Indian custom eating food is one of the spiritual activity for which lot of importance is attached. After the child is weaned off by the mother and the child has to start eating the food, such an occasion is celebrated in hindu families as a religious ceremony. According to Hindu law, there are 16 samskaras which will purify the life of a hindu, in which Annaprashana is one of them when a child is fed with a solid food *'anna'* for the first time. Consuming every meal is also a ritual as per hindu tenets. Lot of religious sanctity is attached even for taking one's meal.

---

[6]'Annabhavantibhutani' as stated in *Bhagavad Gita* 3.4.

### 2.3.3 Vedas

Vedas, one of the sources of Hindu law are believed to be the revelations of lord Brahma. There are four Vedas, namely *Rigveda, Atharvanaveda, Yajurveda* and *Samaveda.* These Vedas include elements such as liturgical material which means the performance of a religious ritual as well as mythological accounts, poems, prayers, and formulas considered to be sacred by the Vedic religion. The *Rigveda* emphasizes this discipline in even stronger terms, saying:

*moghamannam vindate apracetah. satyam bravimi vadha itsa tasya.*
*naryamanam pusyati no sakhayam. kevalagho bhavati kevaladi.*

Which means, food that comes to the one who does not give is indeed a waste. This is the truth. I, the *risi*, say it. The food that such a one obtains is not only wasted, in fact it comes as his very death. He feeds neither the devas, the upholders of various aspects of creation, nor the men who arrive at his door as friends, seekers and guests. Eating for himself alone, he becomes the partaker of sin alone. Thus *Rig veda* also advocates and casts a duty of sharing of food among the needy and not a lone consumer. This discipline of growing an abundance of food and sharing the same among the needy is taught in Smriti texts such as Mahabharata, Ramayana and many puranas and Dharma Shastras. It was realised that life arises from food and the world is sustained by food[7]. *Yajurveda* which mainly deals with sacrificial formulas prescribes a meal time prayer, which will have to be chanted before one's meal,

*Aum annapate annasya no dehyanamivasya shushminah.*
*Prapradataramtarisha urjam no dhehi dvipade chatushpade'*[8].

Which means 'oh lord, the giver of food! May you provide us with healthy and energy producing food, grant happiness to those that give food in kindness. May this food give us strength'. This indirectly expresses that the food will have to be safe as it should give energy and happiness to the consumer. It is everyone's knowledge that, if the food is not good in quality and safe, it neither gives us nutrition nor energy nor allows to survive. *Atharvanaveda* which deals with procedures for everyday life goes one step further and mandates that

*Samani prapa saha vonnbhaga*
*samane yoktray saha wo yunism*
*arah nabhimiv abhite*[9]*:*

---

[7] 'Annabhavantibhutani' as stated in Bhagavad Gita 3.4.
[8] Yajurveda (Yajurveda 11/83)
[9] Samjnana Sukta of Atharvanaveda

Which means all have equal rights in articles of food and water. The yoke of the chariot of life is placed equally on the shoulder of all. All should live together with harmony supporting one another like the spokes of a wheel of the chariot connecting its rim and the hub. This was the first scripture which spoke about one's entitlement to food as Right to food which indirectly means Right to life. The Researcher thinks that, may be our Constitution framers were inspired by this verse from *Atharvanaveda* to include Article 21 in our Constitution.

### 2.3.4 Upanishads

Upanishads form the end part of Vedas and hence they are called Vedantas. They represent philosophical developments beyond Vedas. Some of the thinkers have also contributed for the growth of Upanishads. The word 'food' in the Upanishads has much wider connotation than what is normally understood by it. "Food stands for all the inputs required to sustain life both at the individual and cosmic levels"[10]. In the *Chandogya Upanishad* (VII.26.2) occurs this passage 'when there is purity of food, then the mind becomes pure, when the mind is pure then follows firm remembrance (of real self), when the last is secured all knots (that bind the soul to the world are loosened'[11]. Hence the importance attached to right to safe food is not a novelty and the same can be traced from the Vedic and Upanishads period.

Food is the first form of god, the Anna Brahma of the Upanishads. Without honouring this form of divinity, the other forms may not be able to manifest in our lives. Food not only nourishes the body but also the mind and the emotions. The *Chhandogya Upanishad* also describes the process of food and drink. It is said that hunger is absorption of food that is eaten. Food is like fuel to the fire or heat to the body. When food is not eaten the mind loses its strength and the fire in the body is extinguished. The mind is influenced by the food that is eaten. It is strongly opined in this Upanishad that "in purity of food there is purity of mind, in purity of mind there is established memory".

According to *Vedanta*, the personality of a complete man comprises of five layers and referred to as '*Panchakosam*'. Among these the first and the foremost is the

---

[10] Briha Daranyaka Upanishad(4)-Madhu-Kanda chapter I available at www.esamskriti.com/print.aspx?topicid =927&chapter=1 (accessed on 6/5/2014 at 6.00pm)
[11] Dr. Panduranga Vaman Kane, *History of Dharmasastra*, Oriental Research Institute, Poona. ( 2nd edn,1974) p 757

Annamaya Kosam. 'Annam' means food and 'maya' means modifications. The body is the result of modification of food and hence called 'Annamaya'. The food eaten is digested. It's very essence becomes the source of new life. The child grows up and develops in strength and size due to the food eaten. Finally we die to merge into food (Earth). The earth itself becomes the food we eat. So we are born from food (earth) and go back to food(earth). Annamaya Kosam is our Gross Body (Sthula Sariram).

*Isopanishad* says *Athithi devo bhava,* reception of guests and paying attention to them was one of the propitiations of householders, it is after feeding the guests and the dependants in and around home that husband and wife shall eat[12]. India has always been referred to as the land of Anna purna, which means one who feeds. Food and water are not only the elixir of life, but they are worshiped as gods. In the past, Indians had conferred great prominence on growing food in large quantity, abundant enough to feed all its people and sharing the same. Indians, till recently had considered the abundance of food as the primary condition and pre requisite for human civilisation and sharing the available food as the primary discipline of civilised living which we call it as Dharma.

This outlook towards food and sharing of it is dealt in detail in the basic Indian text of *Taittiriyopanisad*, the most revered sruti even today. In *Tattriyopanishad, Anna*, which mean food, in fact, forms the entrance to the edifice of *brahmavidya*, and what is protected at the centre of that edifice is also *anna*. In Tatrriyopanishad, *Brigu maharshi* wrote '*annam bahu kurvita annam na indyaath tat vratam*'. Which means let us create and share food, let all hungry stomachs be fulfilled with food, let us not waste food, no place you can drop food where there is no hunger (either outside or inside) [13]. Tatrriyopanishad, ordained the people to respect food, share it and not waste it in the light of hungry stomachs.

Brigu maharshi stresses that plenty of food is also needed to feed the many who will come to him for the knowledge of the Brahman[14]. *Annam na nindyat Tadvratam,* which means do not look down upon anna i.e food that is the inviolable discipline of life for the one who knows[15], *annam na paricaksita. Tadvratam* which means do not neglect *anna*. That is the inviolable discipline of life for the one who knows, *annam*

[12] Isopanishad at verse No1

[13] http://www.nithyananda.org/sites/default/files/sharings/food-annalaya-project-annam-bahu-kurvita.pdf (accessed on 11/8/2018 at 12.56pm)

[14] Available at ww.esamskriti.com/e/Spirituality/Upanishads-Commentary/Taittiriya-Upanishad~-Petal-9-3.aspx (accessed on 11/8/2018 at 1.00pm)

[15] J.K. Bajaj and M.D. Srinivas, The Indian Tradition of Growing and Sharing Food Available at http://www.indiatogether.org/manushi/issue92/bajaj.htm (accessed on 11/8/2018 at 1.30pm)

*bahu kurvita. Tadvratam* which means multiply anna many-fold, ensure an abundance of food all around, that is the inviolable discipline of life for the one who knows. *na kamcana vasatau pratyacaksita. tadvratam. tasmadyaya kaya ca vidhaya bahvannam prapnuyat. aradhyasma annamityacaksate* which means Do not turn away anyone who comes seeking your hospitality. this is the inviolable discipline of the one who knows. Therefore, obtain a great abundance of *anna*, exert all your efforts to ensure such abundance and welcome all seekers with the announcement that the food is ready, partake of it. This is the discipline that was expected by *Taittiriyopanisad*, to be adhered to by every human being. Such is the discipline of abundance and sharing of food that *Taittiriyopanisad* teaches.

In summary, this shloka means that everyone ought to vow to never to condemn food. The prana is verily, food the body is the eater of food. The body rests on the prana; the prana rests on the body. Thus food rests on food. He who knows this resting of food on food is established, he becomes a possessor of food and an eater of food. He becomes great in offspring and cattle and in spiritual radiance and great in fame. This Upanishad also deals with meditation on food as, it is considered food as the basis for any living being. It is with the help of food that an individual is able to sustain life and only if an individual sustains he is able to attain enlightenment. Thus contemplation of food is ever desired.

According to *Brihadaranyaka Upanishad*[16], food has much more wider connotation than what is normally understood by it. Food stands for all the inputs required to sustain life both at the individual and the cosmic levels. This Upanishad says that the creator, Prajapati, by means of knowledge and contemplation produced seven kinds of foods viz., 1.Solid food, what we generally call as food for consumption by all 2.Milk, liquid food 3.Sacrifice (yajna) 4.Symbolic offerings to god 5.Speech 6.mind and 7.Life breath. The first four 'gross foods' sustain our physical life. The last three, 'subtle foods' are significant for our metaphysical life both at the individual and cosmic levels.

All these seven kinds of food are essential for a human body. This Upanishad opines that people continue to consume food of various types endlessly, for ages, over centuries and yet the food is not exhausted because the desire of the human mind for that matter is inexhaustible, as long as the desire is present, its object will also be present. So, no food is exhausted as long as there is a need for food. This Upanishad while

[16] Madhu-Kanda-Chapter1

dealing with the Meditation on Prana- vital force embodied in a person, compares Prana to the calf and the subtle body in which it is lodged as its abode. The energy of the whole system which maintains it is the peg to which it is tied by means of attachment to the body. The food that we consume by which energy is generated is the rope by which it is tied to the peg. Thus this Upanishad glorifies the importance of food we consume.

*Aitareya Upanishad*, while dealing with the Origin of the universe and the Man[17] says that Brahma after creating the fields for the functioning of the universe and the human beings which were both empowered by the cosmic forces, the third step in the creative process is said to be the creation of food. This is the reason why even today we consider food, clothing and shelter as the fundamental necessities of human beings to sustain life. This Upanishad brings parlance between the different senses urged by hunger and thirst run after food to catch it like a cat running after a mouse. After this, the god intensely meditated and created food in the form of cereals, animals and other living and non living things. The Arunika Upanishad says that "food should be eaten as a medicine". "From food verity all creatures are produced". Food is called Anna because it is eaten by all beings.

The *Maitrayani Upanishad* says that both the mind and the food to be eaten should be purified before eating food, that is indirectly this Upanishad dealt with safety of food for human consumption. This Upanishad equates food with Lord Vishnu and says that energy is the essence of food and exalts the process of eating of eating food to divine worship because the food and the eater of food are forms of divinity. Therefore it is said no one should insult food and one should not despise food.

### 2.3.5 Bhagavad Geeta

The Bhagavad Geeta which means '*a song of the lord*' is an ancient Indian text and is an important work of Hindu tradition in terms of both literature and philosophy. Bhagavad Geeta is the revelation of lord Krishna to Arjuna in the battle field when Arjuna was crest fallen that he had to fight and kill his kith and kin and loses his will to fight. Lord Krishna reminds Arjuna of his duties which constitutes Bhagavad Geeta in the form of shlokas. Chapter 17, Verses 8, 9, 10 of Bhagavad Geeta also deals with the importance of food and says that stale, putrid, decomposed, foul and impure food should not be consumed. In principle, these shlokas also deal with safety of food meant for consumption of the people.

---

[17] Part2

*āyu-sattva-balārogya-sukha-prīti-vivardhanā*
*rasyā snigdhā sthirā hidyā āhārā sāttvika-priyā*

*kav-amla-lavaāty-uha-tīkha-rūkha-vidāhina*
*āhārā rājasasyehā dukha-śhokāmaya-pradā*

*yāta-yāma gata-rasa pūti paryuhita cha yat*
*uchchhihṭam api chāmedhya bhojana tāmasa-priyam*

These shlokas mean Persons in the mode of goodness prefer foods that promote the life span, and increase virtue, strength, health, happiness, and satisfaction. Such foods are juicy, succulent, nourishing, and naturally tasteful.

Foods that are too bitter, too sour, salty, very hot, pungent, dry, and chiliful, are dear to persons in the mode of passion. Such foods produce pain, grief, and disease. Foods that are overcooked, stale, putrid, polluted, and impure are dear to persons in the mode of ignorance.

### 2.3.6 Mahabharata

Mahabharata is one of the two major Sanskrit epics of ancient India, In Mahabharata also one can find glimpses of the same idea, where Sri Krishna summarizes the teachings of Bhishma, 'Annenadharyatesarvam'[18]. "*Hunger of even one person in a kingdom renders the life of the king forfeited*"- Bhishmacharya warned the king Yudhistra in the Anushasana parva of Mahabharat and instructed, "the king should look after the welfare of the helpless, the aged, the blind, cripple, those suffering from diseases and calamities, pregnant women by giving them food, lodging clothing and medicines according to their needs[19]. In Mahabharatha, the value of food is emphasised in the teachings of Bhishma in a long discourse running to about 25000 verses, the major part of it deals with dharma and which forms a quarter of the epic.

The king Yudhisthira requests Srikrishna to let him know the essence of the teaching of Bhisma. Srikrisna, in response, utters just 15 verses, the first ten of which lay down the centrality of *annadana*, the giving of food, in the life of a disciplined householder and the next five celebrate the greatness of food, its emergence out of the vital essences of the earth and its intimate connection with all life. The first verse Srikrishna utters while summarizing the teachings of Bhisma for Yudhisthira is:

[18] Anushasanaparva
[19] Mahabharata Shanti Parva

*annena dharyate sarvam jagadetaccaracaram*
*annat prabhavati pranah pratyaksam nasti samsayah,*

which means, both animate and inanimate world is sustained by food, life arises from food, this is observed everywhere and there can be no doubt about it and Krishna concludes his discourse on *annadana* with:

*annadah pranado loke pranadah sarvado bhavet*
*tasmadannam visesena datavyam bhutimicchata*

which means, the giver of food is giver of life. Hence, one who is desirous of well being in this world and beyond, ought to give food. Recounting the same incident, the Bhavishyapurana, in its chapter dealing with *Annadanamahatmya*, has rendered the teachings of lord Sri Krishna in the following verse,

*dadasvannam dadasvannam dadasvannam yudhisthira*

which means O Yudhisthira! Give food, Give food and, keep giving.

Bhishma in Mahabharata, reminds Yudhishtira repeatedly of the importance of feeding the others in general but specially of the duty of the king to make sure that agriculture in his domain is taken care of well and farmers are not disgruntled by unreasonable exaction and irrigation is taken care of by the king and not left to the mercy of rain god, so that there is always plenty of food so that nobody has to sleep with empty and hungry stomach. In tune with the advice of Bhishma, in Indraprastha - the kingdom of Pandavas, a hundred thousand women attendants of Yudhisthira used to carry pots of food in their hands and were engaged in feeding the guests day and night. When Yudhisthira travelled out of Indraprastha, he was followed by a hundred thousand horses and a hundred thousand elephants with food articles to be given away to his people on the way. This is how things were when Yudhisthira, residing in Indraprastha.

From these teachings of Bhishma, the Researcher infers that he always advised to concentrate on the permanent solution to the problem of hunger and not resort to temporary measures. This is almost the same advice that lord Sri Rama offered to Bharata, his younger brother, while enquiring after the welfare of Ayodya when Bharata visits him at Chitrakoota during the early phase of Sri Rama's long sojourn in the forests. The vision of Ramarajya is where there is abundance of crops and absolute absence of hunger, thirst, diseases, scarcity and errors on the whole earth. The contrary of Ramarajya is Yugaksaya, when food becomes so scarce that the people will be reduced to selling food and those seek food. water and shelter will be turned down and will be forced to lie hungry and thirsty on the roads, which means the end of times.

### 2.3.7 Manusmriti

In the Indian context, every householder is a king in his own domain and it is expected of him to make sure that no one under his care suffers from hunger and want. In fact, an Indian may partake of food only after the gods representing different aspects of nature and his ancestors are propitiated, bhootas/departed souls have been offered their due share, the guests and the seeker of food at the door have been satisfied and servants and other dependents have been fed. The same is being discussed in Manusmrti and has laid down this daily discipline of feeding and taking care of others before eating for oneself in more than two hundred verses, summarizing the discipline the text says:

*devanrsinmanusyamsca pitrngrhyasca devatah*
*pujayitva tatah pascadgrhasthah sesabhugbhavet*
*agham sa kevalam bhunkte yah pacatyatmakaranat*
*yajnasistasanam hyetatsatamannam vidhiyate*

which means, the householder must eat only what is left after making reverential offerings to the *deva*s, *rishi*s, ancestors, the *boota*s and the men under his care and those who come seeking at his door. A householder who cooks for himself alone does not partake of food, but partakes merely of sin. For the wise one the leftover of what has been shared with all of the above alone is proper food. According to Manu, only after all have been fed, it is indeed time for the *grhastha-dampati*, i.e, the husband and the wife, to sit down to eat for themselves, for their greatness is in eating what is left after feeding others. This establishes the obligation of the head of the family and the head of the State to feed his dependants with safe food.

### 2.3.8 Kautilya's Arthashastra

The classical Indian literature is replete with repeated advice to the kings and emperors to always do what pleases the people most. As Kautilya enunciates

*prajasukhe sukham rajnah prajanam ca hite hitam*
*natmapriyam hitam rajnah prajanam tu priyam hitam*

which means happiness of the people is the happiness of the king, the welfare of the people is the welfare of the king and what is good for the king is not what pleases him but what pleases the people is what is good for the king. The people in the kingdom can be happy only when the basic necessities are met.

In South India, the tradition of feeding the poor and sharing the food with the needy has been strongly rooted. In a Tamil Buddhist text called Manimekalai, there is a reference to Aputran. He on being left alone in an uninhabited island with an inexhaustible pot of food in his hands which is known as akshayapatra, prefers to die of hunger rather than eating alone without sharing the food with anybody[20]. This is the backing that is attached to providing of food to the people by the state or by the administrators or by the king supporting one's right to food. Our ancient scriptures have categorically laid lot of emphasis on providing food to the needy and hungry which though did not use the term 'right to food', in principle in meant the same.

## 2.4 Tradition of Providing Food for Others in Modern India

In the modern India, British administrators in India and early European observers who visited India repeatedly encountered the Indian practice of offering food to the needy and hospitality to those who happened to be at their door. One of the prominent among such observer was Abbe JA Dubois representing the French Missionary who came to India in 1792 and lived here for 31 years enjoyed the fairy-tale Indian hospitality in the villages in and around Mysore. Abbe, is revered as the first sociologist of India for his observations on what he called the "*Hindu manners, customs and ceremonies*". While documenting the practices of Indians, he writes extensively about the strict discipline that the Indians followed in the matter of eating and sharing their food. Abbe has documented the elaborate ceremony linked with taking of a meal by grihastha i.e. a householder, recording the keen and cautious attention that the householder paid to hygiene and piousness in the matter of eating.

## 2.5 Historical Instances of British Denouncing the Tradition of Providing Food

In the beginning of the 19th century, the kings of Tanjavur were popular for having cared deeply about minimising the hunger of the people in their kingdom. In a letter written by Raja Sarfoji, the then king of Thanjavur, in 1801to the British, he had highlighted about the chatrams that had dotted along the road to Rameshwaram, a popular pilgrimage. He had written that these chatrams were running from the time of his ancestors and food was served in these chatrams all through the day who ever came in hungry. At the close of the day that is at midnight, bells were rung to call upon those

[20] The Indian Tradition of growing and sharing food: Annam Bahukurvita available at http://indiafacts.org/the-indian-tradition-of-growing-and-sharing-food/ (accessed on 3/12/2018 at 6.30pm)

who had not received the food to rush and receive the same. This shows the commitment of the kings to see that even the last person in his kingdom does not go hungry and is being fed. When the British took over the kingdom of Tanjavur in 1801, the Raja Sarfoji urged the British to continue this noble tradition of feeding the people. 'The running of the chatrams, the Raja felt, was what gave Thanjavur the title of 'Dharmarajya', and this was the title, the Raja told the British, he valued above all other dignities of his office and he implored the British to ensure that whatever else might happen to his state, this tradition of providing for the hungry was not abridged or eliminated'[21].

History evidences that from the 10th century till the beginning of the 19th century, farming was the main occupation and agricultural produce was in abundance and the same has been reported by the European observers from various parts of the country. Thus, for productivity of food grains in the region around Allahabad, one such observer in 1803 reported a value of 7.5 tons per hectare, and another reported a yield of 13.0 tons of paddy from Coimbatore in 1807[22]. With the coming of the British the abundance of the lands disappeared almost overnight[23] as they did not encourage agriculture but were mostly into trading. As a result, by the end of the British regime in 1947, average productivity of paddy was less than one tonne/hectare and that of wheat was approximately around 700kg/hectare and the productivity of coarse grain was much below this. The cumulative effect of all this was, the traditional plenty was reduced to a scorching scarcity which has persisted even today. They forcibly deflected the Indian polity away from its deep rooted tradition of growing and sharing the food.

The institutional arrangements like chatrams,that our kings had made for providing food to the needy were disliked by the British and gradually they withdrew the resources that funded the noble cause of feeding the poor, which in the opinion of the Researcher, the deprivation of the Right to food thus began. Their insistence on withdrawal of the resources was so stern that in 1799, at the time of the conquest of Mysore, Richard Wellesely, the then Governor General of the East India Company warned Diwan Purnaiah of dreadful consequences in case he continued to alienate the state funds to such purpose. "Bowing to such undisguised threats, Diwan Purniah - who was re-installed as the chief minister of the newly conquered State of Mysore, and who administered the state in the name of the king but on behalf of the British, brought down

21*Ibi*
22 *Ibid*
23 *Ibid* 'A Wasteful Habit'

the resources assigned to what we have been calling the institutions of hospitality and learning from 2,33,954 to 56,993 controy pagodas, in the very first year of the new administration"[24].

Thus began the induced decline of the state's obligation to feed the people. In addition to giving us the barren lands, the British polluted the minds of the Indians by discouraging us from the discipline of giving food to the needy before we eat and compelled us to turn callously indifferent to the hunger and grave need of want of others. The sharing that the Indians practised as a matter of the inherent discipline of being human, was disdained by the British as a wasteful habit and their disdain had such a blow on the newly emerging elite of India that already in 1829, William Bentinck, the then governor-general of the Company could write that, "...much of what used, in old times, to be distributed among beggars and Brahmins, is now, in many instances, devoted to the ostentatious entertainment of Europeans; and generally, the amount expended in useless alms is stated to have been much curtailed..."[25].

The Indians who came under the influence of British and their judgements were swayed by their thought on the Indian discipline of sharing the food and feeding the poor and the needy. The maiden issue of Keshub Chandra Sen[26]'s '*Sulabh Samachar* dated 15th November, 1870 published a write up against the vice of giving alms, the write up declared that "Giving of alms to beggars is not an act of kindness because it is wrong to live on another's charity". 'And the article went on to suggest that incapacitated beggars should instead be trained to do "useful things for society'[27]. This approach of demanding work from those who do not have sufficient food to eat has become over a period of time a cliche among the comparatively well off Indians mainly among those who aver to have acquired a rational and modern consciousness.

But as the Researcher thinks 'habits die hard', despite the efforts of the Britishers to curb the practice of feeding the hungry and sharing the food before eating, the habit

---

24 "How the British Systematically Destroyed Indian Food Production" available at https://dharma dispatch.in/how-the-british-systematically-destroyed-indian-food-production/ (accessed on 20/1/2019 at 6.00pm)

25 Report from the Select Committee of the House of Commons on the Affairs of the East India Company Dated 16th August 1832 available at https://books.google.co.in/books?id=WmlUAAAAYAAJ&pg=RA1-PA377&lpg=RA1-PA377&dq=much+of+what+used,+in+old+times,+to+be+distributed..(accessed on 20/12/2018 at 5.18pm)

26 Hindu philosopher and social reformer who attempted to incorporate Christian theology within the framework of Hindu thought. Available at https://www.britannica.com/biography/Keshab-Chunder-Sen (accessed on 1/12/2018 at 7.30am)

27 "How the British broke our Charitable Institutions and Created a Nation of Beggars" available at https:// dharmadispatch.in/how-the-british-broke-our-charitable-institutions-and-created-a-nation-of-beggars/(acc essed on 21/12/2018 at 6.00pm)

remained widespread that the Famine Commission of 1880 was worried about its impact on the famine administration that such care and concern by the people themselves may destabilise the authority of the majesty and the sovereignty of the state and thus realised that native society of India is popular for its charity and recommended for the continuance of such charity. But the Famine Commission also advocated that when the situation is brought under control by the government by providing reliefs for those who need it, street begging, beggars and public distribution of alms to strange applicants should be shunned and if possible may be entirely stopped. The reliefs were in the form of survival wages which was just adequate to keep oneself alive that is sufficient for maintenance and nothing more than that in return for hard labour for a day at designated work sites.

For those labourers who suffered from ill health and due to which they could not engage themselves in work and earn their bread, the Famine Commissioners recommended for the provision of 'dole of grains' after a careful inspection by the inspecting officers. If a person who availed the provision of 'dole of grains', looked fit enough to work for the inspecting officer, the provision of 'dole of grains' was to be stopped. By and large in India women were not traditionally allowed to appear in public but the Commissioners expected them to work, in return for the dole of grains that was provided, in the form of spinning cotton for the state. "Such was the horror that the British administrators felt for the 'gratuitous' giving out of food, which for the Indians is the very essence of being human. Giving food without demanding work in return seemed to somehow violate the British sense of ethics and morality: they insisted on elaborate controls on the distribution of food in times of great distress, even though they noticed "the reluctance which the people exhibit to accept public charity, and the eagerness with which at the earliest opportunity they recur to their own unaided labour for support ..."[28].

In spite of all these developments, Indians continued to retain the same sense of sharing the food before partaking of it oneself. The Researcher thinks that may be due to the continuous scarcity of food grains and also the changed mindset of the people from the traditional public life to the western ways, our minds are so befuddled that our Indian traditions and practices are fading from the lives of the people and the Researcher

[28] "How the British broke our Charitable Institutions and Created a Nation of Beggars" available at https:// dharma dispatch.in/ how- the- british-broke-our-charitable- institutions-and-created-a-nation- of- beggars/ ( accessed on 21/12/2018 at 6.10pm)

is apprehensive that it might fade from the minds of people even. It is shameful that even among the most affluent class there is no sense of responsibility to curb the prevalence of extreme hunger of the people in India. All of us are a party to this sin and a nation with such a sin in its stomach cannot attain its fullest unless it first expiates the same.

We can liberate ourselves from the shackles of this sin only if we renew our allegiance to the theory of annam bahu kurvita and consider the agrarian sector seriously on par with other sectors thus growing more food again. The evil of scarcity can be wiped off the face of India only when fertile arable idle lands are cultivated as the Researcher thinks that hungry people can exhaust all virtues of a nation. Our Vedas have instructed us kevalagho bhavati kevaladi which means that whoever eats without sharing, eats in sin. We have to resolve ourselves to make a commitment to care for the hunger of our people. Indians need not be taught as to how to share food or how to perform annadaana as we have been taught by our forefathers the greatness and importance of anna it has come to us naturally.

**2.6 The Enactment of the National Food Security Act, 2013 (NFSA)**

The first case concerning the right to food was filed in *Kishen Patnaik* v. *State of Orissa*[29], followed by the Right to food petition in '*PUCL* v *Union of India*'[30], before the Supreme Court and the Apex Court ordered the State to initiate various measures to curb hunger and malnourishment among the needy. Despite the recognition of Right to food by the Indian judiciary as a justiciable right in plethora of cases, many impoverished families were not the beneficiaries of any programme or schemes of the government. The authorities were in pretended deep slumber. The benefits of the schemes meant for the downtrodden never reached the needy. Till 2013, no concrete efforts were made to recognise this as a Right.

In 2013, a need was felt for a legislation supporting the right to food which ultimately metamorphosised in the form of a central legislation, National Food Security Act, 2013 (NFSA) to provide for food and nutritional security of the people who are living below poverty line, by ensuring access to adequate quantity of quality food at subsidised prices to people to live a life with dignity. As has been propounded, the right to food is one of the basic rights. "*The right to food is not an individual Right but a*

[29] 1989 Supp(1) SCC 258
[30] (1997) 1 SCC 301

*rather a broadly formulated programme for governmental policies in the social and economic fields*"[31]. It took more than a decade and a half to the authorities to wake up from the pretended deep slumber to give the right to food a concrete shape in the form of a legislation and make the right a justiciable one.

This Act has benefitted 75% of the rural population & 50% of the urban population, the eligible households i.e., people who are below poverty line, guaranteeing them fixed quantity of food grains on regular basis at the subsidised price, under the targeted public distribution system which made the beneficiaries of this Act a Rights holder. Nutritional security of pregnant ladies, lactating mothers, children below the age of 14 years and emancipation of women were some of the key areas that this legislation has addressed. The Act deals with food security allowance being provided to the eligible households in the event of the state unable to provide grains at subsidised price. This Act also advocated for laying of new railways or extending the railway tracks to transport the food grains from surplus area a to the areas that need them for human consumption.

'*Economist Jean Dreze, former National Advisory Council member and said to be the original architect of the bill, had written(2013) that the bill would be a form of investment in human capital, bringing security to people's lives and making it easier for them to meet their basic needs, protect their health, educate their children, and take risks*'[32]. The then Minister of Consumer Affairs, Food and Public Distribution K V Thomas stated in an interview: '*the day is not far off, when India will be known the world over for this important step towards eradication of hunger, malnutrition and resultant poverty*'[33]. '*The food security scheme gives statutory recognition to right to food under Article 21 of the Constitution of India. Since the scheme rests on the principle of co-operative federalism, it must be able to secure co-operation from various state governments despite their differences in political ideology and must willingly and whole heartedly implement in its true spirit of their legal obligations*'[34]. NFSA has a laudable objective of eradicating hunger and malnutrition from India in the shortest possible time and marked a paradigm shift in addressing the problem of food security from the erstwhile welfare approach to the present Rights based approach. It is therefore

---

[31] J N Saxena, "The Right to Food", Delhi Law Review, [1990], Vol 12, p 126-132
[32] C K Mathew, "Food Security Act in Sleep Mode", Economic and Political weekly, [October 24, 2015] Vol no 43, p20-22
[33] *Ibid*
[34] Umesh Kumar Bhandari, "Fundamental Rights to Life vis-a-vis Food Security: A Study of National Food Security Act", Andhra Law Times, [November & December 2013] Vol 6, p 46-48

important to get it right not just in terms of making it a legal entitlement under the Rights approach but making it successful. The rationale for the Act is derived from Article 47[35] of the Indian Constitution.

### 2.6.1The Objective of NFSA

The objective of the Act is '*to provide for food and nutritional security in human life cycle approach, by ensuring access to adequate quantity of quality food at affordable prices to people to live a life with dignity*'. In line with the stated objective, the Act provides a legal entitlement to receive food grains at subsidized prices by persons belonging to priority households and general households under Targeted Public Distribution System (TPDS). The entitlement has benefitted up to 75% of the rural population and up to 50% of the urban population which sums to be one third of the total population of India. Further, in order to improve the nutritional security, the NFSA has brought various other ongoing welfare schemes of the government under its ambit.

### 2.6.2 The Salient Features of NFSA

- The Act created a statutory entitlement for the eligible households and its obverse is a legal obligation on the government.
- Coverage and entitlement under TPDS upto 75% of the rural population and 50% of the urban population with uniform entitlement
- Food grains supplied at subsidised prices under TPDS
- Nutritional Support to pregnant women and lactating mothers. Children in the age group of 6 months to 14 years will be entitled to meals
- To achieve women Empowerment, eldest woman of the household not below the age of 18 years to be the head of the household for the purpose of issuing of ration cards.
- The Act provides for a Force Majeure clause under Section 5236. It provides immunity to both the Centre and the States against any claim by beneficiaries

[35] "*Article 47 of the Constitution, inter alia, provides that the State shall regard raising the level of nutrition and the standard of living of its people and the improvement of public health as among its primary duties*".

[36] "The Central government, or the State governments, shall not be liable for any claim by persons belonging to the priority households or general households or other groups entitled under this Act for loss/damage/compensation, arising out of failure of supply of foodgrains or meals when such failure of supply is due to conditions such as, war, flood, drought, fire, cyclone, earthquake or any act of God".

entitled under this Act for loss, damage or compensation arising out of failure of supply of food grains or meals in situations like droughts and floods.

- Grievance redressal mechanism will be set up at the District and State levels.
- Provisions have been made for transparency and accountability for disclosure of records relating to PDS, social audits and setting up of Vigilance Committees
- Provision for food security allowance to entitled beneficiaries in case of non-supply of entitled food grains or meals.
- Provision for penalty on public authority/servant in case of failure to comply with the relief recommended by the District Grievance Redressal Officer.

### 2.6.3 Implementation Hiccups of NFSA

Though NFSA has a laudable objective of providing food to the needy, yet it is fraught with many challenges in its implementation

- ❖ Section 16 of the Act requires the constitution of the Food Commission whose duty is to monitor and evaluate the implementation of this Act, advice the state government for the effective implementation of the Act, either suo motu or on receipt of complaint inquire into violations of entitlements, hear appeals against orders of the District Grievance Redressal Officer and prepare annual reports which shall be laid before the State Legislature by the State Government. The State Commission, which is one of the implementation agencies under the Act has not been constituted in many states including Karnataka.
- ❖ The Act is not clear as to the quantum of the food security allowance payable by the state to the beneficiaries in the event of the failure of the state to provide subsidised food grains. If the state contemplates to pay the food security allowance equal to the price at which they are procuring food grains from the farmers for example, which is approximately about Rs 26 per kilo of rice, the Researcher is apprehensive if we can buy a kilo of rice at Rs 26 from the open market. This provision of food security allowance is only an eyewash.
- ❖ The food subsidy in coming years will inflate due to the rise in the population thus increasing the number of entitled beneficiaries and the need to keep raising the Minimum Support Price to cover the rising costs of production and to incentivize farmers to increase production. The existing food security complex of procurement, stocking and distribution would further increase the operational

expenditure of the Act given its creaking infrastructure, leakages and inefficient governance. This entire setup can be replaced by food coupons/vouchers or cash transfers as is done in Brazil[37] through the Bolsa Familia programme. It is imperative that India learns from these social safety net experiences and evolves an innovative strategy that is based on more effective and appropriate policy instruments to enhance social and economic welfare.

- The Act gives a legal sanction to a highly centralized procurement and distribution model. All guidelines and rules will be prescribed by the Centre including criteria for priority households, exclusion criteria, reforms in TPDS, price at which the State Government is required to sell the food grains to the entitled persons. It leaves no room for experimentation/customization for the States suited to their specific choices, institutional strengths and weakness. Section 40(2)[38] of the NFSA allows State Governments to design their own schemes but it is rendered practically ineffective as it essentially imposes an obligation on the State Government to procure food grains from Food Corporation of India only for TPDS instead the state should be free to buy from the open market by calling for tenders depending on the local needs as food habits are varied in our country.
- Section 19(1)[39] of the NFSA deals with Women Empowerment. In the opinion of the legislators this provision goes a long way in the empowerment of women. If women empowerment can be a reality by merely compelling the ration cards to stand in the name of the women, then the Researcher opines that the state can mandate all documents to stand in the name of women. This provision is just a farce and far from reality as we live in a patriarchal society.

**2.6.4 Challenges in the Implementation of NFSA**

The provisions of NFSA are being implemented with its own set of authorities in co-ordination with the public distribution system. But implementation has not been

[37] The Bolsa Familia programme, world's largest conditional cash transfer program, has lifted more than 20 million Brazilians out of acute poverty and also promotes education & health care.
[38] Section 40(2) Notwithstanding anything contained in this Act, the state government may continue with or formulate food or nutrition based plans or schemes providing for benefits higher than the benefit provided under this Act, from its own resources
[39] Section 19(1) the eldest woman who is not less than 18 years of age, in every eligible household, shall be the head of the household for the purpose of issue of ration cards.

satisfactory owing to various inevitable circumstances. In the operation of the Act, the following challenges are encountered-

- The large-scale subsidized grain distribution to almost two-thirds of the country's population of 1.2 billion implies massive procurement of food grains and a very large distribution network entailing a huge financial burden on the already burdened fiscal system.
- NFSA aims at improving the nutritional status of the population especially of women and children. But malnutrition is a multi-dimensional problem and needs a multi-pronged approach. Women's education, access to clean drinking water, availability of hygienic sanitation facilities are the prime prerequisites for improved nutrition
- The implementation of the Act requires additional storage capacity. The cost of procurement by Food Corporation of India has been increasing steadily with rising procurement levels. Currently, the economic cost of FCI for acquiring, storing and distributing food grains is about 40 percent of the procurement price which will add to the cost of the project.
- Periodic overstocking by the public sector has huge implications on the fiscal side, apart from distorting the free functioning of food grain market. Higher level of buffer stock carries the risk of higher wastage of food grains along with higher cost of maintaining the buffer owing to the ill equipped godowns of Food Corporation of India. Overstocking only results in rotting and wastage of food grains.
- All guidelines and rules will be prescribed by the Centre including criteria for priority households, exclusion criteria, reforms in TPDS, price at which the State Government is required to sell the food grains leaving no room for experimentation or customization for the concerned states.
- The NFSA provides Force Majeure clause exempting the state from its obligation of supplying food grains in case of floods and natural calamities. But, it is pertinent to note that in these conditions of failure of market forces, volatility in prices and distress the poor and vulnerable would depend on government to ensure their food security. "The Act would also need to include provisions that would come into effect when faced with emergencies and

disasters such as floods, earthquakes, riots"[40]. "Does not the right to life under Article 21 of the Constitution include the right to food? Does not the right to food, which has been upheld by the apex court, imply that the state has a duty to provide food especially in the situations of drought to people who are drought-affected and are not in a position to purchase food"[41].

- NFSA has tried giving the existing PDS and procurement system a new lease of life in an "as is where is" condition despite its established chequered history of failure and leakages. The estimated leakages from the TPDS go as high as 40 percent[42]. The existing system of TPDS needs to be reformed for efficient delivery of food grains but the norms and types of reforms are to be decided by the Central Government. Before the implementation of the Act, in-limine the state should have updated the godowns. Currently, FCI is facing an acute storage crisis with respect to the infrastructure to store the food grains as we have witnessed that FCI godowns are feeding the rodents more than feeding the hungry citizenry. The additional procurement as a result of NFSA will put enormous pressure on the existing infrastructure which is grossly inadequate to handle the current procurement norms. The modern silo storage and bulk handling is required for preservation of quality of food grains which at present not all FCI godowns have.
- "A food credit card system could be a superior alternative to the prevalent system of specialized fair price shops and perhaps even to a food stamp system. To buy subsidised food grains from the market, the customers could use food credit/debit cards over which retailers can claim the subsidy from the government"[43]. Though the issue costs of a food credit card are likely to be higher than for existing ration card, the running costs may be lower than for specialised fair price shops as the credit card can be used in any existing retail shops that accept such cards.

[40] Reethika Khera, "Right to food Act: Beyond cheap promises", Economic & Political weekly, [July 18-24,2009] Vol XLIV No 29, p40-44

[41] The Hindu Online edition of India's National Newspaper, Tuesday, August 21, 2001 available at https://www.thehindu.com/thehindu/2001/08/21/stories/01210001.htm(accessed on 22/8/2018 at 3.16pm)

[42] Ashok Gulati, jyothi Gujral, T Nandakumar in National Food Security Bill-Challenges and Options in Discussion paper no 2 of Commission for Agricultural Costs and Price

[43] Sudipta Bhattacharjee and Hetal Doshi, "An Evaluation of the Indian Food Security Infrastructure Vis – a-Vis the Right to Food", Social Action, [January-December 2004] Vol 54 No4, p 279-291

- India's tryst with TPDS has not been very encouraging. Many studies have shown that implicit subsidy has drained the exchequer and targeting effectiveness is very poor due to rampant corruption and lack of awareness on the part of consumers. Leaving aside the procurement of food grains for TPDS, the selection of BPL households has been fraught with difficulties as the criteria used for identification of BPL is widely debated across the nation. Efforts to revise and fine tune the selection process have not yet yielded satisfactory results. There have been lot of arbitrariness and manipulations in the implementation of BPL surveys. Even if the perfect selection criteria were implemented genuinely, it would still fall short of accuracy. This is because poverty is not a static phenomenon. Mainly in rural areas, household can fall into and come out of poverty any time.
- On the other hand BPL surveys are conducted only once in 5 years and the status of the households are fixed as BPL or APL. This results in TPDS suffering from large inclusion and exclusion errors as the case may be, whereby the BPL cards may be held by non poor and many poor households may have APL cards. As Amartya Sen, Indian economist and philosopher aptly remarked, 'benefits for the poor remain poor benefits'. "A flagship government food subsidy scheme is failing and millions in India remain hungry despite years of economic boom, a UN report showed. The Targeted Public Distribution System, meant to sell food essentials to India's poorest people at subsidized prices, has excluded large numbers because of poor data and lack of adequate definitions of hunger, the report said. About 20 per cent of the world's 1 billion hungry poor live in India. Apart from failing to serve the intended goal of reduction of food insecurity, the TPDS also led to greater food insecurity for large sections of the poor and the near-poor,"[44].
- "For a few years now the decay of public distribution has been in political debate in Parliament and in public without any serious measure to reverse it"45. "The functioning of the public distribution system in India has come under scrutiny because of rising burden of subsidy and storage cost and meagre

---

[44] Report by U N World Food Programme "Indian food policy failing, millions hungry :UN" dated 1st April 2009 available at https://www.wfp.org/news/hunger-in-the-news?page=4&tid=211 (accessed on 10/8/2018 at 8.50pm)

[45] Kamala Prasad, "Politics of Food" Mainstream, [July5, 2008] Vol XLVI Issue no 29, p121-132

coverage of the poor and the actual benefits received by them"46. "One of the major impediments for functional public distribution system is the lack of effective grievance redressal mechanisms. There is no satisfactory feedback mechanism to monitor whether and how much reaches the ration card holder after the grain is delivered to the ration dealer"[47].

### 2.7 Live Instances of People Dying of Starvation Even After NFSA

All these shortcomings of NFSA on one hand and the instances of people still dying of starvation on the other hand, stares boldly in our face. Despite NFSA, on September 28th, 2017 an 11-year-old girl Santoshi Kumar, in Simdega district of Jharkhand, died because of starving for nearly 8 days[48]. Her family had not received any food grains for months as their ration card was cancelled after being struck off the public distribution system for not linking it with Aadhaar. The system of Aadhaar has been under continuous criticism for 'depriving the most vulnerable people of their grain entitlements'. A set of public interest litigations has been decided by the Supreme Court urging the government from mandating the compulsory possession of Aadhaar.

Ashamedly, there were reports[49] about three sisters aged between two and eight years having starved to death in east Delhi's Mandavali, barely 12 km from Parliament house. Is this incident not a testimony to the fact that the benefits of NFSA has not trickled to the really needy. The post mortem report[50] of these girls revealed that they may not have eaten anything for three days since there were no traces of food or water in their stomachs. The officials tried to hush up the matter saying that they were natural deaths but the post mortem report let the cat out of the bag. The father of these girls was a rickshaw puller and had gone out for work. It was reported that the father was out of work for the past two weeks because his rickshaw was stolen. The family did not have ration card as they were migrants from Bengal but living in Delhi for many years. This incident only led to a war of words between the ruling and the opposition party, opposition cornering the ruling party. All of us need to hang our head in shame. We

---

[46] V M Rao and R S Deshpande, "Food Security in Drought prone Areas – A Study in Karnataka", Economic and Political Weekly, [June 29 2002] Vol XXXVII No 26, p3677-3681

[47] Reethika Khera, "Right to food Act: Beyond cheap promises", Economic & Political weekly, [July 18-24,2009] Vol XLIV No 29, p40-44

[48] ANURAG BHASKAR "Death By Starvation, Right To Food & Indian Democracy" available at: http://www.livelaw.in/death-starvation-right-to-food-indian-democracy/(accessed on 11/8/2018 at 11.00pm)

[49] "3 sisters, aged between 2 and 8, die of starvation" Deccan Herald, Bangalore Edition, dated 26/7/2017

[50] "No trace of food in girl's stomach" Deccan Herald, Bangalore Edition, dated 27/7/2017

speak about availability, access and right to food, the Researcher thinks that, in the light of these incidents all these concepts are dead in spirit and have lost their meaning. It is unfortunate that these incidents will be hushed up by the government and in turn media will encash the situation. The public memory is very weak and short lived, we tend to forget things unless we bear the brunt of it.

In some parts of Bangalore, malnourished children studying in government schools get their free breakfast served by Upahaar- a collaboration between Inner Wheel Club of Bangalore IT corridor and Rotary Bangalore IT corridor[51]. It is reported that these children suffer from stunted growth, have sunken eyes, breathing problems and white patches on their faces- all signs of malnourishment. The Researcher thinks that, this report again holds testimony to the inefficacy of NFSA. Only when the State has failed in its duty of providing safe and standard mid-day meals under NFSA in this government school, the private individuals have donned the role of saviours. What disturbs the Researcher is that, how can the authorities turn a blind eye and shirk away from their obligation.

These instances are not isolated incidents. They have been the consequences of manmade starvation, caused by the negligence and mis-governance by the authorities. The godowns of the Food Corporation of India are brimming with enough stock of food grain to meet the requirement of Public Distribution System and guaranteeing strategic reserve despite drought situation. Despite being one of the world's largest grain producers and its self-sufficiency in food availability, about one-fourth of Indians go to bed without food. Lakhs of tons of food grain have often been wasted away because there are not enough warehouses for storage. Being a food surplus country, India has not been able to curb food grain wastage on one hand and hunger & malnutrition on the other. Enormous funds are invested in various poverty alleviation schemes, yet the benefits don't reach the needy. Being a welfare state, the responsibility and accountability of superior and subordinate authorities must be fixed and be made answerable. Only then this basic natural and socio–economic justice would be available to the needy which is one of the most essential promissory provision of our Preamble.

## 2.8 History of Emergence of Adulteration of Food and the Laws Relating to it

In the history of mankind, struggle for food has always been an unpleasant part and consumed much of everybody's time, energy and efforts. In common parlance, the

[51] "Malnourished kids in govt. schools get free breakfast", Bangalore Edition, dated 3/8/2018

word denotes anything edible by a human being and fit for consumption, which indirectly connotes food that is safe for human consumption. All of us are that what we eat should act as building material for a healthy and robust body. Thus the safety and quality of food is crucial for healthy body & mind and life of a consumer. Owing to the advancement in science and technology and the advent of the modern & fast living, new kinds of articles of food which were unknown in the past, were added to the palate of a man. Hence, it became tough for a common man to identify the nature, safety and quality of the food that one consumed. This development gave rise to the modern states to come up with new definitions with renewed standards for the term 'food' through legislations.

Due to industrialisation and thus ensuing urbanisation, majority of the people no longer live within the easy reach of sources of food. Consequently, traditional modes of preserving, processing and packing of food became impossible and thus preservatives and anti microbial agents were used in the preparation/manufacture and preservation of articles of food. An article of food may be nutritious but might fail to look attractive and appealing to the eyes of a consumer. To make it look appealing, the traders began to use food colour which would enhance their colour and to render better taste, flavouring agents were added generously. Whatever began as a taste enhancer or making the food more appealing to eyes, over period of time gave rise to rampant and unscrupulous use and rendered the articles of food unsafe for human consumption.

Mixing the article of lower quality with that of good quality has been the mode of adulteration from ancient times. The vice of adulteration is not of recent origin but, dates back to ancient India. May be the ways to do it and substances empolyed to adulterate may have varied from time to time. In ancient India, adulteration was only a matter of deception, but, with the passage of time this deception has been constantly resorted to conceal the flaws of the article of food and paved the way for adulteration. The Researcher thinks that the history of adulteration is inextricably mixed with the history of deception and the history of deception is as old as the history of mankind because of which it is not possible to separate them from each other. Age old means of adulteration has been replaced by new methods and new articles, thanks to the development in science especially chemistry.

Fraudulent and devious means of adulteration were prohibited even by earlier codes. While passing the Hebrew laws[52], Moses[53] declared "You shall not steal, nor deal falsely, nor lie to one another". This prohibition which is in the nature of a negative command, treats false dealers as thieves. He thinks thieves rob people in the night whereas the traders indulge in day light robbery through deceitful means. When Moses propounded this, he had not foreseen adulteration in quality and he thought only about cheating in terms of quality. Initially adulteration was present only in quantity and an attempt was made to curb only that. But, with the passage of time, adulteration both in terms of quality and quantity became rampant which demanded the splitting of adulteration into various components.

While, adulteration involved the wilful concealment of defects of the article from the purchaser, an effort was made to compel the trader to disclose the blemishes of the article to the intending purchaser which led to openness in the sale. The celebrated Twelve Tables[54] of the Romans, which enjoyed the same repute as that of Magna Carta, infused the rule of openness in sale by specific disclosure of defects in the merchandise. Table IV of the Twelve Table prescribed a penalty of double damage to the vendor if there were to be any defect in the goods which the seller knew but had specifically denied which reads as the following: "It is sufficient to make good such faults as have been named by word of mouth, and that for any flaws which the vendor had expressly denied he shall undergo penalty of double damage"[55]. This casts a liability on the vendor to honestly reveal the defects of the goods which he was aware of. In concealing the truth, the seller commits fraud and is therefore liable to an increased punishment. But this dictate was not clear as to the latent or blatant defect in the goods. But later on, historians dismissed the obligation of revealing the blatant defects as they were apparent and did not need any disclosure.

In tune with this, Greeks also recognised the idea of openness in sale through their Athenian Law, which demanded that every statement made in the course of

---

[52] Body of ancient Hebrew Law codes found in various places in the old testament similar to the earlier codes of Hamurabi.

[53] Moses was a prophet according to the teachings of the Abrahamic religion. The jews were living as slaves in Egypt. Their leader was Moses. He led the jews out of slavery in Egypt.

[54] Set of laws inscribed on 12 bronze tablets created in ancient Rome in 451 and 450 BCE. They were the beginning of a new approach to laws where they would be passed by government and written down so that all citizens might be treated equally before them. Available at https://www.ancient.eu.Twelve.... (Accessed on 24/1/2019 at 2.00pm.)

[55] Lewis, Naphatali & Reinhold, Meyer: *Roman Civilisation: The Republic* Vol 1Columbia University Press, New York (1951)

business transactions should be true[56]. This idea of openness was also followed by Muslims under their personal Law. Since adulteration is the off shoot of fraud and deception, Muslim law prohibited its practice in any form. Muslim law very clearly dictates its people that "whenever you enter into a transaction, there should not any attempt to deceive", which indirectly ordained the traders to reveal the flaws of the goods if any. Muslim law terms the concealment of defects affecting the goods as dallas in Arabic which means nasty. Prophet never allowed the mixing of dates of different quality, this was considered as usury by him. Massive increase in deceitful and fraudulent activities will result in moral derailment of the society at large. Such outcome is unjustifiable under in Islam.

While, it was fraud to conceal the defects of the goods, it was also illegal to sell defective goods without revealing the defects to the intending buyers. The trader was *ipso jure* bound to deliver the article devoid of any defect or reveal the same in expressly or impliedly. When the disclosure about the defects of the goods by the vendor was in express terms, the seller's act amounted to a warranty guaranteeing the quality and wholesomeness of the article. When the assertion of the quality of goods was done impliedly, it would act like an implied warranty of the merchandise.

### 2.8.1 Contribution of Smritikaras for the Growth of Anti – Adulteration Laws

Literature of ancient India which is more ancient than Muslim Law is replete with laws containing adulteration. Smritis[57] played a major role in shaping the anti adulteration laws.

#### 2.8.1.1 Manusmriti

Ancient history shows that it was Manu, the first smritikara, i.e the writer of smriti who raised cudgel against adulteration for the first time in his Manusmriti. He segregated the thieves as open thieves and concealed thieves[58]. According to him open thieves are those who deprive the people of their property, whereas, the concealed thieves are those who act secretly in the shadow of darkness and Manu opined that the adulterators are the concealed thieves who cheat the public in broad day light by deception. Against the adulterators Manu ordained that "*One thing should never be sold mixed with another, nor (should anything be sold) damaged, deficient, far away or*

---

[56] Jones, J. Walter: *The Law and Legal Theory of the Greeks,* Clarendon Press, Oxford University (1956)
[57] Smritis means that which is remembered from the precepts of almighty and they are one of the sources of un codified Hindu law.
[58] Muller, F.Max: *The Sacred Books of the East Val.*XXV 1965, Motilal Banarsidas, NewDelhi

*concealed"*[59], which means that the obnoxious practice of mixing one substance with the other has been there from the times of Manu. He also spoke about sale of damaged goods. Another Smritikara Brihaspati also called adulterators as open thieves.

### 2.8.1.2 Yajnavalkya Smriti

Yajnavalkya in his Yajnavalkya smriti threatened adulteration of drugs, oils and fats, salts scents,sugar and paddy with penal consequences[60]. Though another smritikara Narada also spoke against adulteration, he had not rightly appraised the gravity of the evil[61]. But Narada also dealt with the principle of *Caveat Emptor* in his own words, when he stated "*the intending purchaser shall first examine an article(before purchasing it), in order to find out its good and bad qualities. That which has been approved by the purchaser after close examination, cannot be returned to the vendor*"[62]. This doctrine allowed the unscrupulous traders to take shelter under it to cheat the innocent and ignorant consumers on quality, thus, restricting the scope of consumer sovereignty. Narada also prescribed lighter penalty for adulteration.

## 2.8.2 Kautilya's Arthashastra's Dictum on Adulteration of Food

*Kautilya's Arthashastra,* which was written about 321 BC during the regime of Chandra Gupta Maurya of Maurya dynasty also made a major contribution to the administration of laws dealing with adulteration in general and adulteration of articles of food in particular. "*The evil of food adulteration is not only present in the society to a great extent, but its history can be traced back to the times of Kautilya*[63]". According to Kautilya who is also known as Chanakya, the king was not a mere protector of the person and property of his subjects but was the fountain of the welfare of the people. This is apparent from the use of the Sanskrit word *Yogakshema* which means welfare that he had used while describing the duties of the king towards his people. Kautilya had chosen to use the word *Yogakshema* instead of palana or rakshaka which means only the protector. In principle, thus there is no anti thesis between Kautiya's *yogakshema* and the modern public welfare doctrine[64], while modern states had taken centuries to

---

[59] *ibid*
[60] Thakur, Upendra, *Corruption In Ancient India* 1st Edn1979, Abhinav Publications, N.Delh)
[61] Jolly Dr. Julius ' *Naradiya Dharmasastra,* 1st edn, 1981, Takshila Hardboullds, New.Delhi
[62] *ibid*

[63] Dhulia, Anubha, "Laws on Food Adulteration: A Critical Study with Special Reference to the Food Safety and Standards Act, 2006", ILI Law Review, (April 30, 2010), p 56-67
[64] Kangle RP, The Kautiya Arthashastra Part III (1965), University of Bombay

conceive the idea of public welfare in their administration. But, Kautilya's *yogakshema* was also severely threatened by anti social elements like deceitful and corrupt merchants for which Kautilya did not budge and he called these anti social elements & adulterators as Kantakas.

The act of suppression, removal and weeding out these anti social elements, he called it as Kantaka Sodhana which in principle had the same effect as modern administration of justice. According to Kautilya, it is the duty of the king or his officials to bring the culprits and adulterators to book. Kautilya in his *Arthashastra* has dealt in detail about the sale of spurious, adulterated commodities in the market and has prescribed the fines for all these wrongs, what catches the attention of the Researcher is that he has dealt with the adulteration of the salt which is a raw material for any food prepararion. He has prescribed punishment for unfair trade practices by the merchants like misrepresenting the goods as belonging to a particular region or bears a particular quality or bears a special property.

Adulteration of salt shall be punished with the highest amercement. Middlemen who cause to a merchant or a purchaser the loss of one eighth of a pana by substituting with tricks of hand, false weights or measures or other kinds of inferior articles, shall be punished with a fine of 200 panas. Fines for, greater losses shall be proportionally increased commencing from 200 panas. Kautilya declared that "*Adulteration of grains, oils, alkalies, salts, scents, and medical articles with similar articles of no quality shall be punished with fine of 12 panas*"[65]. Kautilya not only dealt with the quality of the goods and the market reforms but he also spoke voluminously about the consumer interest. He further ordained "*When a trader sells or mortgages inferior as superior commodities, articles of some other locality as the produce of particular locality, adulterated things, or deceitful mixtures or when dextrously substitutes other articles for those just sold (samuta parivartinam) he shall not only be punished with fine of 54 panas but also be compelled to make good the loss*"[66].

When a trader sells articles of some other locality, as the produce of a particular locality which in modern times we call it as geographical indication of goods, adulterated things, or deceitful mixtures, or when he dexterously substitutes other articles for those just sold, he shall not only be punished but also be compelled to make

---

[65] R.Shamasastry *Kautilya's Arthasastra* (ed), Mysore Printing & Publishing House, Mysore(7th edn,1961)
[66] *ibid*

good the loss. A crucial examination of the Kautilyan code as propounded in his Arthashastra unveils a well formulated law on adulteration of goods in general articles of food in particular. We can break down his idea of adulteration into many components and one of them is sale of goods of inferior quality with false description of being of good quality. This is an widely practiced age old unfair trade practice among unscrupulous traders to market their merchandise of inferior quality with enticing and alluring descriptions.

Another component of the Kautilyan code is the mode of deceiving the consumer by misrepresenting that the goods are sourced from a particular locality, but not in reality. It is natural that some goods are of good quality owing to the natural conditions that are prevailing in the place of origin compared to similar goods from other localities. It is common knowledge that people who are aware of the fact that the natural conditions prevailing in a locality makes a lot difference to the quality of the goods will not buy the same goods from a different place which in essence amounts misbranding. The traders would misbrand the product as products of accredited locality. The third component that can be inferred from the Kautilyan code is that deceitful mixtures which could trick the people to buy these mixtures presuming it to be what they wanted to buy but, truly they were not what they seemed to be. Such mixtures were made either by adding or substituting the foreign ingredients or with the subtraction of essential ingredients. More often, thus added or substituted foreign ingredients were injurious to the health of the consumers. Rashastrakutas attached lot of importance for purity of food[67].

On an elaborate study of our ancient history dealing with adulteration and consumer interest, it is clear that these laws formed a foundation for the laws to develop further. These laws were based on the premise of fraud and deception. In course of time many significant allied concepts were developed from them and later on the struggle to wipe out the menace of adulteration was centred on openness in sale/transaction. This is how the law regulating the evil of adulteration blossomed in ancient India.

### 2.8.3 Laws Regulating Adulteration of Food in the Medieval India

In the Medieval India Alauddin Khilji's Market reforms were well known. Rulers of Medieval India believed that necessities of life, especially food grains should be available to the folk at reasonable prices thus ensuring food security to all its

---

[67] V D Mahajan, '*Ancient India*', S. Chand & Company Ltd, New Delhi, (13th edn, 2000)p 234

citizenry. Allaudin fixed the market prices for all the commodities including the food grains and set up a separate market for the food grains. He controlled not only the supply of food grains from the producers and villages and its transportation to the city by the grain merchants but also its proper distribution to the citizens thus preventing the hoarding of the essential commodities. An officer-Shehna was in charge of the market to see that no one violates the royal orders.

Barids who were intelligence officers and Munhiyan who were secret spies were appointed. Allauddin also tried to ensure that there were sufficient stocks of food grains with the government so that the traders did not hike up prices by creating artificial scarcity or indulge in profiteering. All the food grains were to be brought to the market-Mandis and sold only at official prices[68]. The emperor Ashoka, the Great is also known for market reforms. He took measures to cut down the middlemen in trade. Took extensive measures to make sure that lot of food grains are available in the market and no scarcity of the same. Any trader selling the adulterated food or anything meant for human consumption was spurious he was punished severely.

The laws relating to adulteration and consumer interests are one part of the right to safe food. The other part is the paucity of food grains after the advent of British rule in India. "Hunger and malnutrition is not a new affliction and they have been persistent features of human history"[69]. Though, India had a rich history in terms of agricultural abundance, it was during the period of British administration that agriculture started declining in most parts of the country and the country's per capita availability of food decreased to about 200 kilo gram/year, which is sufficient only to maintain the bare physical existence which the British administrators considered to be the minimal requirement for staving off droughts. India reached the state of almost near famine within few decades of British administration and we have remained so in that state ever since. After being liberated from the clutches of British about 7 decades ago, efforts were made to put agriculture in the front burner, yet per capita availability of food or articles of food has almost remained unchanged.

[68] http://www.universityofcalicut.info/SDE/BA_his_medieval_india_society.pdf (accessed on 12/2/2017 at 6.24pm)

[69] Dr Vishwanath. M, "Right to life in devoid of Right to food: Is it governance? In search of answers", Indian Bar Review, [October-December 2011],Vol XXXVIII(4), p135-160

Consequently, a large proportion of Indian people have continued to remain hungry and malnourished[70]. Today, for a population of 900 million, India produces around 180 million tons of food grains which implies an average of 200kg per capita/year. Out of this, a portion of it has to be reserved for seeds, a small portion of it goes waste which is inevitable, and another portion goes towards fodder for the animals. After all these deductions, it is estimated that the quantity of food grains available for human consumption in 1990 was only around 180kg per capita/year which was less than what had been estimated to be the bare minimum by the Famine Commission appointed by the British administrators in 1880.

Hence, in reality the average consumption of staple food in India is below the ordinary standards of the world by one third. Apart from India, there are only a few countries where the average consumption of staple food is at this level. The statistics and the day today images that we witness on the streets is a testimony to the fact that hunger is stalking India.

### 2.8.4 Advent of Prevention of Food Adulteration Act, 1954 (PFAA)

In India, as early as 1919, when the portfolio of health was transferred to the provincial governments, most of the provincial authorities made special provisions in municipal acts for the prevention of food adulterations. There was, however, little uniformity either in the field of standards or in the mode of enforcement. To reconcile the divergent laws of the various provinces and to fix uniform standards of purity of food articles, the Department of Health of the Government of India in 1937 set up a Central Advisory Board of Health.

The Board appointed a Committee to go into the question of food adulteration in the country, with particular reference to the varying food standards and legislations then in force. This Committee was the Food Adulteration Committee. In pursuance of the recommendations made by this Committee, a Central Committee for Food Standards was formed in 1941, under the aegis of the Central Ministry of Health. This Committee functioned as an advisory body along the lines of the Society of Public Analysts in the UK. After India's independence in 1947, more serious thought was given to the problem of food adulteration. It was soon realized that provincial food Acts were not only outdated for India's purpose, but they also hampered trade and industry. To ensure the

[70] J. K. Bajaj (edited) "Food For All", Centre for Policy Studies, Chennai, (1st ed 2001) Available at http://www.indiatogether.org/manushi/issue92/bajaj.htm (accessed on 11/8/2018 at 2.30pm)

purity of articles of food sold throughout the country, the central government enacted the Prevention of Food Adulteration Act of 1954.

## 2.9 Conclusion

The grave problem that needs the attention of the legislators is the conundrum of the evil practice of adulteration which can be traced to antiquity and so is the legal remedy that has been in place. With the passage of time one remedy was replaced by another remedy but without there being any change in the incidence of the problem. In fact as on today, the mammoth evil of adulteration has grown by leaps and bounds in different forms and has become overwhelmingly insurmountable, invincible and undetectable even by a man of prudence. In fact the history of food adulteration has been a history of legislations. Going back in time, Kautiya's Arthashatra contained rules to fight out this evil.

During the British Indian period, the Indian Penal Code also contained provisions to combat food adulteration. Thereafter, from 1912, the individual state laws began to impose strict liability. Consequently came the central legislation, Prevention of Food Adulteration Act, 1954 and the Rules under it and then a number of orders relating to fruits, meat, and vegetable oil. In 1986, Consumer protection Act, 1986 was enacted which provided the remedy in the form of compensation.

Now, finally Food Safety and Standards Act, 2006, subsumed eight different legislations and has come to stay to curb the menace of adulteration of food & the availability of unsafe food in the market and to fix scientific standards for food. National Food Security Act, 2013 has been promulgated by the central government to give effect to the Right to food of the general public and Food Safety and Standards Act, 2006 has been enacted to save the people from unsafe food. These two are the major food laws that are operating in our country today. Though, we have rich culture of making the food safe under ancient Indian scriptures, today making the food safe is a perennial challenge faced by the authorities.

Tthe Right to food has been recognised in the form of NFSA, 2013. In the modern world, food safety is a scientific discipline describing handling, preparation and storage of food in ways that prevent food borne illness[71], which requires a number of routines to be followed to prevent the potentially severe health hazards. The food safety

---

[71] Seth & Capoor's Commentary on '*The Food Safety and Standards Act, 2006 with Rules 2011*', Delhi Law House, Vol1(9th edn, 2017), p 21

more than often overlaps the food defence to avoid the damage that may be caused to the consumers. This line of thought makes the food safe between industry and the market and the market and the consumer. In dealing with industry and market practices, food safety considerations comprise the origins of food involving the practices pertaining to labelling, hygiene, residues of pesticides, food additives and also policies on biotechnology and food and requisites for the management of import and export inspection by the government.

When we consider market to consumer practices, the standard requirement is that the food will have to be safe in the market stressing on the need to make safe production and delivery of the same to the consumer. Food is not only a medium of growth for bacteria, which can turn the food to poison but also can transmit disease among the people. When developed countries are known for stipulating and implementing stringent standards for food preparation, in developing countries making the safe food and drinking water available to the common man itself is a challenge faced by the authorities.

# CHAPTER – III

# CONSTITUTIONAL PERSPECTIVE AND JUDICIAL DICTUM ON RIGHT TO FOOD IN INDIA – AN EVALUATION

## 3.1 Introduction

December 10 is being celebrated as the Human Rights Day with pomp and gaiety all over the world but the most basic and inalienable right to life is elusive to a very large section of the people in India. Adam Smith, an economist and a philosopher who authored what is considered as bible of capitalism stoically remarked *"No society can surely be flourishing and happy, of which the far greater part of the members are poor and miserable"*. In '*A Tryst with Destiny Speech*' of India's first Prime Minister, Jawaharlal Nehru emphatically observed that *"It means the ending of poverty and ignorance and disease and inequality of opportunity. The ambition of the greatest man of our generation has been to wipe every tear from every eye"*.

Even after 70 years of independence, about one third of population still lives below poverty line. This is the contribution of the successive governments. They may not have contributed anything towards the value addition to the nation but certainly they have contributed their might in increasing the number of people who fall below poverty line. Soon after the making of our constitution, our then President Rajendra Prasad gave a pledge saying that *"To all we give assurance that it will be our endeavour to end poverty and squalor and its companions, hunger and disease; to abolish distinctions and exploitation and to ensure decent conditions of living. We are embarking on a great task"*[1].

But successive governments have successfully failed to redeem this pledge. The principal need to secure to the people the right to life had been felt and articulated long back by Swami Vivekananda which he reflected in his popular address to the Parliament of Religions on September 20, 1893 at Chicago and said *"But it is bread that the suffering millions of burning India cry out for with parched throats. They ask us for bread but we give them stones. It is an insult to a starving man to teach him metaphysics"*. In another occasion he expressed his anguish by saying, *"my heart aches to think of the conditions of the poor and the low in India. So long as millions live in hunger and ignorance, I hold everyone traitor who has been educated at their expense*

---

[1] B N Arora, "Human Rights: Right to Food Eluding Millions in India", Mainstream, New Delhi, [December 13, 2003] Vol XLI No 52, p8-10.

*and pays not the least heed to them. No amount of politics can be of any avail until the masses of India are well fed, well educated and well cared for*"[2].

The Researcher thinks that 'what poverty is to violence, hunger is to crime against humanity' and 'mere schemes without implementation is of no use'. The Researcher is also convinced that the Constitution has ample provisions to take justice to the door step of the poor[3]. Indisputably, widespread poverty is the off shoot of hunger, which is the negation of the right to life, despite the rhetoric of successive governments and their schemes for poverty alleviation, during the last 70 years, the ground reality about hunger is very grim. Hunger among children is detrimental to the existence of any society, for it stunts the mental as well as the physical development of the future of the nation, thus perpetuating the vicious circle of poverty. The UNDP Human Development Report released on 24th April, 2003 stated that India is home to the largest number of hungry people and India has come third among the poor countries even among the poverty stricken South Asian countries, after Maldives and Sri lanka. On 9th January, 2003 at a public hearing held at Delhi University, Nobel laureate Amartya Sen, endorsed this fact and said that India had more endemic hunger and regular under nourishment than any part of the world. A joint study by UN World Food Programme and MS Swaminathan Research Foundation[4] revealed that in India, about one third of people starve for want of food and some 160 million go without a square meal, while granaries overflow.

"Living a life with dignity should not be considered to be a dream but a reality, but in many major developing and underdeveloped countries it is still considered to be a dream for many families. Human right of an individual is compromised on many grounds and food is one such ground depriving a person of his human right"[5]. It has been propounded globally that Right to Food is a basic Human Right. However, the obligations to protect and employ the same in the municipal sphere are primarily

---

[2] *Ibid*

[3] Article 32(3) of the Constitution of India states that "Without prejudice to the powers conferred upon the Supreme Court by clause (1) and (2) Parliament may by law empower any other court to exercise within the local limits of its jurisdiction all or any of the powers exercisable by the Supreme Court under clause (2)". Therefore, the Parliament can by adopting appropriate legislation bring justiciability of the right to food at the village level by empowering the Munsiff Courts or the Nyaya Panchayat.

[4] 'Report on the state of food insecurity in urban India' available at https:// www. wfp. org/ sites/ default/files/Report%20on%20Food%20Insecurity%20in%20Urban%20India.pdf (accessed on 10/8/2018 at 8.55pm).

[5] Anubhav Pandey, "People's Union For Civil Liberty Vs Union Of India – Right to Food" available at https://blog.ipleaders.in/peoples-union-for-civil-liberty-vs-union-of-india/ (accessed on 13/8/2018 at 1.24pm).

domestic, in the relationships between respective states and their own people where the major obligations of national governments are towards the people living under their jurisdictions. It is only the Constitution that plays a fundamental role in the realization of the Right to food as it is the supreme law of the land and the source of all political power within a nation. The Constitution of India commands justice, liberty, equality and fraternity as supreme virtues/values to escort in the egalitarian social, economic and political democracy. Social justice, equality and dignity of a person are the major cornerstones of democracy. Hunger is incompatible with democracy. The Constitution is not a mere document faltering in the still waters of jurisprudence, but a means to a larger socio-economic end. "*The Constitution should not remain merely as a document of admiration. It must truly serve as an effective, meaningful and purposeful instrument of social transformation to bring cheer and human dignity to all citizens in particular to the excluded, deprived, downtrodden and vulnerable sections of the people so as to integrate them in the main stream. In other words 'we the people'......without any exclusion, should really feel that they live dignified lives and get justice as stated in the Preamble of the Constitution*"[6].

The fundamental rights and directive principles are not ends in itself but are the means which enable us to achieve a society characterised by liberty and freedom. Hence, the Constitution cannot be subjected to narrow juristic limitations but must be interpreted in a broader socio economic context. The Constitutionality of every law and act of government is one of the most important political principles of democracies and universally accepted rule of law norms. The logical consequence of the superiority of the Constitution is that it supersedes all acts of the legislature contrary to it. Consequently, such acts will not bind either the courts or the citizens. Constitutional provisions are also binding for the executive, so all administrative authorities are equally limited by its provisions. Any executive or administrative act that contravenes the provisions of the Constitution must be considered void and the courts must invalidate it. 'A constitutional right cannot be thwarted by any concession of counsel'[7]. The existing system of guaranteeing and implementing the right to food for the welfare of the people will have to be examined under the three inter related frame works namely legal, administrative and enforcement mechanism. "*Owing to the General comment*

[6] P .Ishwara Bhat, *Law and Social Transformation*, Eastern Book Company, Luknow( 2009) p 176
[7] *Election Commission of India* v. *St.Mary's School,* (2008) 2 SCC 390

*No12*[8]*, it is no longer possible for states to evade the responsibility for failure to provide for adequate food and in fact, there has been a positive obligation cast ensuring the adoption of national strategy to ensure food and nutrition security for all, which is based on human rights principles*"[9]. "*In recent years, the battle against hunger has been placed at the centre of the development discourse in India. This has come about mainly due to the efforts of the Right to food campaign and as a direct result of a writ petition filed in the Supreme Court*"[10]. The realisation of basic right to food should be at the centre of the country's overall development programme, but if the situation is such that the state cannot afford food, there can hardly be any option of judicial or other recourse. Safe food is one of the most significant components of right to food. From production to consumption, it is the bounden responsibility of national governments, consumers themselves and food industry to ensure that the food is safe.

### 3.2 Jurisprudential Analysis of Right to Food

The doctrine of *Ubi jus ibi remedium* means, where there is a right, there is a remedy. In order to study if right to food is justiciable, we need to study different kinds of Rights. Rights are of two types, substantive and procedural. Recognition of a right provides the substantive basis and access provides the procedure that leads to full realisation of the right. '*The right to food is not about charity, but about ensuring that all people have the capacity to feed themselves in dignity*'[11]. It is a basic human right which belongs to everybody irrespective of one's citizenship, nationality, race, ethnicity, language, political affiliation. '*No right has a meaning or value once starvation strikes. It is an ultimate deprivation of rights, for without food, life ends, and rights are of value only for the living*'[12].

From time immemorial, traditionally, the basic needs of a man are thought to be food, shelter and clothing. When this was propounded, there was no concept of right to food or shelter or clothing. Our forefathers must have thought that they are basic and shall be provided. The right to safe food as an enforceable claim to a minimum quantity of food of a certain quality, carries with it correlated duties on the part of others,

---

[8] Right to adequate food.
[9] Eide.Asbjourn, "The Right to Food: From Vision to Substance", Oxford University Press.
[10] Priyanka Rathi and Vinayak Mishra, "Right to food and development", All India Reporter, Punjab and Haryana Section[2008], Vol 95, p 61-64.
[11] Dr Manish Kumar Chaubey, "Right to Food in India", Civil and Military law Journal[January-March 2012] Vol 48 No 1, p 277-284.
[12] GOROVITZ , a Professor at Syracuse University, New York, United States.

particularly the state. Despite great rhetoric various governments both at the centre and states and various schemes for poverty alleviation during the last 70 years, the scourge of hunger and malnutrition pervades, haunting the large sections of people of the country. In 1996, at the World Food Summit, the countries took upon the responsibility of halving the hungry from 800 million in 1995. But, the set target is yet to be achieved though we are way past the dead line. The right to food obligates the states to ensure equitable distribution of food supplies. This right also implies the right to live with dignity and the people have natural right to the basic minimum quantity of food necessary for survival, which includes the quality food as well.

### 3.3 Nature of the Right to Food

In order to study the right to food in a proper perspective, at the outset, we need to examine the nature of this right. Its nature can be categorised into the following forms:

#### 3.3.1Right to Food is a Natural Right.

"*The natural right proponents say that natural right is a right that rests ultimately neither on law nor on the good of the society but on the nature of man*"[13]. Natural rights are sometimes construed as rights that were based on and which a man possessed as an individual before they entered into a state of political society. They belonged to a man in a state of nature before the advent of civilization. But TH Green[14] clarifies by stating that it is not that they actually existed when a man was born or have been as long as the human race, but that they are necessary for and arise out of a moral capacity without which a man could not be a man. Thus it can be inferred that adopting either argument, right to food becomes a natural right. They are certain inalienable rights. The American Declaration of Independence stresses '*it is self evident truth that all men are endowed by their creator with certain inalienable rights and among these are life, liberty and pursuit of happiness*'.

#### 3.3.2 Right to Food Is a Justiciable Right

The concept of right to food has traversed a long way from being non-justiciable to the justiciable one, though our ancient scriptures dealt with the aspect of food as a basic requirement to one's life. The German philosopher *Immanuel Kant* defined justiciability as the power to award to each person that which is due to him under the

---

[13] Sheeba Pillai, "Right to Safe food: Laws and Remedies", The Banaras Law journal, [2012], Vol 41, p 119-135.

[14] Thomas Hill Green (7 April 1836 – 15 March 1882) was an English philosopher, political radical and temperance reformer, and a member of the British idealism movement.

law. In the light of this definition, the right to food becomes justiciable only if it is legally recognised in the country. Subsequently in the modern times for the first time, right to food was expressly guaranteed as basic human right under Article 25 of the Universal Declaration of Human Rights in 1948. As it was only a declaration, this right was not enforceable. Our Indian Constitution also does not make a specific reference to right to food expressly which consequently rendered the right to food a non justiciable one. But, thanks to the judicial activism in the second half of the 19th century, where the judiciary donned the role of a saviour and brought relief to almost one third of malnourished and impoverished population of India. Though, Right to food is the outcome of judicial pronouncements, yet a justiciable one as on today. Heads of different states and governments, at the invitation of FAO assembled at Rome for the World Food Summit on 13th November, 1996 and reaffirmed the right of everyone to have access to safe and nutritious food consistent with the right to adequate food and the fundamental right of everyone to be from hunger.

### 3.3.3 Right to Food is a Socio – Economic Right

Right to food is one of the basic socio economic rights, the fulfilment of which is important for the accomplishment of economic democracy and in turn social democracy, without which the Indian political democracy will become hollow. Dr Ambedkar was prophetic in his final speech delivered to the Constituent Assembly on 25th November, 1949 where he had clearly explained the essence of the Indian democracy in the following words:...

"*On the 26th of January 1950, we are going to enter into a life of contradictions. In politics we will have equality and in social and economic life we will have inequality. In politics we will be recognising the principle of one man one vote and one vote one value. In our social and economic life, we shall, by reason of our social and economic structure, continue to deny the principle of one man one value. How long shall we continue to live this life of contradictions? How long shall we continue to deny equality in our social and economic life? If we continue to deny it for long, we will do so only by putting our political democracy in peril. We must remove this contradiction at the earliest possible moment or else those who suffer from inequality will blow up the structure of political democracy which this Assembly has laboriously built up.*"[15]

---

[15] Available at http://www.livelaw.in/death-starvation-right-to-food-indian-democracy/...(accessed on 18/6/2018 at 7.00am).

The above speech by Dr B R Ambedkar, portrays his concern for the gap between the rich and the poor and how the socio economic rights have been deprived to the people. He has also expressed how such inequalities breed contempt and how the disgruntled can blow up the political democracy. Hence he urged for the bridging of the gap between the haves and have nots. The question arises, only if the courts can enforce socio economic rights. To analyse this predicament of the Researcher, we need to differentiate between socio economic rights and civil & political rights. Civil & political rights are negative rights, which embody the obligations of the state to not to interfere with the enjoyment of certain freedoms. Article 14 guarantees that the state shall not deny to any person 'equality before the law or the equal protection of laws'. Likewise, Article 21 ensures that no person shall be deprived of his life or personal liberty except according to procedure established by law, the effect of these two provisions on the state is negative. But, socio economic rights are inherently positive rights, which are to be implemented by the state. Even the International Covenant on Economic, Social and Cultural Rights, 1966 calls upon all the state parties to ensure the progressive realisation of socio economic rights through appropriate means.

Hence, civil & political rights are negative rights and socio economic rights are positive or programmatic rights which require affirmative state measures and also involves difficulty in the implementation. '*The failure to enact right to food as a legal right as opposed to a social standard or an obligation is often rationalised on the ground that it would be prohibitively expensive to the state*'[16]. The lack of resource is a valid defence for the states but the courts can examine the measures undertaken by the states for realisation of the right in the context of the available resources. As, in India starvation deaths are not due to paucity of resources, the justiciability of the right to food cannot be denied on the ground of lack of resources. In *Kapila Hingorani* v. *State of Biha'*[17], a public spirited person was allowed by the Supreme Court to take up the cause of nearly 250 employees suffering from hunger and starvation due to non payment of their salaries for a long period of time. Unable to suffer and to put up a bold front, many employees committed suicide. The court called upon the state to uphold the constitutional mandate and to discharge its international obligation and observed that economic rights of the people are often denied citing financial stringency and the states

[16] Raghav Gaiha, "Does the Right to Food Matter?", Economic and political weekly, [October 4-10, 2003] Vol XXXVIII No 40, p 4269-427.3

[17] (2003) 6 SCC 1.

take up the plea of helplessness to address/redress the problem of the people. Hence, the Apex Court emphatically observed that incapacity cannot be an excuse for justifying the violation of fundamental right to food. The right to food can be made justiciable by promulgating a suitable legislation, as on today we have National Food Security Act, 2013 which has made the right to food justiciable, though this Act suffers from some of the shortcomings. Independent judiciary is crucial for the justiciability of any right.

### 3.4 Provisions of the Constitution of India Dealing with Right to Food

Since its inception, the United Nations has identified access to adequate food as both individual right and a collective responsibility. The right to food is written into the Constitutions of over 20 countries and about 145 countries have ratified the International Covenant on Economic, Social and Cultural Rights, 1966 which expressly mandates the signatory states to legislate for the right to adequate food. In 1987, the Special Rapporteur, Ashbjorn Eide, to the Economic and Social Council of the United Nations on the right to food gave the following meaning- '*Everyone requires, food which is (a) sufficient, balanced and safe to satisfy the nutritional requirement (b) culturally acceptable (c) accessible in a manner which does not destroy one's dignity as a human being*'[18]. In 2008, Dr. Olivier De Schutter, the special Rapporteur on the right to food stated that the right to food entails '*the right to have regular, permanent and uninterrupted access either directly or by means of financial purchases to quantitatively and qualitatively adequate and sufficient food, corresponding to the cultural tradition of the people to which the consumer belongs and which ensures a physical and mental, individual and collective, fulfilling and dignified life free of fear*'[19].

Constitution of India, being a written Constitution is a fundamental law of the land, provides a sound framework for the right to food as it is one of the basic economic and social rights that is crucial to accomplish and attain 'economic democracy' without which the political democracy cannot be at its best. The Researcher believes that mass hunger is fundamentally incompatible with democracy in any meaningful sense of the term. Though, our Indian Constitution does not deal with the right to food directly or in express words, we do have some fundamental rights and directive principles which, only when interpreted implies the right to food. Fundamental rights are not one time

---

[18] E/CN.4/Sub/1987/23 "The Right to Adequate Food as a Human Right," Report prepared by Asbjorn Eide, Final Report available at http://hrlibrary.umn.edu/edumat/IHRIP/ripple /app/ appendixb.html (accessed on 16/8/2018 at 11.30pm).

[19] A/HRC/&/5, A Report of the special Rapporteur Jean Ziegler on the right to food at http://undocsrg/en/ A/HRC/7/5 (accessed on 16/8/2018 at 11.11pm).

right but continues throughout the life of an individual and directive principles aim at the establishment of a welfare state based on socialistic principles and the following are some of them which deals with right to food.

### 3.4.1 Values Enshrined in the Preamble of the Constitution of India

A careful examination of the Preamble of the Indian Constitution reveals that India is not a police state[20] anymore but a welfare state[21], which strives to achieve the overall development of all its citizens and is based on the principles of equality of opportunity, equitable distribution of wealth, and public responsibility for those unable to avail themselves of the minimal provisions for a good life. The concept of welfare state is further reinforced by the directive principles which enunciate the economic, political and social goals of our Constitution. These directive principles bestow certain rights on the people which are non justiciable and the government is obligated to achieve and maximise social welfare and basic social values, namely, education, health and employment. The Constitution strives to achieve a Socialistic form of society[22] that tries to secure Social justice to all its citizens.

From the statement of objects and reasons of the 42nd Amendment Act 1976, it appears that the word Socialism[23] was inserted to 'spell out expressly the high ideals of Socialism'. What many do not know is that it is not the 42nd amendment of the Constitution which made our Constitution Socialist. The amendment only made explicit what was implicit. The Preamble of the Constitution among other things pledges to ensure social, economic and political justice and equality of status and opportunity and they cannot be enforced except under a socialist management of economic affairs[24]. Social security is a facet of socio-economic justice to the people and a means to livelihood. The primary motto of a welfare state is the attainment of substantial degree of social, economic and political equalities and to accomplish self expression in one's work as a citizen. As viewed by Justice K Krishnaswamy, "*social justice is an integral*

---

[20] A nation in which the police, especially a secret police, summarily suppresses any social, economic, or political act that conflicts with governmental policy. dictionary.reference.com/browse/police+state accessed on 24/12/2014 at 7.00am.

[21] A welfare state is a concept of government in which the state plays a key role in the protection and promotion of the economic and social well-being of its citizens.

[22] Constitution (42nd Amendment) Act 1976.

[23] Socialism is a political and economic theory of social organization which advocates that the means of production, distribution, and exchange should be owned or regulated by the community as a whole. Available.atwww.google.co.in/search?source=hp&ei=ScpzW7ydK5G99QOCnqBo&q=what+is+socialis & oq (accessed on 15/8/2018 at 12.11noon.

[24] Justice P B Sawant, Socialism under the Indian Constitution, Society for Community Organisation Trust, Madhurai (1994) p 4.

*part of justice in the generic sense. Justice is the genus, of which social justice is one of the species. Social justice is a dynamic devise to mitigate the sufferings of the poor, weak, dalits, tribals and deprived sections of the society and to elevate them to the level of equality to live with dignity of person......it is an essential part of complex social change to relieve the poor from handicaps, penury.........to make their life liveable, for the greater good of the society at large*"[25]. Socio economic democracy ought to take strong roots and must be expressed in the very fabric of life.

In 1983, the Constitution bench in *Nakara's case*[26] explained the meaning of 'Socialist' in the preamble referring to the foregoing statement of objects and reasons appended to in these words "*The principal aim of the socialist state is to eliminate inequality in income and status and standard of life. The basic aim of socialism is to provide a decent standard of life to the working class and especially to provide security from cradle to the grave…………*" Thus inclusion of the word "socialist" in the preamble was clearly *to set up a vibrant, throbbing welfare society in the place of a feudal exploited society*[27]. *Democratic Socialism aims to end poverty, ignorance, disease and inequality of opportunity*[28]. The court also propounded that socialism means distributive justice which brings about the allocation of material resources of the community so as to sub serve the common good.

It was held in *Sodan Singh* v *New Delhi Municipal Committee*[29] that greater concern must be shown to improve the condition of population of the country and every effort should be made to allow them as much benefit as may be possible after insertion of words Socialist in the preamble. Addition of the word socialism is to alienate in equalities in income and status and standard of life. The emphasis is on economic equality in a Socialist welfare society. In order to achieve the essence of Socialism[30], each individual must be guaranteed of their basic human Right of Right to Food. Providing safe food becomes the primary obligation of the state which is Socialist in nature.

---

[25] P .Ishwara Bhat, Law and Social Transformation, Eastern Book Company, Luknow ( 2009), p 178.
[26] AIR 1983 SC 130 (para33).
[27] The dominant social system in medieval Europe, in which the nobility held lands from the crown in exchange for military service and lower orders of society held lands from and worked for the nobles. Oxford Dictionary Thesaurus 2001, Indian edition, Oxford University Press.
[28] MP Jain, *Indian Constitutional Law,* Lexis Nexis (7th edn, 2014), p32.
[29] AIR1989 SC1988.
[30] Socialism is a social and economic system characterised by social ownership of the means of production and co-operative management of the economy. Oxford Dictionary Thesaurus 2001, Indian edition, Oxford University Press.

In *Air India Statutory Corporation* v *United labour Union*[31] the Supreme Court also laid emphasis on social justice so as to attain substantial degree of social, economic and political equality and opined that social justice and equality are complementary to each other. In *Municipal corporation* v. *Jan Mohammad*[32], while dealing with Article 19(g) the expression 'in the interest of general public' has been defined as 'is of wide import comprehending public order, public health, public security, morals, economic welfare of the community and the objects mentioned in Part IV of the Constitution'.

In *Peerless General Finance and Investment Co Ltd* v. *Reserve Bank of India*[33], it was held that right to economic empowerment is a fundamental right. Accordingly the Apex court has held that stability of political democracy hinges upon socio economic democracy. In Indian Constitution, certain rights are fundamental for the human beings and are recognised as justiciable. These are treated as basic rights as they are norms of treatment that every human being must enjoy, in order to be able to live a life of dignity and pursue opportunities to realise one's potential.

In *Ahmedabad Municipal Corporation* v. *Nawab Khan Gulab Khan*[34], the judiciary reminded the government of its duty towards the hungry and held that socio economic justice is the goal of the Preamble and the nation state must promote socio economic justice and fulfil the basic human and constitutional rights of the public so as to make their life more meaningful. '*The framers of the Constitution were conscious of the fact that political democracy alone is not enough, hence they were endeavouring to promote the concept of a welfare state by laying down these fundamental principles of social and economic order which fearing the wrath of the people, the legislators and executives could not easily ignore*'[35]. '*The preamble of our Constitution guarantees to all citizens justice, social, economic and political, an equal opportunity to all to develop and thus resulting to lead a dignified and decent life. But we have failed miserably to achieve it that is even to provide a morsel of food, forget other things*'[36].

---

[31] AIR1997 SC 645.
[32] AIR 1986 SC 1205.
[33] AIR 1992 SC 1033.
[34] (1997) 11 SCC 121.
[35] Kamal Nayan Kabra, "Right to Food in National and International Legislations and Instruments", Social Action, [January-December] Vol 54 No 4, p 231-245.
[36] P S Lathwal, "Right to Food: A Human Rights Approach",M.D.U law Journal, [2004],Vol 9, No 1&2, p 99-107.

### 3.4.2 Article 14 of the Constitution of India

Article 14 of the Constitution[37] which demonstrates Right to equality has been declared to be the basic tenet of the Constitution by the Supreme Court. Preamble of the Constitution of India emphasizes principle of equality as basic to the Constitution. Equality before law means that among equals the law should be equal and should be equally administered, that like should be treated alike. Equal protection of the laws is now being read as a positive obligation on the state to ensure equal protection of laws by bringing in necessary social and economic changes. Article 14 though establishes equality before law, yet it can lay reasonable restrictions in the form of Protective discrimination and make reasonable classification by making suitable legislations in order to conform to the Socialistic principle that the Preamble of the Constitution has embodied. In the light of the essence of Article 14, the state can make legislation to provide safe food to a particular class of the society i.e. people who are living below poverty line, thus striving to bring them to the mainstream.

### 3.4.3 Article 19(1) (g) of the Constitution of India

Article 19 (1) (g)[38] *of* the Constitution guarantees Freedom of Trade and Occupation, which means we have the right to carry on, indulge in trade and occupation and the vocation that we desire. But, at the same time, this right is not absolute, but subject to restrictions laid by the state from time to time. At the same time, trade in liquor or other contraband articles cannot be demanded as a matter of right. In *Sodan Singh* v *New Delhi Municipal Committee*[39], *Justice Kuldip Singh* defined the four terms in Article 19(1) (g)[40]. Everyone is endowed with fundamental right to pursue any trade, occupation or profession that one wants, but this right is not absolute and it was ruled in *Municipal Corpn of the City of Ahmedabad* v. *Jan Mohd. Usmanbhai*[41], that the freedom guaranteed under this Article is not unrestrained but authorizes legislation which imposes reasonable restrictions on this Right in the interest of the general public, prescribes professional or technical qualification necessary for carrying on any profession, trade or business and enables the state to carry on any trade or business to

[37] 'Equality before law and equal protection of laws'.
[38] Article 19(1) (g) guarantees that all citizens have the right to practice any profession or to carry on any Occupation that all citizens have the Right to practice any profession or to carry on any occupation or trade or business.
[39] AIR1989 SC1988.
[40] Profession, trade, occupation, business.
[41](1986) 3 SCC 20, 31.

the exclusion of private persons wholly or partially. For long, India has denounced lasseiz faire and has practiced a regulated and planned economy, reasonable restrictions laid by the Constitution on trade, commerce or business have to be considered in the light of this aspect. The right to carry on any trade is very much regulated in India and the courts have hailed social control over private enterprises.

Article 19(5)[42] permitted the imposition of reasonable restriction in the interest of general public and will include public order, public health, morality etc in its ambit. Despite Article 19(1) (g), by and large the government enjoys power to regulate and monitor the economy in the way it pleases within the purview of the Constitution. But the reasonableness of the restriction is to be ascertained in an objective manner from the standpoint of general public interest and not from the interests of the persons on whom the restrictions have been laid. A restriction cannot be termed to be unreasonable, merely because it operates harshly. To ascertain if there is any unfairness involved, the nature of the right, the underlying principle of the restriction imposed, the scope and the exigency of the menace sought to be remedied, the existing condition at the relevant time will be considered for judicial verdict. Hence, the reasonableness of the justifiable expectation needs to be determined with respect to the circumstances pertaining to trade and business in question. Therefore, the restrictions to be reasonable, must not be excessive or arbitrary in nature which involves balancing of private interest against public interest.

In this course, the courts have leaned towards consumers interests. Consequently, while far-reaching restrictions have been laid on trade and commerce, only hardly ever will a restriction be held as unreasonable. In *Sivani* v. *State of Maharashtra*[43], the Supreme Court laid down the standard to assess the reasonableness of the restriction under Article 19(6)[44] and held that the court must consider whether the

---

[42] 'Nothing in sub-clauses (d) and (e) of the said clause shall affect the operation of any existing law in so far as it imposes or prevent the state from making any law imposing, reasonable restrictions on the exercise of any of the rights conferred by the said sub clauses either in the interests of the general public or for the protection of the interests of any scheduled tribe'.

[43] AIR 1995 S C 1770.

[44] Article 19(6) in The Constitution Of India 1949.

(6) Nothing in sub clause (g) of the said clause shall affect the operation of any existing law in so far as it imposes, or prevent the State from making any law imposing, in the interests of the general public, reasonable restrictions on the exercise of the right conferred by the said sub clause, and, in particular, nothing in the said sub clause shall affect the operation of any existing law in so far as it relates to, or prevent the State from making any law relating to,

(i) the professional or technical qualifications necessary for practising any profession or carrying on any occupation, trade or business, or

law has struck a fair and proper balance between social control and the right of the individual on the other. The regulatory measures for better efficiency, conduct and behaviour in the public interest are desired. In the light of this Article, the Right of the 'food business operators[45]' is not absolute but they have to conform to the rules and regulations as set by Food Safety and Standards Act, 2006 with respect to safety and standards of food articles and the FBOS cannot allege that it is an infringement of right guaranteed under 19 (1) (g).

**3.4.4 Article 21 of the Constitution of India**

One who is born into this world has an inherent and inalienable right to life. The right to life guaranteed under Article 21[46] of the constitution was initially interpreted in a narrow, literal and restrictive sense only to mean a protection against arbitrary deprivation of life[47]. But, over a period of time, the expression 'life' in Article 21 has been interpreted by the Supreme Court in myriad of cases liberally. While doing so, the Supreme Court has often quoted the observation of Justice Field in an American case *Munn* v. *Illinois*[48], "*By the term 'life' as here used something more is meant than mere animal existence. The inhibition against its deprivation extends to all those limbs and faculties by which life is enjoyed.*". The right to life guaranteed by Art 21 encompasses within its sweep not only the physical existence but also the quality of life and if any statutory provisions are contrary to this right, it shall be held unconstitutional. Right to food is an important component of right to life in its broad conceptualisation. It is common knowledge that the life would be meaningless without adequate nutrient food. Efforts to eliminate malnutrition and epidemics are an important aspect of state's duty to sustain and guarantee the right to life. Various expressions like 'right to eat', 'right to be saved from starvation', 'freedom from hunger' 'right to nourishment' are used interchangeably for right to food. Generally it is assumed that the measures assuring right to work will automatically take care of the right to food.

"The Right to be free from Hunger is fundamental, which means that the state has an obligation to ensure, as a minimum, that people do not starve. The right is closely

---

(ii) the carrying on by the State, or by a corporation owned or controlled by the State, of any trade, business, industry or service, whether to the exclusion, complete or partial, of citizens or otherwise

[45] Food Safety and Standards Act,2006 Section 3(1) (O) Food business operator in relation to food business means a person by whom the business is carri4ed on or owned and is responsible for ensuring the compliance of this Act, rules and regulations made there under;

[46] Protection of life and personal liberty.

[47] *AK Gopalan* v *State of Madras*, AIR 1950 SC 27.

[48] 94 U.S 113(1877).

linked to the right to life itself. A series of judicial interventions and interpretations have further deepened the normative content of Right to life. Indeed, several times, the Supreme Court has explicitly stated that the right to life ought to be interpreted as a right to live with human dignity which includes the right to food and other basic necessities. In addition, however, state should also take all the necessary steps possible towards the goal of full enjoyment of the right to adequate food"[49]. The term 'freedom from hunger involves several interpretations: getting two square meals a day, meeting the dietary requirements and specific calorie intake, avoiding nutrition related ailments which the Researcher, ideally means that the right to food should be seen as a right to nutrition as enunciated in Article 47 of Indian constitution.

"State parties to the ICESCR are required to adopt, inter alia, the legislative measures necessary to realize the right to an adequate standard of living, including the right to food. Several countries already have provisions on the right to food in their national constitutions, but there is still a lack of experience in designing and using national legislations to implement those provisions"[50]. Article 21, must be interpreted in the light of Article 25 of Universal Declaration of Human Rights 1948[51], Article 27 of the Convention on the Rights of the Child[52] and Articles 11[53] &12[54] of the International Covenant on Economic, Social and Cultural Rights 1966.

---

[49] Yogendra Kumar Srivastava, "Right to Food: A Human Right", The PRP Journal of Human Rights [June-March 2001],Vol 5, Issue 1, p15-23.

[50] Yogendra Kumar Srivastava, "Right to Food: An international Social Problem", The PRP Journal of Human Rights,[October-December 2004]Vol8, Issue No1, p 30-37.

[51] Article 25 1. Everyone has the right to a standard of living adequate for the health and well-being of himself and of his family, including food, clothing, housing and medical care and necessary social services, and the right to security in the event of unemployment, sickness, disability, widowhood, old age or other lack of livelihood in circumstances beyond his control.
2. Motherhood and childhood are entitled to special care and assistance. All children, whether born in or out of wedlock, shall enjoy the same social protection.

[52] Article 27 (Adequate standard of living): Children have the right to a standard of living that is good enough to meet their physical and mental needs. Governments should help families and guardians who cannot afford to provide this, particularly with regard to food, clothing and housing.

[53] Article 11.
1. The States Parties to the present Covenant recognize the right of everyone to an adequate standard of living for himself and his family, including adequate food, clothing and housing, and to the continuous improvement of living conditions. The States Parties will take appropriate steps to ensure the realization of this right, recognizing to this effect the essential importance of international co-operation based on free consent.
2. The States Parties to the present Covenant, recognizing the fundamental right of everyone to be free from hunger, shall take, individually and through international co-operation, the measures, including specific programmes, which are needed:(a) To improve methods of production, conservation and distribution of food by making full use of technical and scientific knowledge, by disseminating knowledge of the principles of nutrition and by developing or reforming agrarian systems in such a way as to achieve the most efficient development and utilization of natural resources;(b) Taking into account the problems

*"There are two main ways of depriving the right to life, firstly by execution, torture and other various forms of killing and secondly, by starvation and lack of medical care, the second seem to be more prominent in depriving the right to life"*[55]. *"The judges at the Supreme Court during the post emergency period, found themselves free from its previous jurisprudence and took an extensive view of life and liberty to include in the meaning of right to life, every aspect that made life meaningful and worthwhile"*[56]. In Article 21 of the Constitution of India, the expression 'Life' has been judiciously interpreted to mean a life with human dignity and not mere survival or animal existence[57]. In the light of this, the State is obliged to provide for all those minimum requirements which must be satisfied in order to enable a person to live with human dignity, such as education, health care, just and humane conditions of work, protection against exploitation, the Right to Safe Food is inherent to a life with human dignity.

The obligation of the state is not only to provide food but safe food which is fit for human consumption as the decision in '*Bandhua Mukthi Morcha* v *Union of India*'[58] states '*it is the fundamental right of everyone in the country... to live with human dignity……… it must include protection of health and strength of workers………..these are the requirements which must exist in order to enable a person to live with human dignity and no state …… has the right to take any action which will deprive a person of the enjoyment of these basic essentials*'. The Apex court disposing of the Writ petition in '*Centre for public interest litigation* v. *Union of India*[59]', held that any food article which is hazardous or injurious to public health is a potential danger to the fundamental

---

of both food-importing and food-exporting countries, to ensure an equitable distribution of world food supplies in relation to need.

[54] Article 12

1. The States Parties to the present Covenant recognize the right of everyone to the enjoyment of the highest attainable standard of physical and mental health.

2. The steps to be taken by the States Parties to the present Covenant to achieve the full realization of this right shall include those necessary for:

(a) The provision for the reduction of the stillbirth-rate and of infant mortality and for the healthy development of the child;

(b) The improvement of all aspects of environmental and industrial hygiene;

(c) The prevention, treatment and control of epidemic, endemic, occupational and other diseases;

(d) The creation of conditions which would assure to all medical service and medical attention in the event of sickness.

[55] Dr Vishwanath. M, "Right to life in devoid of Right to food: Is it governance? In search of answers", Indian Bar Review, [October-December 2011],Vol XXXVIII(4), p135-160.

[56] Ramesh, "Right to Food: A need for changes in the present policy", Indian Socio-Legal Journal, [2003]Vol XXIX Nos 1&2.

[57] *Maneka Gandhi* v. *Union of India*, AIR 1978 SC 597.

[58] (1984)3SCC 161, AIR 1984 SC 802.

[59] (2013) 16 SCC 279.

right to life guaranteed under Article 21 of the Constitution. A paramount duty is cast on the states and its authorities to achieve an appropriate level of protection to human life and health which is a fundamental right guaranteed to the citizens under Article 21 read with Article 47 of the Constitution.

### 3.4.5 Articles 32 and 226 of the Constitution of India

These two provisions of Indian Constitution confers writ jurisdiction on the High Courts and the Supreme Courts to enforce one's Fundamental Rights in case of violation by the state. These two articles mainly act as the custodians of the Fundamental Rights that are enshrined in the Indian Constitution. These two provisions guarantee the availability of the food at the outset and also the safe food in particular even to the last person in the society by the state. In the event of infringement of right to safe food these two provisions aid the affected to use these provisions as the first and last weapon in his armour.

### 3.4.6 Articles 39 and 47 of the Constitution of India

The citizen's right to be free from hunger enshrined in Article 21 is to be ensured by the fulfilment of the obligation of the State set out in Articles 39(a) and Article 47 of Indian Constitution, which are Directive Principles but fundamental in the governance of the country, requires the State to direct its policy towards securing that the citizens, men and women equally, have the right to an adequate means of livelihood. The reading of Article 21 together with Articles 39(a) and 47 places the Right to Food in the correct perspective, thus making it a guaranteed Fundamental Right which is enforceable by virtue of the constitutional remedy provided under Articles 32[60] and Article 226[61] of the Constitution.

Article 37 of the Indian Constitution states that the state shall in particular, direct its policy towards securing that the citizens, men and women equally, have the right to adequate means of livelihood. Further Article 47 of Indian Constitution calls upon the state to raise the level of nutrition and the standard of living and to improve the public health. It shows that these Articles of Part IV of Indian Constitution mandates that the right to food is important for right to live with dignity. 'The specific instruments on children indicate clearly the right of every child to be adequately nourished as means of

---

[60] 'Remedies for enforcement of rights conferred by part III of the Constitution'.

[61] 'Power of High Courts to issue certain writs'.

attaining and maintaining health. By nutritious food, it is meant proper intake of all principles of food-protein, carbohydrates, fats, minerals and vitamins'[62].

Articles 39 and 47 are "Directive Principles of State Policy", which are not supposed to be enforceable in the court of law. But Article 37[63], enunciates that the directive principles are nevertheless fundamental in the governance of the state. However, it is possible to argue that Articles 39(a)[64] and 47[65] are enforceable in Court as expressions of the fundamental Right to life as National human Rights Commission in its report stated that "there is a fundamental Right to be free from hunger"[66]. In continuation of the principles contained in the preamble of the constitution, Article 38[67] enjoins the state to strive to promote the welfare of the people by securing and protecting as effectively as it may, a social order in which justice, social, economic and political shall inform all the institutions of the national life.

There cannot be a social justice in the state unless the basic requirement of Right to Safe food is guaranteed to its citizenry. Article 39(e)[68], mandates the state to secure the health and strength of general public. Article 39(f)[69], ordains that the children be given opportunities and facilities to develop in a healthy manner. This provision indirectly instructs the state to take care of the children in terms of the food and calorie intake. Different clauses of Article 39 deal with the obligations of the state to take care of the health of the people at large. It is understood that, unless one consumes safe food which is good in all respects in terms of quantity, quality, nutrition and safety, he cannot enjoy good health. Safe and nutritious food is crucial for one's health.

---

[62] Dr Harpal Kaur Khehra, "Legal Regulation of Infant foods in India", M.D.U law Journal, [2004] Vol 9 No 1&2,p 57-76.

[63] Application of the principles contained in this part(Part IV).

[64] Art 39(a)The State shall, in particular, direct its policy towards securing (a) that the citizens, men and women equally, have the right to an adequate means to livelihood.

[65] Article 47, Duty of the State to raise the level of nutrition and the standard of living and to improve public health The State shall regard the raising of the level of nutrition and the standard of living of its people and the improvement of public health as among its primary duties and, in particular, the State shall endeavour to bring about prohibition of the consumption except for medicinal purposes of intoxicating drinks and of drugs which are injurious to health.

[66] Report dated17th January 2003.

[67] 'The state to secure a Social order for the promotion of welfare of the people'.

[68] Art 39(e) The State shall, in particular, direct its policy towards securing that the health and strength of workers, men and women, and the tender age of children are not abused and that citizens are not forced by economic necessity to enter avocations unsuited to their age or strength;

[69] Article 39(f) The State shall, in particular, direct its policy towards securing that children are given opportunities and facilities to develop in a healthy manner and in conditions of freedom and dignity and that childhood and youth are protected against exploitation and against moral and material abandonment.

### 3.4.7 Article 41 of the Constitution of India

Article 41[70], though does not deal with the right to food, it dictates the states to provide work to the citizens. Unless the people are provided with employment opportunities, they may not have access to food though it is available. This is another dimension of right to food. By providing employment opportunities, the state can make the people self reliant and not depend on the state to feed them. The provisions dealing with right to food and that fall under directive principles of state policy are the conscience of the Constitution to achieve socio-economic justice. The Constitution can work vibrantly only if there is a proper balance between part III and Part IV of the Constitution as propounded in *Minerva Mills* v. *Union of India*[71]. Though, the Directive Principles of State policy are not justiciable, the Supreme Court subsequently held in *State of Tamilnadu* v. *Abhi*[72] and *Bandua Mukthi Morcha* v. *Union of India*[73] that "*the court should make every attempt to reconcile the fundamental rights with the Directive principles remembering that the reason why the Directive Principles were left by the fathers of the Constitution as non-enforceable in the courts was to give the government sufficient latitude to implement these principles from time to time according to capacity and circumstances that might arise. A fundamental right may be read together with the Directive Principles in order to enforce the directive principles through the fundamental rights*".

This ruling means that the state in order to achieve the ideals mentioned in the preamble of the Constitution must strike a balance between the fundamental rights and Directive principles. The state cannot altogether neglect the directive principles just because they are not enforceable but depending on the economic affordability, the state should take measures to implement directive principles. One important provision of Indian Constitution dealing with the primary duty of the state to raise the level of nutrition, the standards of living and to improve public health which is the hard rock foundation of right to food has still not received the adequate attention of the legislators, civil society and the courts, that it should have. The first part of Article 47 is based on the recommendations of the UN conference on Food and Agriculture, 1945 which

---

[70] Article 41. Right to work, to education and to public assistance in certain cases The State shall, within the limits of its economic capacity and development, make effective provision for securing the right to work, to education and to public assistance in cases of unemployment, old age, sickness and disablement, and in other cases of undeserved want.

[71] AIR (1980)SC 1789.

[72] 1984 SC 326.

[73] 1984 SC 802.

remarked that the primary responsibility lies with the respective nation for ensuring that its own people have enough food required for life and health.

Unfortunately even today the Directive principles of state policy are accorded a secondary importance when compared to fundamental rights with a very limited judicial role in the implementation. Even the framers of the Constitution appear to have thought that in respect of economic rights such as food, shelter and health, that the concept of public duty is more appropriate than the notion of individual rights. Few Directive principles dealing with right to an adequate means of livelihood, education, right to work, the principle relating to the duty of the state to secure social order for the promotion of welfare, public assistance in cases of unemployment, old age , sickness, living wages for workers are thought to be important to ensure right to food.

### 3.4.8 Article 246 of the Constitution of India

Article 246[74] gives powers to the state as well as to the union to make Rules with respect to the subject matters contained in VII schedule. The matter of food being in the List III of the Concurrent List, the Constitution authorises both the parliament and the state legislatures to legislate concurrently[75]. The subject matter of food finds a mention in the List III in the entry 18[76] and 33(b)[77] and price control in entry 34, hence the Union government has passed central legislations, which are the two major food laws in the country, 'Food Safety and Standards Act 2006' and 'National Food Security Act 2013'. An elaborate set of control orders have been promulgated by both the Central and State government under the enabling provisions of Essential Commodities Act 1955 for the regulation of trade and commerce and price control. The central and state governments have enacted and enforced different control orders and food laws for better distribution of available food grains. The main objectives of these control orders and food laws have been to check the undesirable activities of the traders, to curb adulteration of food stuffs, to check the escalating prices of food stuffs and to ensure the supply of available food grains to the consumers primarily to the low-income vulnerable sections of the community at subsidised prices.

---

[74] 'Subject-matter of laws made by Parliament and by the Legislature of States'.
[75] Article 246(2).
[76] Adulteration of food stuffs and other goods.
[77] Trade and commerce in and the production, supply and distribution of foodstuffs including edible oil seeds and oils.

### 3.4.9 Article 253 of the Constitution of India

Right to safe food has become a global concern because of which United Nations Organisation is constantly endeavouring to secure this Right globally because of which it has held various conventions and entered into treaties and agreements with all its member nations. India being a member nation of the United Nations Organisation is obligated to adopt the recommendations of the conventions and to implement the obligations under the treaties and other international agreements. Article 253 [78] empowers the state to make legislation for giving effect to international agreements and treaties and for adopting the spirit of the International conventions. Unless the state makes a suitable municipal legislation for adopting the International conventions or treaties or agreements, no citizen can make use of these instruments.

It is interesting to note that, Justice Krishna Iyer took cognisance of international covenants on civil and political rights in *Jolly George* v. *Bank of Cochin*[79], and observed that "*India being a party to the covenant, it must respect the same*". Subsequently in *Chairman Railway Board* v. *Chandrima Das*[80], the Supreme Court stated that "*The International Covenants and Declarations as adopted by the United Nations have to be respected by all signatory states and the meaning given to the above words in those declarations and covenants have to be such as would help in the effective implementation of those rights. The applicability of the Universal Declaration of Human Rights and the principles thereof may have to be read, if need be into the domestic jurisprudence*". In *Kapila Hingorani* v. *State of Bihar*[81], it was well settled that *a statute ought to be interpreted in the light of international treaties and conventions.*

On the same analogy, India being a signatory to International Covenant on Economic, Social and Cultural Rights 1956, the Researcher submits that India should have accorded legislative recognition to the right to food and must have made it enforceable long back. In *Chameli Singh* v *State of UP*[82], referring to Article 11 of the International Covenant on Economic, Social and Cultural Rights, held that the state parties should recognise "*the right of everyone to an adequate standard of living for himself and for his family including food, clothing, housing and to the continuous*

---

[78] 'Legislation for giving effect to international agreements'.
[79] AIR 1980 SC 70.
[80] 2000(2)SCC 465.
[81] 2003 6 SCC 1.
[82] 1996(2) SCC 549.

*improvement of living conditions*". 'The Limberg principles[83] state that "*legislative measures alone are not sufficient to fulfil obligations under the covenant*". '*Recognition of a right involves the acknowledgement of such a right by the state party*'[84]. Hence in order to give effect to various International conventions and covenants relating to Right to food, for which India is a party, Union government has enacted the Food Laws.

### 3.5 Judicial Approach to Right to Food

This basic human right to food is never expressed in our Constitution. May be our Constitution framers never thought that such a need would never arise as it is a very basic necessity for the existence of life and that the same would be taken care of by the authorities without making any bones about it. Or on the flip side, we have always witnessed that India has been suffering from poverty and the resultant malnutrition. Hence, perhaps the framers wanted to give importance to other aspects than to this basic right or they did not want to further burden the scrawny exchequer by making the people rights-holder, as they must have known that our financial resources cannot bear the brunt of feeding almost one third of the population. Thanks to the judges mainly the Apex court, whose timely intervention and interpretation salvaged many uncared for lives. The Apex court has reiterated in several of its decisions that the right to life in its true meaning includes basic right to food, clothing and shelter. Due to this the Researcher thinks that '*Right to food is the product of judicial decisions*'. But for the timely intervention by the apex court, the one third population of our country which was poverty stricken would have been wiped out with hungry bellies.

Way back 1954 itself, the Supreme Court established the justiciability of the right to food through various case laws ruling that the wages paid to a labourer should be adequate enough for him to feed himself and his family. At this juncture, the Supreme Court also termed the low wages paid to the labourers with which they were not able to make both ends meet as the 'starvation wage'. The Supreme Court condemned and prohibited the starvation wages and ruled that the minimum wage

---

[83] In June 1986 a group of distinguished experts in international (human rights) law convened in Maastricht, the Netherlands, to deliberate the nature and scope of state parties' obligations under the International Covenant on Economic, Social and Cultural Rights (ICESCR). This meeting resulted in the agreement and adoption of the so-called Limburg Principles on the Implementation of the International Covenant on Economic, Social and Cultural Rights. Available at https://www.maastricht university .nl/ ev ents/ imburg-principles-and-migration (accessed on 24/8/2018 at 3.52pm).

[84] Rajat Khosla, "Recognition and Access : An Experience with the Right to Food", Cochin University Law Review [2003] Vol XXVII, p 224-232.

which enables the labourers to maintain themselves and their family should be paid. The Apex Court in *Crown Alluminium Work* v. *Their Workmen*[85], held that the unorganised sector due to poverty, may at times be available at starvation wages, but India being a welfare state cannot exploit the labourers. The nation is required to pay a fair and living wages to its labourers so that they can afford a minimum level of comfort and decency in their lives.

Though, this was the maiden step towards establishing the principle that wages ought to guarantee the bare minimum comfort in the life of the poverty stricken famished labourers, yet a remarkable one at that. The same view point was echoed in *Express Newspaper Private Ltd* v. *Union of India*[86]. Despite these progressive judgements, the authorities had not initiated any concrete efforts to put an end to the practice of hiring labourers on starvation wages. Quite a few instances that were brought before the judiciary reflected the payment of wages which were starvation wages in exchange for a day's hard labour. Though, initially the judiciary exercised self restraint, but its compassion towards the starvation of the poor and needy was reflected in *U. Unichoyi and others* v. *State of Kerala*[87]. In this it was observed that in underdeveloped countries like India, poverty may compel the people to work for starvation wages but, the welfare state should ensure a minimum wage which can buy the bare necessities of life.

Thus for the first time guidelines were laid down for the fixation of fair minimum wages, this decision was again reiterated in *Hydro (Engineers) Pvt. Ltd* v. *Workman*[88], by the Supreme Court. In continuation, in *Jaydip Paper Industries* v. *Workmen*[89] and *Gujarat Agricultural University* v. *Rathod Labhu Bechar*[90], the Supreme Court held that the state should provide the labourers with minimum wages and also protection against starvation. Another landmark case dealing with the war against hunger was decided by the Supreme Court i.e. *Bandhua Mukti Morcha* v. *Union of India*[91], and held that the duty of the state does not end with the mere release of the labourers from bondage and it has the duty to rehabilitate the released labourers by providing them with necessities of life, so that the released labourers do not get

[85] AIR 1958 SC 30.
[86] AIR 1958 SC 578.
[87] AIR 1962 SC 12.
[88] AIR 1969 SC182.
[89] (1974) 4 SCC 316.
[90] AIR 2001 (SCW) 351.
[91] (1984) 3 SCC 115.

embroiled in another cycle of starvation. In *Neeraja Choudhury* v. *State of Madhya Pradesh*[92], reiterating its previous decision, the court called upon the government to draw up suitable plans to rehabilitate the released bonded labourers so that they don't come out of the bonded labour to enter into the bondage of hunger and struggle.

There cannot be food security unless there is adequate supply of the same in the market. This can be ensured only when there is total cultivation of all agricultural lands in the country. The state should make sure that the lands are not left barren by the farmers for whatsoever reasons. The Supreme Court in *Dasaudha Singh and others* v. *State of Haryana*[93], upheld the government policy of compelling the Zamindars and farmers to cultivate all the agricultural lands. The court upheld the legislation which in effect contributed to the increase of food grains production of the country, thus transforming India from food importing country to food export country. The increase in food production is vital to ensure self sufficiency and food security in the country. By examining all these decisions it is evident that the Supreme Court has gone all out, to erase even the traces of hunger from welfare and a civilised society like ours.

In *Keshavananda Bharati* v. *State of Kerala*[94], the Supreme Court while interpreting the Constitution in the light of municipal law and the Charter of United Nations, held that the object of the Constitution should be to promote the social and economic justice. The onus of protecting the people from starvation is on the wings of the legislature, executive and judiciary. It further held that protection from starvation is a fundamental right of every citizen and it also laid down the role to be played by different organs of the state in eliminating hunger. The Supreme Court carved a niche in creating new jurisprudence of right to food in the Indian Legal system. But, sadly international law supporting the right to food was not fully established at that point of time and India had still not ratified the Conventions relating to food, and international obligations for providing these basic rights were also not very rigid in application. In this scenario, it is the far sightedness and compassionate judiciary which felt for the hungry and starving population of India. Due to all these factors, the judgement failed to specify the relevant article of the Constitution, which could be invoked in cases of violation of fundamental right to food and freedom from hunger and malnutrition. Thus the judgement fell short of enforceability to protect the right to food.

---

[92] (1984) 3 SCC 243.
[93] (1973) 2 SCC 393.
[94] (1973) 4 SCC 225 (paragraph 1700).

The post emergency judicial activism encouraged the judges of the Supreme Court to come up with a new kind of sensitivity to the cause of the poor, the bonded labourers, wage earners and other weaker sections of the society. This highly valuable judicial sensitivity is abundantly reflected in several notable cases such as the Beggar's Case[95], the Pavement Dweller's case[96], Rickshaw pullers case[97], the Asiad Games case[98], Bonded Workers case[99] and a host of other cases.

For the first time in the Indian judicial history, the Apex court in *Maneka Gandhi* v. *Union of India*[100] gave a wide interpretation to Article 21 and ruled that *right to life is not restricted to mere physical existence but includes the right to live with human dignity*. The Supreme Court in *State of Maharashtra* v. *Chandrabhan*[101], held that *right to food is a component of the right to life guaranteed under Art 21 of the Constitution*. In *Vincent* v. *Union of India*[102], the Supreme Court held that *right to maintenance and improvement of public health is included in the right to life with human dignity*. The justiciability of the right to food was upheld by the Supreme Court in *State of Uttar Pradesh* v. *Uptron Employees Union CMD*[103]. The court ruled that "*the State cannot escape the liability when a human rights problem of starvation deaths and suicides by the employees takes place as an offshoot of non payment of salaries of the employees for a long period*".

The compassion of the judiciary to the sufferings of hungry and starving populace is again evident from the judgement in *Indian Council of Legal Aid and Advice & others* v. *State of Orissa*[104]. The court in this case requested the Human Rights Commission to look into the different schemes initiated by the government for the realisation of the right to food, thus preventing starvation and death. In the *Board of Trustee of the Port of Bombay* v. *Dilip Kumar Raghavendranath Nadkarni*[105], the Supreme Court held that "*people cannot be allowed to lead a life of continued drudgery and they have a right to livelihood which makes life worth living*".

---

[95] *Kalidas* v. *Government of Jammu and Kashmir*(1987)3 SCC 430.
[96] *Olga Tellis* v. *Bombay Municipal Corporation*(1985)3SCC 545.
[97] *Azad Rickshaw Pullers Union* v. *State'*(1980) SSC Supp 601.
[98] *PUDR* v. *Union of India* (1982) 2SCC 161.
[99] *Bandhua Mukthi Morcha* v. *Union of India* (1984) 3 SCC 4,161.
[100] AIR 1978 SC 597.
[101] (1983) 3 SCC 387.
[102] AIR 1987 SC 165.
[103] (2006) 5 SCC 319.
[104] (2008) (1) SCC Supreme 421, Writ Petition (Civil) 43 of 1997.
[105] (1983) 1 SCC 124.

In *Francis Coralie Mullin* v. *The Union Territory of Delhi*[106], the Former Chief Justice of India P N Bhagavathi, noted "*we think that the right to life includes the right to live with human dignity and all that goes along with it, namely, the bare necessities of life such as adequate nutrition, clothing and the like. Further the objective of the Directive Principle is to achieve a welfare state (among others, ensuring basic needs) by supplementing fundamental rights*". He also conceded that *"the magnitude and extent of the components of this right would depend upon the extent of development of the country*".

In *Shantistar Builders* v. *Narayan Khamalal Totame*[107] and *P.G. Gupta* v. *State of Gujrat*[108] the Supreme Court held that the *basic need of a man is food, clothing and shelter*. In *Olga Tellis* v. *Bombay Municipal Corporation*[109], Chief Justice Chandra Chud said "*deprive a person of his right to livelihood and you will have to deprive him of his life*". Justice Mukherji in Prabhakaran Nair's case[110] among other things, observed that "*basic needs of man have traditionally been accepted to be three-food, clothing and shelter, the right to life is guaranteed in any civilised society that would take within its sweep the right to food, the right to clothing and the right to clean environment and a reasonable accommodation to live in*".

Justice Sawant, implying that the right to life also includes the right to work in *Delhi Transport Corporation* v. *DTC Mazdoor Congress*[111], emphatically noted "*income is the foundation of many fundamental rights and when work is the sole source of income, the right to work becomes as much fundamental*". "*Overtime, the right to work moved closer to the centre of the stage, as it came to be seen as an essential precondition of the right to food. Social security for those who are unable to work is another abiding concern of the campaign*"[112].

*C.E.S.C Ltd.* v. *Subhash Chandra Bose*[113], in this case it was held that right to health is a part of right to live and also considered the Universal Declaration of Human Rights, International Conventions of Economic, Social and Cultural Rights which have included right to food, fair wages, decent working conditions etc as a part of right to

---

[106] (1981) 4 SCC 494.
[107] (1990) 1 SCC 520.
[108] (1995) Supplementary 2 SCC 182.
[109] AIR 1986 SC 180.
[110] AIR 1987 SC 2117.
[111] AIR 1991 SC 101: Supp (1) SCC 600.
[112] Jean Dreze, "Bhopal Convention on the right to food and work: Brief report and personal observations", Social Action,[January –December, 2004],Vol 54,No4, p 237-247.
[113] (1992) 1 SCC 441.

life. It was held in *Ahmedabad Municipal Corporation* v. *Nawab Khan Gulab Khan & Ors*[114], that "*Right to food is an inbuilt and inalienable part of right to life which cannot be compromised on any ground. Right to live guaranteed in any civilised society implies the right to food, water, decent environment, education, medical care and shelter*". Justice K. Ramaswamy, in *Samatha* v. *State of Andhra Pradesh*[115], while recognising the right to socio economic empowerment held that "*right to life enshrined in Article 21 means something more than survival of animal existence. The right to live with human dignity with minimum of sustenance and shelter and those rights and aspects of life which would go to make a man's life complete and worth living would form part of it*".

Food security not only implies availability but also access to quality and a particular quantity of food, which means that right to work is also a fundamental right. The Researcher thinks that unless a person enjoys the right to work, he cannot earn money and cannot have access to food, though it is available.

### 3.5.1 Case Analysis of *Kishen Pattanaik* v. *State of Orissa*

The first case concerning specifically the Right to food that went up to the Supreme Court in 1986 was the case of *Kishen Pattanaik* v. *State of Orissa*[116], in which the petitioner wrote a letter to the Supreme Court invoking the epistolary jurisdiction about the extreme poverty of the people of Kalahandi[117] in Orissa where hundreds of people were dying due to starvation and where several people were forced to sell their children to make both ends meet. In this case the court converted a letter addressed to the Chief Justice of India as a public interest litigation. The letter prayed that the State government should take immediate action.

While allowing the writ petition, *P.N. Bhagwati, CJ* observed "*No one in this country can be allowed to suffer deprivation and exploitation particularly when social justice is the watchword of our Constitution. It is the bounden duty of the state government to see that everyone is provided with bare necessities of life and no one is driven to a position where he is compelled to sell his sweat and labour for a pittance*". The court directed the state to take immediate action in order to ameliorate this miserable condition of the people in Kalahandi.

---

[114] AIR 1997 SC 152.
[115] AIR 1997 SC 3297,3330.
[116] 1989 Supp(1) SCC 258.
[117] Kalahandi, is a district of Odisha in India. The region had a glorious past and great civilisation in ancient time. Archaeological evidence of stone age and Iron Age human settlement has been recovered from the region.

In its judgment, the Supreme Court took a pro-government approach and directed the state to take macro level measures to address the starvation problem. But none of these measures actually affected the immediate needs of the petitioner and the benefits did not reach the needy. Even at this juncture, the Supreme Court did not recognize the specific Right to Food. This was the first case specifically taking up the issue of starvation and lack of food.

Justice VR Krishna Iyer, aptly described the new found judicial concern for the oppressed class of the society and for poor in the following words:

*"The challenge in these petitions compels us to remind ourselves that under our constitutional system, courts are havens for the toiler, not the exploiter, for the weaker claimant of social justice not the stranger pretender who seeks to sustain the status quo ante by the judicial writ in the name of fundamental rights. This right to food was further strengthened due to landmark decisions of the Supreme Court in the popular known Kalahandi Food Petitions[118]. It was alleged in these petitions that the suffering and misery of the habitants were due to utter negligence and callousness of the administration of the government of Orissa".*

The Supreme Court after the lapse of about 3 years finally disposed off the petition accepting the relief measures initiated by the Orissa government. "It was believed by some people that some events narrated in the petition would compel the court to come out with an extraordinary response wherein the court would critically examine the social reality and government response to hunger and starvation and award bold and humane consequential relief for the people facing hunger and starvation. Unfortunately there was no reference to the basic human needs involved in the famine and starvation conditions nor any legal or constitutional recognition of right to food or right to satisfaction of basic needs of the citizens"[119]. However, the court recognised the obligation on the part of the state under such situations but it was not clear as to the nature and extent of such an important and crucial obligation. It was very unfortunate that the court turned a blind eye on the real issues presented in the petitions by accepting the government's version of the situation and honouring the submissions made by it.

From the above cases it can be inferred that the Supreme Court has indirectly conceded right to food as an aspect of right to life, which includes right to health, work,

---

[118] *Kishen Patnaik and others* v. *State of Orissa, India People Front* v. *State of Orissa,* AIR 1989 SC 677.
[119] Ramesh, "Right to Food: A Need For Changes In The Present Policy", Indian Socio-Legal Journal, [2003] Vol XXIX Nos 1&2, p 55-62.

environment. But such implied recognition may not be of much help to protect the interest of vulnerable section of the society from the pangs of hunger and malnutrition. Even after *Kishen Patnaik's* case, in 2001, 48 people were reported to have died of starvation in the Kashipur block of the Rayagada district of the State of Orissa[120]. The right to food campaign has also confirmed starvation deaths in Jharkahand in 2002'[121].

As discussed above, the Supreme court impliedly recognised the right to food mainly with the help of Article 21, Right to life and personal liberty but this was not sufficient to protect the interests of vulnerable groups of people suffering from hunger and malnutrition, the turning point in accepting right to food as right to life was in 2001 when the People's Union for Civil Liberties filed a Writ petition in the Supreme Court praying for directions to the central government to release food grains from Food Corporation of India godowns to the people who were starving to death in the State of Rajasthan.

### 3.5.2 Case Analysis of *PUCL* v. *Union of India*

The justiciability of the specific Right to food as an integral Right under Article 21 had never been articulated until Right to Food Petition filed in '*PUCL*[122] v. *Union of India*[123]'. In 2001, 60 million tonnes of food grains were in the godowns of Food Corporation of India, whereas the buffer stocks required to be maintained was only 20 million tonnes. The Government had 40 million tonnes of food grains above the buffer stock but still people were dying of starvation. In 2001, Starvation deaths had occurred in the State of Rajasthan, 47 tribals and dalits were starved to death in south-eastern Rajasthan, despite excess grain being kept for official times of famine, and various schemes and despite India's food stocks having an excess of around 40 million tonnes of food grain that year, activists and organizations had approached the apex court to secure food security. In 2001, the godowns of Food Corporation of India which was situated about 5 kms away from Jaipur city were overflowing with food grains. The food grains were rotting owing to fermentation as they were stocked in the open area, rainwater had percolated down to the grain stock. In a village which was near the godown, people were eating in rotation, which can be classically termed as "rotation

[120] '48 died of starvation', The Hindu, September 4, 2001, Available at https://www.thehindu.com / the hindu/ 2001/09/04/stories/0204000c.ht (accessed on 17/8/2018 at 3.25pm).
[121] Rajat Khosla, "Recognition and Access : An Experience with the Right to Food", Cochin University law Review[2003] Vol XXVII, p 224-232.
[122] People's Union for Civil Liberties.
[123] (1997) 1 SCC 301.

eating", or "rotation hunger" where some members of the family eat on one day and the remaining persons eat on the other day. Natural disasters, lack of purchasing power, massive unemployment and other factors led to starvation and death of many poor people. In response to this, the PUCL sought recognition of the right to food under Article 21 in the Supreme Court in 2001.

The PUCL petitioned the court for enforcement of both the food schemes and the Famine Code[124]. They grounded their arguments on the right to food, deriving it from the Right to life. The Petitioner requested the Supreme Court to intercede and issue a writ of mandamus and direct the administration 1. To implement the famine code in the famine hit states 2. To release the surplus stored food grains in order to meet out the shortage of food grains 3. To revise the PDS framework and frame a fresh scheme and policy of public distribution for scientific and reasonable distribution of grains to the poor people at subsidised rates. It was contended on behalf of the FCI, that it had no powers to release food grains to States on its own. Food grains would be released to the States only after the allotment of food grains quota to them by the Centre on the basis of requisitions made by the states. "Nevertheless the court expressed unhappiness over the bureaucratic functioning of the FCI and the inaction of both the Central and State Governments to come to the rescue of the starving people"[125].

On 23rd July, 2001 the Supreme Court directed the petitioner to amend the petition and make all the states and the union territories, parties to the petition. The Supreme Court took up the matter very seriously, over two years, various interim orders pertaining to nutrition-related schemes were made by the court, but with meagre implementation by the national and state governments. In 2003, the court issued a strong judgment which found the Right to life was imperilled due to the failure of the schemes. The Court noted the paradox of food being available in granaries but that the poor were starving and it refused to hear arguments concerning the non-availability of resources given the severity of the situation and ordered that all individuals without means of support (older persons, widows, disabled adults) are to be granted an *Antyodaya Anna Yozana*[126] ration card for free grain and State governments should

---

[124] A code permitting the release of grain stocks in times of famine.

[125] J. Venkatesan, "Food must reach the starving, not rot: S.C", The Hindu, Online Edition of Indian National News Paper. Dated Tuesday, August 21, 2001 available at https://www.thehindu.com/thehindu/2001/08/21/stories/01210001.htm (accessed on 16/8/2018 at 10.26am).

[126] Antyodaya ration cards are issued to such families who have the income of less than Rs. 250/- per capita per month. Ishwar Bhat's '*Fundamental Rights*' 1st edn 2004, Eastern Law House Pvt Ltd, Kolkata.

progressively implement the mid-day meal scheme in schools. The Supreme Court observed in the case "in our opinion, what is of utmost importance is to see that food is provided to the aged, infirm, disabled, destitute women, destitute men who are in danger of starvation, pregnant and lactating women and destitute children, especially in cases where they or members of their family do not have sufficient funds to provide for them. In case of famine, there may be shortage of food, but here the situation is that amongst plenty there is scarcity. Plenty of food is available, but distribution of the same amongst the very poor and the destitute is scarce and non-existent, leading to malnourishment, starvation and other related problems.......by way of an interim order, we direct the states to see that all the PDS shops, if closed, are reopened and start functioning within one week from today and regular supplies made".

On August 20th 2001, the Supreme Court made the following order:

"*The anxiety of the court is to see that the poor and the destitute and the weaker sections of the society do not suffer from hunger and starvation. The prevention of the same is one of the prime responsibilities of the government – whether central or the state. How this is to be ensured would be a matter of policy, which is best left to the government. All that the court has to be satisfied and which it may have to ensure is that the food grains which are overflowing in the storage receptacles, especially of FCI godowns and which are in abundance, should not be wasted by dumping into the sea or eaten by the rats. Mere schemes without any implementation are of no use. What is important is that the food must reach the hungry*".

The Supreme Court criticised the lax attitude of the government in failing to take adequate and appropriate measures for storage, transportation and distribution of food grains from FCI godowns which resulted in wastage of the food grains meant for food related schemes benefitting the poor and the needy. On 28th November 2001, the court focussed on eight food related schemes, the Public Distribution System, Antyodaya Anna Yojana, the National Programme of Nutritional Support to Primary Education which is popularly known as Mid Day Meal Scheme, the Integrated Child Development Services, Annapurna, the National Old Age Pension Scheme, the National Maternity Benefit Scheme and the National Family Benefit Scheme. Essentially, the interim order of 28th November 2001 converted the benefits of these eight schemes into legal entitlements[127]. Subsequent to this in 2002, the court set up institutional mechanisms in the form of Commissioners to monitor, co-ordinate and report on the implementation of the court order and also to suggest the ways to promote food security rights of the poor, which is free of the authority of the executive. On finding the approach of the government distressing and being not satisfied with the government's action in this case, on May 2nd 2003, the Supreme Court holding that the right to life would encompass right to food stressing that right to food is an integral part of Article 21 of the Constitution, held as follows:

"*Article 21 of the Constitution of India protects for every citizen a right to live with human dignity. Would the very existence of life of those families who are below poverty line not come under the danger for want of appropriate schemes and implementation thereof, to provide requisite aid to such families? Reference can also be made to Article 47 which inter alia provides that the State shall regard the raising of the level of nutrition and the standard of living of its people and the improvement of public health as among its primary duties*".

The Supreme Court also expanded the list of beneficiaries for the Anthodaya Anna Yojana Scheme providing highly subsidised food grains for the poor. The Apex Court also directed the inclusion of six classes of people in the AAY Scheme namely,

1. Aged, infirm, disabled, destitute men and women, pregnant and lactating women;
2. Widows and other single women with no regular support
3. Old persons(aged 60 or above) with no regular support and no assured means of subsistence
4. Households with a disabled adult and assured means of subsistence
5. Households where due to old age, lack of physical or mental fitness, social customs, need to care for a disabled or
6. Other reasons, no adult member is available to engage in gainful employment outside the house.

This Writ Petition was filed on16th April, 2001 and since then 427 affidavits and 71interlocutory applications have been filed by both the respondents and petitioners. This petition resulted in the significant consolidation and expansion of the National Campaign on the Right to Food. This petition was instrumental in highlighting the new

---

127 Yamini Jaishankar and Jean Dreze, "Supreme Court orders on the right to food: A tool for action" available at http://escrj.southasianrights.org/front/view_document/58(accessed on 11/8/2018 at 10.50pm).

facet of Article 21 for the first time and addressing the grievances of poor and downtrodden section of the society and also pioneered the right to food movement. In one of the interim orders dated November 28, 2001, the Supreme Court converted most of the food and employment related schemes to "legal entitlements". This petition exposed the corruption, lack of transparency, incapacity and accountability of the government. Though the final decision of the case has not yet been passed, regular hearings of the writ petition have been taking place in the Supreme Court and the authorities and the government are constantly reminded of their duty to ensure the implementation of the right to food. "*The central organisational challenge for the right to food campaign is to develop ways of working together that are both effective and consistent with our basic values-including democracy, equality and transparency*"[128]. "*This case is perhaps the longest continuing writ of mandamus in the world on the right to food issue*"[129].

'Mihir shah, designated Advisor to the National Commissioner appointed by the Supreme Court in the Right to Food case, while questioning the distinction of poor as below poverty line' and 'above poverty line' observed that "*while official data themselves show that 50 percent of Indians are chronically malnourished, less than 30 percent belong to the government's BPL category*'[130]. This reflects the standoff in fixing the criteria for BPL.

### 3.6 Limitations of a Judicial Process

The Researcher thinks that the limitation suffered by a judicial process is that it needs/lacks resources to carry out sociological survey in identifying the needy and hungry. It has been suggested by the World Food Summit, 2009 that every state ought to have national food insecurity and vulnerability data and mapping system to find out who is hungry and the reasons for hunger[131]. Most of the times, the court will be compelled to rely on the information supplied by the authorities. Interestingly, the Supreme Court while disposing off the public interest litigation filed by the *PUCL* was anguished by the starvation deaths in some states. The Supreme Court remarked that the

[128] Jean Dreze, "Bhopal Convention on the right to food and work: Brief report and personal observations", Social Action,[January –December, 2004],Vol 54, Issue No4, p 237- 247.
[129] Dr KV Ravikumar, "Right to food in India-whether a protection under fundamental rights", Indian Bar review, [July-September, 2012] Vol XXXIX No3, p 39-48.
[130] Seema Singh, "Menace of hunger vis-à-vis Right to food: A Constitutional perspective", Civil & Military Law Journal, [2009], Vol 45, p 70-79.
[131] FAO Right to Food Theory and Practice (Rome, FAO, United Nations 1998).

primary responsibility of the government both central and state to ensure that the food grains overflowing in the godowns of Food Corporation of India reached the starving and not wasted by being dumped in the sea or eaten by rats[132]. But, unfortunately no machinery is in place to ensure the implementation of the guidelines especially if it is for the public good. It is a case of 'who will bell the cat'. '*A right to food law is meaningless if it neglects people living with and dying of starvation. Current, scarcity and drought codes, inherited and modified from the colonial famine codes, contain no binding duties on states to prevent and address starvation*'[133].

When the courts begin to implement socio economic rights, they enter into a realm of administration, policy and budgetary allocation. The wafer thin line between the executive and judiciary diminishes and the courts will be accused of usurping the powers and functions that are not theirs originally. For instance, ameliorating the nutritional status in the country might involve a direction to the executive that the government should increase the amount of expenditure on nutrition in the country. But such judicial activism adversely affects the healthy functioning of the democracy. But, in principle in a nation beleaguered by populist executive action, judicial recourse is the most effective option for replenishing the Constitutionally fundamental and universally essential socio economic rights.

### 3.7 Law Commission Reports Supporting the Right to Food

Though, none of the law commission reports dealt with right to food directly, yet some law commissions have attempted to ensure that there is adequate safe food in the market meant for human consumption. Some law commission even concentrated on early child hood development measures that ought to be undertaken, to have healthy citizenry. Here the Researcher has undertaken a limited study of law commission reports pertaining to consumer interests, early child hood development and providing for have not's.

#### 3.7.1 105th Report on '*Quality Control and Inspection of Consumer Goods*' of 10th Law Commission of India.

The 10th Law Commission was chaired by K K Mathew and was constituted in October 1984 to deal with various provisions of Sale of Goods Act and the need to

[132] The Hindu, Bangalore Edition, Dated August 21, Tuesday 2001, available at https://www.thehindu.com/thehindu/2001/08/21/index.htm (accessed on 10/8/2018 at 3.47pm).
[133] Ankita Aggarwal, Harsh Mander, "Abandoning the Right to Food", Economic and Political Weekly[February 23, 2013] Vol XLVIII Issue No8, p 21- 23.

protect the interest of the Consumers. The Commission while dealing with other aspects, in Chapter V in Para 5.12 (2) has dealt with 'Recalling' which means a system where recalling of the dangerous goods from the reach of the public is introduced. The Commission opined that this step may be initiated either by the manufacturer or by the Government agency. In Chapter VI in Para 6.7 the Commission has dealt with Prevention Food Adulteration Act, 1954 under which it has discussed about the appointment of Public analysts & Food analysts and their power to take samples and subject the same to the laboratory test.

### 3.7.2 223rd Report on '*Need for Ameliorating the Lot of the Have-Nots–Supreme Court's Judgments*' of 18th Law Commission of India

18th Law Commission of India was constituted in April 2009 under the Chairmanship of Justice A. R. Lakshmanan. The Commission was constituted for a period of three years from 1st September, 2006 by Order No. A.45012/1/2006-Admn.III (LA) dated the 16th October, 2006, issued by the Government of India, Ministry of Law and Justice, Department of Legal Affairs, New Delhi. The Commission has borrowed the definition of poverty from United Nations Committee on Economic, Social and Cultural Rights and also studied the factors responsible for the same. The Commission unanimously opined that poverty has been and remains a constructed social and economic reality and the poor are not poor simply because they are less human or because they are physiologically or mentally inferior to others whose conditions are better off. On the contrary, their poverty is often a direct or indirect consequence of society's failure to establish equity and fairness as the basis of its social and economic relations. The Commission adopted the concept of '*parens patriae*' which means a duty is cast on the State under the Constitution to take up for its citizens when they are not in a position to assert their rights or defend themselves. The Commission speaks about the Universal Declaration of Human Rights and has enunciated that every man and woman has the human right to a standard of living adequate for health and well-being, to food, clothing, housing, medical care and social services. The Commission has remarkably stressed that the human right to live in dignity, free from want, is itself a fundamental right, and is also essential to the realization of all other human rights. With the help of

*Olga Tellis* v *Bombay Municipal Corporation'*[134], the Commission held that the right to life includes the right to livelihood.

### 3.7.3 259th Report on '*Early Childhood Development and Legal Entitlements*' 20th Law Commission of India.

20th Law Commission of India was constituted in August 2015 and chaired by Justice Ajith Prakash Shah. The main objective of this Commission was to build a healthy citizenry, which is one of the National powers. The Report deals with the nutritional requirements of the child, especially between the age of 1 and 6years termed as 'formation years' where the need for nutrition is critical. The Commission has dealt with a few declarations which strengthen the claim of children for better nutrition. Apart from this, the Commission has dealt with various provisions of Indian Constitution which speaks of Right to food. Though in the context of this report it aims at only the rights of the children for better nutrition and their right to claim the same, in essence the Researcher believes that they are all basic human rights which should be afforded for everybody irrespective of one's age.

## 3.8. Conclusion

During the pre-independence drought, the people were starving not because there was no food but because there was no distribution system. So, the state had to work to progressively achieve the full realisation of the rights. The Indian Constitution and its underlying ideas provide a sound framework for thinking about the right to food and is one of the basic economic and social rights that are essential to achieve "economic democracy", without which political democracy is at best incomplete[135]. A press report[136] states that a whopping 6 lakh tonnes of food grain has been rendered unfit for human consumption as the same was rotting in open storage spaces and plinths at various places for long time in Punjab.

The grains deteriorated significantly for want of adequate storage in ware houses and the stock has been certified as fit for cattle feed. This is the situation in spite of the Supreme Court observing in August 2010 that in a country where people are starving, it

---

[134] AIR 1986 SC 180.

[135] Jean Dreze, "Democracy and the Right to Food", Economic and Political Weekly[2004], Vol 39,Issue 17,p 1723-1731.

[136] '6 lakh tonnes of grain goes waste in Punjab', Deccan Herald Dated 11th March, 2015, Bengaluru Edition.

is a crime to waste even a single grain of food. This is not a lone instance but has only seen the light of the day, thanks to the probing press. If this is the amount of wastage of food grains in one state, how much it amounts to if we aggregate all the wastage in the rest of 28 states of the country- disheartening & demoralising!. Ware houses of the state and godowns of Food Corporation of India should be repaired, updated and modernised to ensure that the food grains are properly stocked without being wasted, so that the state could feed the worthy citizens but not the rodents or the cattle.

*'In fact, it is doubtful if a sincere attempt to attain this objective is even under contemplation. This becomes all the more perplexing in so far as one finds it really hard to visualise how the objective of livelihood security can conflict with any palpable, long term economic interests of the presently powerful sections*"[137]. *'Hunger is not a consequence of under-production, but the result of inequality and mal distribution. The situation is such that, if all the sacks of grain lying in warehouses were lined up in a row, the line would stretch for a million kilometres or so more than twice the distance from the earth to the moon. Starvation deaths, in such a scenario, cannot by any stretch of imagination, be justified*'[138].

Despite the efforts of the state to boost the agricultural sector, sharp fluctuation in agricultural output still persists. This can be attributed to several factors, firstly, agricultural output is highly dependent not on stable component of area but on the yield. Secondly, while only 30% of India's cultivated area is under irrigation, a major portion of the cultivable area depends on tanks and wells which again depend on rains. Thirdly, concentration of output growth in limited pockets also contributes to fluctuations in agricultural output. Improving yields per crop and multi cropping must be stressed. "*Effective food security analysis should direct the government to put in place those policies, investments and institution building efforts which are consistent with long term reduction of food insecurity*"[139].

Legal protection is a necessary step towards the realisation of the right to food as a Fundamental right. Food security can be achieved and secured only if it is made legally enforceable right. It is questionable whether the Constitution that fails to incorporate the right to food can ensure the implementation of the right in future. The

---

[137] Kamal Nayan Kabra, "The Right to food: Reappraising the concept and its implications", Social Action, [January-December 2004] Vol 54 No 4, p249-263.
[138] Abhinav Chandrachud, "Malnutrition and the Writ of Continuing Mandamus: The Remedy Befitting the Right", All India Reporter [2006] Rajasthan Section Vol 93, p19-28.
[139] Nira Ramachandran, "How Sustainable is our Food Security", The Administrator, [April-June, 1995], Vol XL, No1, p 131-155.

Researcher is intrigued as to, if such a Constitutional law can provide the human right to food with the required degree of security and protection from future legislative encroachment.

Water management also plays a pivotal role in evolving the sustainable food production strategy. The development of irrigation projects, the exploitation of ground water in high rainfall areas where the water table is favourable, evolving new techniques for water harvesting, rain harvesting, and moisture conservation are most likely to bring about increase in food production resulting in sustainable food security. Lack of access to food can be due to either a decline in the capacity of the people to purchase food or a sharp increase in the prices of food without a corresponding increase in the purchasing power. The state should initiate measures to deal with both these primary causes of hunger. "*India has a number of food and nutrition programmes to reduce the impact of food insecurity. Corruption and lethargy are standing in the way of implementation of these programmes*"[140]. Employment generation schemes for those who can work and relief measures for those who cannot, should be profusely undertaken by the state.

"*Food crisis is obviously built into the domestic and global economic systems. The global system can stand only on the basis of a sustainable, national food security system. And grassroots security for the local community is the foundation of such building blocks. India's old system was a good model. The global system that the 1974 World Food Conference recommended took partly from our system as the most suitable one for the developing world*"[141]. 'Swaminathan's[142] message at the New Delhi Summit 2007 was clear: from green revolution to ever green revolution to ensure that nobody goes hungry in India after 60 years of independence. As a scientist, he has done his bit. It is for the politicians and the governments at the centre and the states to make the slogan of a hunger free India by 2007 a reality. We have the technical competence and the expertise within the country, but what we lack is good governance'[143]. Food security issues are best dealt with as a subset of poverty issues and long term economic growth is the solution to both poverty and hunger.

---

[140] *Worrying reversal in war on hunger,* Deccan Herald, Bangalore edn dated 23/11/2018.
[141] Kamala Prasad, "Politics of Food" Mainstream, [July5, 2008] Vol XLVI no 29, p 1234-1241.
[142] He is an Indian geneticist and international administrator, renowned for his leading role in India's Green Revolution, a program under which high-yield varieties of wheat and rice seedlings were planted in the fields of poor farmers.
[143] Ashish Bose, "Hunger-free India by 2007", Economic and Political weekly, [12th March, 2004], Vol XXXIX No 12, p 1196-1198.

A nation may require to achieve self sufficiency of food grains at any given point of time, but that will not solve the problem. The concept of food security necessitates that, timely, reliable and nutritionally adequate supply of food should be ensured on a long term basis. This implies that the nation has to make sure of the growth rate in food supply so that it can take care of the increase in population, increase in the demand resulting from increase in the income of the people.

The Researcher thinks that if the right to food is to be achieved, it needs to be linked with other economic and social rights relating to education, work, health and information. Food security must be viewed as an approach to development planning. Hordes of frustrated, hunger stricken people are a stigma weakening the society and may cause an increase in the incidence of crime and violence. Prosperity of a country is impossible when it's people are engaged in unending cramping struggle for bread and sustenance. 'The 2018 Global Hunger Index(GHI) report published by Welt Hunger Wilfe and Concern Worldwide depicts a depressing picture of the extent and severity of hunger in India despite all these boasts about its high and fast growing economy. It notes that the levels of hunger in 51 countries, including India are still serious or alarming......but what is appalling is that some African and Asian countries fare better than India'[144]. Researcher honestly thinks that 'democracy and hunger cannot go together'.

Eliminating hunger and resultant malnutrition is well within the reach of mankind. What we need is Political resolve and well-conceived policies and concerted actions at both national and international levels which can have a dramatic impact on these nutrition problems. Many nations, including some of the poorest countries, have adopted and implemented measures to strengthen education & health, agriculture, food, nutrition and family welfare programmes which have noticeably decreased hunger and malnutrition. But, the current challenge we are facing is to build upon and further accelerate the progress already accomplished.

"As per the media reports from outside India, the reason for the hard stance of developed countries is the burgeoning food stocks built by Food Corporation of India. They are apprehensive that India may unload them in the international market at cheaper prices. It must be admitted that India's policies are haphazard. It is scandalous that India sits on a mountain of food grains when hunger and malnutrition continue to

[144] KN Ninan, "*Hunger Amidst Prosperity: Of What Use is Fast Growth*?", Deccan Herald Dated 29/10/2018 Bangalore Edition.

be serious problems. Often we hear that these grains are sold at cheap rates in international markets'[145]. It is time for us to set our house in order so that we may regain our international credibility.

The right to food is commonly misconstrued as a right to be fed and a right to a minimum ration of calories, proteins and other nutrients. According to the Food and Agriculture Organisation of the United Nations, the right to food does not mean and imply that the respective governments are obligated to handover free food to everyone. The right to food is, being guaranteed the right to feed oneself that each household either has the means to produce or to buy the food. But if anyone is deprived of access to food due to the reasons beyond one's control, like in case of armed conflict, drought, physical incapacity and natural disasters, then the state is obliged to provide them with sufficient food for their survival and which also meets their dietary requirement.

International Covenant on Economic, Social and Cultural Rights declares that when anyone is in fact without adequate food the state must proactively create an enabling environment where people become self-reliant for food or where people are unable to do so, must ensure that it is provided but has no where mandated that the state en masse should provide cheap food at their doorstep. The Researcher strongly opines that instead of providing food to the able bodied and physically fit adults and youths at subsidized price and end up with a superficial feeling of ensuring food security, the state should create more employment opportunities which encourages them to earn their livelihood instead of allowing them to languish and become a bane on the already scrawny state exchequer or on the alternate, after providing food, the state should mandate them to serve the country in some capacity as it is done by the Singapore government[146] as 'No international treaty or convention mandates the member nations to supply the food to its citizenry at their door step.

Global experience on social security programmes have made it evident that food coupons/vouchers, cash transfers are better alternatives than physical handling of food as handling charges are more onerous than the procurement cost of the food grains. It may be noted that this Act is enacted at such a time when internationally, conditional cash transfers rather than physical distribution of subsidized food, have been found to be more efficient in achieving food and nutritional security which has been adopted

---

[145] A Jayagovind, "India and Food Security: WTO Perspective", Indian Journal of International Law, [January-June 2014] Vol 54 Issue 1-2, p 505-513.

[146] In Singapore every male citizen and male permanent residents after attaining 18years are required to compulsorily serve 'National Services Liability' at least for two years.

successfully by many countries across the world. Brazil is a classic example of this - the Bolsa Familia programme, world's largest conditional cash transfer program, has lifted more than 20 million Brazilians out of acute poverty and also promotes education & health care. It is imperative that India learns from these social safety net experiences and evolves an innovative strategy that is based on more effective and appropriate policy instruments to enhance social and economic welfare.

One instance that the Researcher would like to quote at this juncture is - during the period preceding the World War II, annually, India was importing about 5.9 million tonnes of food grains which was largely rice from Burma. With the war, food imports were disrupted and fell to just about 1.16 million tonnes in 1948-52. Adding insult to injury, India was devastated by drought/famine which is popularly known as Bengal famine in 1943. In 1965 India was having severe shortage of Rice and basic cereals, though people had money, there were no food grains to buy. In this heart wrenching situation, US stepped in and entered into an agreement called PL 480[147] and helped with a program of funding, through which rice from US was supplied to India through 600 ships over a year. This consignment of rice was distributed through the public distribution system. This means to say that the state should provide food, when people have access to it but there is non- availability of food. Food security does not mean the state should provide food at the doorstep of the people en masse. The state should venture out creating more employment opportunities which can be a long term solution to the problem of food security. As the famous saying goes "Give a man a fish and you feed him a day, but teach a man to fish and you feed him for his life". Creating a dependent-on-government-for-food economy is not the answer to the problem, creating an independent-self-sustaining one is.

[147] Food for Peace Programme

# CHAPTER – IV

# LEGAL REGULATION OF FOOD SAFETY AND STANDARDS IN INDIA-AN OVERVIEW

## 4.1 Introduction

"Dining was a very risky business in the nineteenth century, you could never be quite sure exactly what you were eating. Of course, the adulteration of food was nothing new; it had probably existed in some form or another as long as people had sold or bartered food. In fourteenth century London, for example, people were fined for selling, amongst other things, stale fish, stinking pig meat and wooden nutmegs. There were laws to deal with deception, but any admixtures were difficult to detect before scientific testing, which only became possible some five hundred years later. Medieval testing was along the lines of sprinkling powdered unicorn's horn into wine to test for poison; if present, it would change the colour.

The scale of the problem increased dramatically in the eighteenth and nineteenth centuries, with the rapid growth of the towns. Supplying enough food to meet the demand was a problem, so the obvious answer to many traders was to cheat. Bulk was added to basic foodstuffs, with inferior substances: mustard powder was mixed with wheat flour and turmeric, cheese was coloured with red lead, coffee was found to contain burnt beans, burnt sugar, acorns and mangel-wurzel. Flour, perhaps the most important staple, was adulterated with plaster of Paris, pipe clay and even ground up bones. It was claimed in an anonymous pamphlet of 1757 that: "The charnel houses of the dead are raked to give filthiness to the food of the living". This was never proved with certainty. What was well known, and accepted by some, was the use of mineral salt called alum to whiten the bread and improve the volume. Alum is an astringent, and whilst not especially harmful, it caused serious digestive problems in those whose diet consisted of very little other than bread. The public was first alerted to the state of their food by the German-born Frederick Accum[1], a chemist in London, whose work, '*A*

[1] Frederick Accum was the first to raise the alarm about food adulteration. Accum was a German chemist who had come to London in 1793 and who quickly established himself as a chemical analyst, consultant and teacher of chemistry. By 1820 Accum had become aware of the problem through his analytical work and this led him to publish 'A treatise on adulterations of food and culinary poisons' - the first serious attempt to expose the nature, extent and dangers of food adulteration. The title page of the book bore a skull and a quotation from the Old Testament 'there is death in the pot' (II Kings chap.4, verse 40). The first edition sold out within a month; a US edition was published in the same year and a German translation was brought out in 1822. In his preface Accum remarked that the art of counterfeiting and adulteration had developed in England to such an extent that spurious articles of all kinds could be found

*Treatise on Adulterations of Food and Culinary Poisons*' caused a storm when it was first published in 1820. It sold out in a month and ran through four editions in two years. Accum took the bold step of naming those who had defrauded their customers and was eventually hounded out of the country by the enemies he created. It was to be another fifty years before effective legislation was introduced" [2].

According to Bible, the first ever case concerning the deception of quality of food practised on human being was the deceitful, yet colourful description made by the Serpent to deceive Eve. The book of genesis outlines as to how the Lord god had prohibited Adam and Eve from touching the fruits of a specific tree which was standing in the middle of Eden. But the cunning Serpent with his sugar coated description about the magical qualities of the forbidden fruit, deceived Eve. Here a perilous article was glorified as of higher quality with a sinister motive, though the first ever deception case though did not take place in commercial transaction, deceitful traders subsequently took the place of Serpent and habitually began to deceive innocent people with the only objective of making profit. Consequently, fraudulent and deceitful dealings were prohibited by the earlier codes. We are what we eat' is an old proverb connoting that one's health, physical and mental health and nutritional status depend on what one eats. Since most of the food articles consist of complex biological substance, an exact assessment of its quality is difficult. *'Food safety involves everybody in the food chain'* USDA Secretary Mike Johanns once remarked[3].

Close on the heels of this, due to the development in the science and technology new kinds of articles of food which were unknown in the past, have been constantly getting added. In this scenario, it has become difficult for a common man to gauge the quality and safety of the food that he is consuming. These developments have prompted the states to devise artificial definitions with new standards for the term 'food' through legislations. Safety of food has been a grave concern from the beginning of the civilisation. Food safety implies absence of contaminants or adulterants or even if they are present, they need to be well within the prescribed limit. Food systems in most of the developing countries are not as organised as they are in developed countries.

---

everywhere, but he regarded the adulteration of food and drink as a criminal offence. 'The man who robs a fellow subject of a few shillings on the highway is sentenced to death', he wrote, but 'he who distributes a slow poison to the whole community escapes unpunished'.

[2] Liz Calvert Smith, a food historian, writer, public speaker, historical food stylist and food related exhibition curator available at https://edible-history.com/2013/02/13/there-is-death-in-the-pot-a-brief-history-of-food-adulterati (accessed on 19/1/2018 at 3.43pm).

[3] Available at http://www.azquotes.com/quote/754710 (accessed on 11/7/2018 at 9.00am).

Contributing factors like scarcity of resources, problem of surging population, urbanisation, and environmental problems have worsened the situation. Due to these undesirable factors, food system continues to suffer in developing countries, negatively impacting the quality and safety of the food articles meant for human consumption and people are exposed to a wide range of various potential food hazards. Safe food is that food which is not only devoid of toxins, pesticides and physical and chemical contaminants but is also rid of microbiological pathogens[4]. The main concern of the people working for food safety is microbial food borne illness which is widely present yet goes undetected. The factors which may cause food borne disease may be 1. Pathogens, 2. Vehicle in which food is transported 3.Conditions which are conducive for the pathogen to survive, multiply and to produce toxins 4.The consumer of this chemically contaminated food article which though physically looks to be fit for human consumption.

## 4.2 Definition and Meaning of Adulteration

- *"It consists in the intentional addition to an article, for purposes of gain or deception, of any substance or substances the presence of which is not acknowledged in the name under which the article is sold"*[5].
- Adulteration usually refers to *mixing other matter of an inferior and sometimes ha rmful quality with food or drink intended to be sold. As a result of adulteration, food or drink becomes impure and unfit for human consumption*[6].
- *To corrupt, debase, or make impure by the addition of a foreign or inferior substance or element; especially: to prepare for sale by replacing more valuable with less valuable or inert ingredients*[7].
- *The act of corrupting or debasing. The term is generally applied to the act of mixing up with food or drink intended to be sold other matters of an inferior quality, and usually of a more or less deleterious quality*[8].
- *To make food or drink weaker or to lower its quality, by adding something else*[9].

---

[4] any small organism, such as a virus or a bacterium that can cause disease: available at https://dictionary.cambridge.org/dictionary/english/bacterium (accessed on 22/1/2018 at 10.47am)

[5] Arthur Hill Hassall, '*Adulterations Detected*', Longman, Brown, Green, Longmans and Roberts, 1857 available at https://books.google.com.au/books?id=r0anIiWCa_gC&dq= adulteration &as_brr= 1& source =gbs _navlinks_s (accessed on 27/6/2018 at 3.40pm)

[6] https://legal-dictionary.thefreedictionary.com/adulteration (accessed on 21/1/2018 at12.24pm)

[7] https://www.merriam-webster.com/dictionary/adulterate (accessed on 21.1.2018 at 12.27pm)

[8] https://thelawdictionary.org/adulteration/ (accessed on 21/1/2018 at 12.29pm)

[9] https://dictionary.cambridge.org/dictionary/english/adulterat (accessed on 21/1/2018 at 12.31pm)

From the above definitions and meanings, it is clear that when something of inferior quality is added to another base product knowing fully well that it deteriorates/diminishes the quality of a base product, it amounts to adulteration. This act need not be accompanied by mens rea, the very act is punishable sans mens rea. In this context, when the Researcher is dealing with adulteration of food, the Researcher means that when any article of food, meant for human consumption is mixed with something which results in deterioration of quality but increases the quantity, it amounts to adulteration. It may or may not impact the quality of the food article, it may or may not be obnoxious for human health, yet it amounts to adulteration.

### 4.2.1 Reasons for Adulteration

One most prominent reason assigned for adulteration is that it is practiced to keep in tune with the wishes and tastes of the public. Over a period of time, people get used to the taste of adulterated food stuffs and does not like the taste of unadulterated food. Another widely put forth argument is that addition made to several articles of food, constitute improvements and enhances the taste. Another widely advocated defence is that it does no harm and not hurtful to the health. In some instances addition is made to the articles of food only to increase the shelf life.

Another favourite defence used in extenuation of adulteration is that the substances employed are too inconsiderable in quantities like those used for colouring that it does not result in any harmful results. '*The great cause which accounts for the larger part of the adulteration which prevails is the desire of increased profit; a second cause is excessive and unfair competition. A trader perceiving that his neighbour in the same business is selling his goods at prices at which, if genuine, it would be impossible to realise a profit, knows that this can be done by having recourse to adulteration, and finding that he cannot compete with his unscrupulous fellow trader, at length he himself too often has recourse to the same practice. We thus perceive how difficult it is for many tradesmen who desire to do so to conduct their businesses in a honourable way, and to resist the temptation to adulterate. The main causes of the prevalence of adulteration are, then the desire of increased profit and excessive and unfair competition*[10]'.

---

[10] Supra note 5

## 4.3 History of Adulteration

Food laws can be traced back to times of the earliest societies. Ancient food regulations are referred to in Egyptian, Chinese, Hindu, Greek, and Roman literature[11]. The birth of modern chemistry in the early nineteenth century made possible the production of materials possessing properties similar to normal foods which, when fraudulently used, did not readily attract the attention of the unsuspecting consumer[12]. Many years ago, about 2500 years BC Mosaic and Egyptian laws had provisions to prevent the contamination of meat. Also, more than 2000 years ago, India already had regulations prohibiting the adulteration of grains and edible fats[13].

### 4.3.1 England

Possibly, the most perennial problem in the food and drug industry has been the problem of adulteration which lowers the quality of products by the addition of impure and inferior ingredients. History is abound with instances where the producers of food articles and drugs used to adulterate their product with an ulterior motive of maximising their profits. Water was added to wine, cream used to be skimmed off from milk and chalk added to bread. "In 1202, King John of England proclaimed the first English food law, the Assize of Bread, which prohibited adulteration of bread with such ingredients as ground peas or beans[14]. In 1848, the first national law concerned with regulating food came after the Mexican–American War, and "banned the importation of adulterated drugs." Food inspection was largely thought to be the duty of the consumer, not the government. Peter Collier, chief chemist at the United States Department of Agriculture in 1880, recommended passage of a national food and drug law, following his own food adulteration investigations. The bill was defeated, but during the next 25 years more than 100 food and drug bills were introduced in Congress"[15].

---

[11] History of Food Quality Standards available at http://citeseerx.ist.psu.edu/viewdoc/ download? Doi =10.1.1.564.1348&rep=rep1&type=pdf (accessed on 19/1/2018 at 7.44pm)

[12] Ihegwuagu Nnemeka Edith[1] and Emeje Martins Ochubiojo[2] 1Agricultural Research Council of Nigeria 2National Institute for Pharmaceutical Research and Development Nigeria, '*Food Quality Control: History, Present and Future*' available at https://www.researchgate.net/ profile/Nnemeka_ Ihegwuagu2/ publication/221923609_Food_Quality_Control_History_Pres (accessed on 19/1/2018 at 8.00pm)

[13] *Ibid*

[14] 'The Horrific History of Food Adulteration' available at https://www.facebook.com/notes/junior-morris/the-horrific-history-of-food-adulteration/10201164119735779 (accessed on 22/6/2014 at 7.04pm)

[15] *Ibid*

### 4.3.2 India

It is evident from the history that adulteration of food stuffs and other goods is not alien to Indian sub-continent. Manusmriti and Kautilya's Arthashastra[16] contained prohibitions of adulteration of food. Kautilya explicitly states that 'Adulteration of grains, oils, alkalis, salts, scents, and medicinal articles with similar articles of no quality shall be punished with a fine[17]. After the advent of Arabs in India, The tenets of Hindu jurisprudence underwent drastic changes. In the earlier period, the Muslim rulers applied Muslim criminal law to the inhabitants in place of the then existing Hindu laws[18]. After the British invasion, for a long time, they applied Muslim law to all the people of India with necessary changes. All this resulted in divergent systems of law in British India. The ambition of the British to establish their empire here, compelled them unify the legal system, without which they could not exercise uniform control on all parts of India[19].

This culminated in the appointment of Law Commission under the Chairmanship of Lord Macauly in 1834 for the unification and codification of penal laws. The need to incorporate express provision for guaranteeing the purity and wholesomeness of food and drinks was manifested in Sections 272[20] and Section 273[21] of Indian Penal Code, 1860(IPC). Section 272 deals with the sale of noxious food and drinks which are adulterated whereas Section 273 does not refer to adulteration but it absolutely prohibits the sale of noxious food and drinks with the knowledge of its obnoxious nature of it. These two provisions in effect prohibit the sale of obnoxious food and drink to the public but incorporate the necessity of *Mens Rea* for culpability. Hence, the courts were compelled to quash any prosecution filed under Section 3 of the

---

[16] Economic thoughts of Kautilya '*Contemporary Relevance of Economic Thoughts of Kautilya*' available at http://shodhganga.inflibnet.ac.in/bitstream/10603/15999/15/15_chapter%207.pdf (accessed on 30/1/2018 at 2.00pm)

[17] Summary on Kautilya's Arthashastra: Its Contemporary Relevance Publshied by Indian Merchants' Chamber (2004) available at https://www.esamskriti.com/essays/docfile/11_359.pdf (accessed on 30/1/2018 at 2.06pm)

[18] Dubey, Harihar Prasad, '*The Judicial Systems of India*', Thripathi, Bombay 47(1980).

[19] '*Essays on The Indian Penal Code*', ILI ed. 1962, Tripathi, Bombay 33. P99

[20] IPC Section 272: Whoever adulterates any article of food or drink, so as to make such article noxious as food or drink, intending to sell such article as food or drink, or knowing it to be likely that the same will be sold as food or drink, shall be punished with imprisonment of either description for a term which may extend to six months, or with fine which may extend to one thousand rupees, or with both.

[21] Ibid Section 273: Whoever sells, or offers or exposes for sale, as food or drink, any article which has been rendered or has become noxious or is in a state unfit for food or drink , knowing or having reason to believe that the same is noxious as food or drink, shall be punished with imprisonment of either description for a term which may extend to six months, or with fine which may extend to one thousand rupees or with both.

Bihar and Orissa Prevention of Adulteration Act 1919 since the impugned statute did not include 'knowledge' as required by the IPC[22]. The 'knowledge' in this context meant, the 'noxious nature' of the food and drink intended for sale[23]. This means that if the adulterated food was not noxious to the health of the consumer, neither adulteration nor its sale would attract Sections 272 or 273. This interpretation further leads us to the conclusion that adulteration with harmless ingredients for the purpose of getting more profit is not punishable under Section 272[24].

Their Lordship of the Lahore High Court emphasised this aspect in 1926 while setting aside a conviction imposed by the trial Court on a charge under Section 273[25]. In the instant case, though milk was adulterated with water, the court did not punish the accused as it opined that the adulterant is not noxious to human health. The meaning of the term 'noxious as food' as present in Sections 272 and 273 were for the first time considered in *Ram Dayal's* case[26]. In this case, ghee was adulterated with pig's fat, but the court was of the opinion that adulterant was not injurious to human health though it may be against the religion and feelings of many in the society. As per the observation of the court, the term 'noxious as food' meant 'unwholesome as food' or 'injurious to health'. In the light of many decisions by the courts, it was evident that IPC was not effective in combating adulteration. The United Provinces Prevention of Adulteration Act[27] was passed by the erstwhile United Province in 1912. This was copied mutatis mutandis by the provincial Government of Madras in 1918.

In 1919, when the portfolio of health was shifted to the local provincial governments, most of them made special provisions in their municipal Acts for the prevention of adulteration of food. However, there was no uniformity in standards or modes of implementation[28]. To resolve and to reconcile the contradictory laws of different provinces and to fix uniform standards, in 1937, the Department of Health, Government of India set up the Central Advisory Board of Health. The Board set up a Food Adulteration Committee to probe into issues of food adulteration in India

---

[22] *Awath Prasad* v. *The State* A 1952 Pat 77 (78)
[23] Ratnalal & Dhirajlal; The Law of Crimes 1982 Wadhava Sales Corpon. Nagpur) M3~.
[24] Gour, Dr.Hari Singh: The Penal Law of India Vol.II (1975) Diamond Jubilee ed. Law Publishers, Allahabad.
[25] *Dhawa* v. *Emperor,* 1926 Lah.49
[26] *Ram Dayal & Others* v. *King Emperor* A 1924 All.214
[27] Mathur, Justice G.C: '*A.P. Mathur's Commentary on Prevention of Food Adulteration Act 1954'* Eastern Book Co., Lucknow.(1989) 131
[28] Shenoi.V. Rama, "*Adulteration of Food stuffs and other Goods*" Cochin University Law Review [1983] Vol 7, p43-56

focussing majorly on inconsistent legislations and standards of food under such legislations in force then[29]. Food Adulteration Committee recommended for the enactment of an all India legislation[30]. The Bengal food Act 1919 defined adulteration as "An article of food shall be deemed to be 'adulterated' if it has been mixed or packed with any other substance or if any part of it has been extracted so as it either affects injuriously its quality, substance or nature"[31].

After India's independence in 1947, more serious thought was given to the problem of food adulteration. It was soon realized that provincial food acts[32] were not only outdated, but they also hampered trade and industry in the country. Many a time, persons committing the same offences were liable to different punishments[33]. But the recommendation was not acted upon as it was a state subject, upon which only the state legislature could deliberate. When India became Republic, after the adoption of the Constitution, the subject of food adulteration was included in Concurrent list in Entry 18 of the Seventh Schedule for which both the central and state government are competent to legislate[34]. To ensure the purity of articles of food sold throughout the country, the central government enacted the Prevention of Food Adulteration Act, 1954(PFAA)[35].

---

[29] http://shodhganga.inflibnet.ac.in/bitstream/10603/50306/9/09_chapter%202.pdf (accessed on 29/1/2018 at 5. 32 pm)

[30] Purohit, Shree DhBIjoshi, Kanshi Nath *Supreme Court on Prevention of Food Adulteration Law in India,* Jain Brothers, New Delhi (1973)

[31] Gupta J P *Commentary on the Prevention of Food Adulteration Act, 1976*, Bharat Publishing House, Allahabad ( 1983)

[32] Singh, Jaspal, *Handbook of Socio -Economic offences,* Pioneer, New Delhi, (1993). The following are precursor to the Prevention of Food Adulteration Act, 1954.

a. The Assam Pure Food Act, 1919
b. The Bengal Food Adulteration (Amendments) Act, of 1925 and 1930.
c. The Bihar and Orissa Prevention of Food Adulteration Act, 1919.
d. The Bihar Prevention of Food Adulteration Act, 1948.
e. The Bombay Prevention of Adulteration Act, 1925 and (Amendment) 1935.
f. The Central Provinces Prevention of Adulteration Act, 1919 and its Amendment Act 1928.
g. The Madras Prevention of Adulteration Act, 1918 and Amendment Act of 1928 and 1932.
h. The Orissa Prevention of Adulteration & Control of Sale of Food Act, 1938.
i. The Punjab Pure Food Act, 1929.
j. The United Provinces Prevention of Adulteration Act, 1912.
k. The United Province Act, 1916.
l. The United Provinces Act, XII of 1932.
m.The UP Pure Food Act, 1950.

[33] Mathur, Justice, *A.P. Mathur's Commentary on Prevention of Food Adulteration Act, 1954,* Eastern Book Co., Lucknow.(1989) p 234

[34] MP Jain, '*Indian Constitutional law*', LexisNexis, (7th edn 2014) p 348

[35] Radomir Lasztity, Marta Petro-Turza, Tamas Foldesi, (2004), History Of Food Quality Standards, in Food Quality and Standards, [Ed. Radomir Lasztity], in EOLSS, Developed under the Auspices of the UNESCO, Eolss Publishers, Oxford, UK, available at http://www.eolss.net (accessed on 21/7/2017 at 5.30pm)

### 4.4 Food Safety At the International Level

Food Safety is a global concern, thus almost all the civilised nations made efforts to pass legislations governing regulation of food and articles of food. In many nations regulations were reviewed by professional specialised and trade agencies from time to time to keep pace with the changing times. Nonetheless, these individual efforts restricted the development of international trade and cooperation. Most of the time, the standards fixed/adhered by the countries were contradicting the standards prescribed by other countries for similar food/article of food. To overcome this difficulty, in the beginning of the twentieth century, some of the European countries tried to harmonise the standards, but the political sub divisions, diverse regulations of food, different prescribed standards, and absence of common language contributed to the prevailing imbroglio.

In the light of growing concern towards safe food, various international organisations are constantly deliberating to improve the safety of the world's supply of food. In 1945, United Nations founded Food and Agricultural Organisation with the motto of increasing agricultural production, of raising levels of nutrition thus improving the standard of living of the people the world over. It also aimed to better the lives of people living in rural areas. To achieve all the mottos, food safety is crucial. The food borne illnesses which are the prime causes for reducing the economic productivity are rampant posing a major threat to the health of the people the world over. The World Health Organisation has also set global standards for health and also supports the governments in strengthening the national health programmes in their respective countries. The strategy of World Health Organisation to overcome malnutrition is to protect the consumers from adulteration and contamination of food and to thwart the food borne diseases.

The food safety activities of WHO hovers around the development of national food safety policies and the required infrastructure for the same, legislations relating to food and their effective implementation, food safety education, monitoring of chemical contaminants in food which may not be apparent, formulation & promotion of better food technologies and surveillance of food borne illnesses and food safety in urban scenario. WHO and FAO co-ordinate with each other and collaborate on many issues pertaining to food safety, to attain the mottos set by them at joint committees and conferences.

One such collaboration is the Codex Alimentarius Commisssion[36] (CAC), Centerband Only Detection of Exchange. It is responsible for developing and setting international food standards. Its primary task is to develop the uniform food standards which should be possible for all the nations in the world to follow. This uniform standard food code is known as Codex Alimentarius. The Codex Alimenratius or the Food code is the compilation of all standards, codes of practice, guidelines and recommendation of CAC. The Commission has published a code of ethics for international trade in food. One of the main objectives of this code is to curb the exporting countries and the exporters from dumping poor quality and unsafe food in the international market. The code also contains general principles such as international trade in food should be conducted on the principle that all consumers are entitled for safe and wholesome food and should be protected from unfair trade practices.

'Further the code says that no food should be in the international market which has in it or upon it any substance in an amount which renders it poisonous, harmful or otherwise injurious to health or consist in part or whole of a filthy, putrid, rotten, decomposed or diseased substance or foreign matter or is otherwise unfit for human consumption; is adulterated; is labelled or presented in a manner which is false, misleading or is deceptive or is sold, prepared, packaged/stored or transported for sale under insanitary conditions'[37].

Though the aim of the Codex is to set uniform regulatory standards in the interests of the international trade, yet it strives to improve the food safety standards in most of the countries in the world. As on today 140 member nations including India have agreed to abide by the standards set by it and follow its codes of practice religiously. So far, CAC has adopted more than 200 food commodity standards, more than 40 hygiene and technological practice codes and has set more than 3200 maximum residue limits for pesticides and veterinary drugs. This Code consists of standards for food articles, codes of practice for hygiene and technology, evaluation of the presence of pesticides in food and the limits for their residues, guidelines for the levels of contaminants in food, assessment of food additives and evaluation of veterinary drugs

---

[36] A body that was established in early November 1961 by the Food and Agriculture Organization of the United Nations (FAO), was joined by the World Health Organization (WHO) in June 1962

[37] Dominique Lauterburg, Food Law: Policy and Ethics, Cavendish Publishing House, London (2nd Revised edn, 2004)

residues in articles of food. The code advocates the application of safe biological principles, designing the systems and food plants that use the existing and the latest good manufacturing practices and the adoption of sound Hazard Analysis and Critical Control Points to make the food safe as the food is the only source of healthy nutrition. India became a member of the CAC in the year 1964, when the legislation dealing with food regulation was the erstwhile PFAA. The Researcher commends the efforts of the Indian government to pass PFAA, to tackle the menace of food adulteration much before the efforts of UN in the form of CAC.

## 4.5 Genesis of Food Safety Laws in India.

One of the main concerns of national governments is that any food produced in the country or imported from outside is safe and does not pose a threat to human, animal or plant health. Therefore, national governments have their own mandatory standards and regulations to avoid such threats.

### 4.5.1 Prevention of Food Adulteration Act, 1954

The laws regulating the quality of food have been in force in the country since 1899[38]. Since the early 1950's Parliament has endeavoured to craft an anti-food adulteration law that can be fruitfully enforced. "In a social welfare state, public health is the main responsibility of the state and so the state should endeavour to protect the health of citizens from insanitation, environmental pollution, malnutrition and adulteration of food items"[39]. Food adulteration has been a constant problem in India. As early as 1950 parliament concluded that food adulteration was rampant[40]. Even after 13 years parliament discovered that the problem of adulteration of food was increasing rather than decreasing even after PFAA had come into being. The Act was only a beginning to curb the socio-economic crime which was posing a threat to the society and also to breaking down the image of white-collar criminals as upstanding citizens[41]. Till 1954 different states had enacted their own laws governing food leading to ambiguities and confusions in the rules and standards for food affecting the inter provincial trade. Adulteration of articles of food was so rampant, widespread and

---

[38] https://archive.india.gov.in/sectors/health_family/food_prevention.php (accessed on 31/1/2018 at 10.55am)

[39] Parkash C Juneja, "Prevention of food adulteration Act and Consumer protection", Central India Law Quarterly, [1988] Vol 1, p 370-386

[40] The Gazette of India, 1950, Part II, Sec 5.

[41] Alan M Katz, "The Law against Food Adulteration: A Current Assessment and a Proposal for an Enforcement alternative", Journal of the Indian Law Institute, [1977] Vol 1, p 68-76

persistent that nothing short of a somewhat drastic remedy in the form of a comprehensive legislation became the need of the hour[42]. The Central Advisory Board appointed by the Government of India in 1937 and the Food adulteration Committee appointed in 1943 jointly reviewed the subject matter of food adulteration and recommended a need for a central legislation. The Constitution of India has conferred on the Central government to make such legislation as the subject of food has been enlisted in the Concurrent List. It is the duty of the Ministry of health and family welfare to ensure the supply of safe food to the consumers.

In the light of this the Prevention of Food Adulteration Act 1954 was enacted with the sole objective of protecting the public from poisonous and harmful foods, to prevent the sale of substandard foods, to protect the interests of the consumers by eliminating fraudulent practices by the traders who were in food industry. 'It is enacted to curb the widespread evil of food adulteration and is legislative measure for social defence'[43]. The object of PFAA has been outlined by the Apex court in *Dinesh Chandra* v. *State of Gujarat*[44]. In the later part of 20th century, the menace of adulteration of articles of food had grown to such bad heights that nothing short of stringent remedy and absolute liability could change the situation arising out of the deep rooted evil. Having realised that only a resolute onslaught of this anti social behaviour could bring relief to the society, the Prevention of Food Adulteration Act was enacted and was applicable to the whole of India including Jammu and Kashmir[45].

The Act came into effect from 15th June 1955, repealing all laws existing in different states concerning adulteration of food. The production, sale, accumulation or distribution of adulterated or misbranded food was prohibited under the Act[46]. Section 5 of the Act[47], prohibited the import of adulterated and misbranded food and the import of

---

[42] http://www.medindia.net/indian_health_act/the-prevention-of-food-adulteration-act-1954 - introduction htm (accessed on 31/1/2018, at5. 31pm)

[43] Prakash C Juneja, "Prevention of Food Adulteration Act and Consumer Protection", Central Law Quarterly [1988] Vol 8, p 370-386

[44] AIR SC 1011 "the object and the purpose of PFA are to eliminate the dangers to human life from the sale of unwholesome articles of food........ it is enacted to curb the widespread evil of food adulteration and is a legislative measure for social defence. It is intended to suppress a social and economic mischief, an evil which attempts to poison, for monetary gains, the very sources of substance of life and the well being of the community".

[45] The words "except the state of Jammu and Kashmir " omitted by Act (4) of 1971, Section 2 w.e.f 25/1/1972

[46] https://lawyerslaw.org/the-prevention-of-food-adulteration-act-1954/(accessed on31/1/2018at 11.43am)

[47] PFA Section 5. Prohibition of Import of certain articles of food.—No person shall import into India— (a) any adulterated food; (b) an misbranded food; (c) any article of food for the import of which a license is prescribed, except in accordance with the conditions of the license; and (d) any article of food in contravention of any other provision of this Act or any rule made there under

any other food articles was allowed strictly in accordance with the terms of the license only. In this respect, the law pertaining to sea customs also applied and the customs officials had the same powers in respect of adulterated and spurious articles of food as they had in respect of goods prohibited under the Sea Customs Act[48].

The PFAA broadly covered standards for food, general procedures for sampling, analysis of food, powers of authorised officers, nature of penalties, parameters relating to food additives, preservatives, colouring agents and packing and labelling of foods. The Act to some extent regulated the consumer –supplier relations. It also ensured that food is prepared, packed and stored hygienically. It gave a comprehensive definition for the term 'adulteration' that it became impossible for the adulterators to escape from the offences as enumerated under the Act. Inter alia, the Act provided for the establishment of a Central Food Laboratory which analysed the samples of food and gave a final opinion in disputed cases. The Act also established the Central Committee for food standards and vested the central government with powers to stipulate the standards of quality for food and other allied articles of food.

#### 4.5.1.1 Meaning of Adulteration under the PFAA

Adulteration connotes mixing of something spurious or of inferior quality to any commodity which lessens its purity and makes it dangerous for use. Any material which is or could be employed for the purposes of adulteration is called adulterant[49]. Section 2 (i a) of the Act lists out the instances where the articles of food may be considered to be adulterated[50]. An article of food, being a primary food, is not deemed to be adulterated

---

[48] Sea Customs Act 1878, Prohibitions and Restrictions of Importation and Exportation. Section 18 No goods specified in the following clauses shall be brought, whether by land or sea, into British India:- (a) any book printed in infringement of any law in force in British India on the subject of copyright, when the proprietor of such copyright, or his agent, has given to the Chief Customs-Authority a notice in writing that such copyright subsists, and a statement of the date on which it will expire: (b) counterfeit coin : or coin which purports to be Queen's coin of India, or to be coin made under the Native Coinage Act, 1876, but which is not of the established standard in weight or fineness : (o) any obscene book, pamphlet, paper, drawing; painting, representation, figure or article : (d) articles bearing any names, brands or marks being, or purporting to be, the names, brands or marks of manufacturers resident in the United Kingdom or British India, and not made by such manufacturers.

[49] The Prevention of Food Adulteration Act, 1954, sec. 2(i).

[50] Under section 2 (ia) of the PFAA, an article of food can be said to be adulterated if:

(1) it contains any other substance which affects, or is so processed as to affect injuriously the nature, substance or quality thereof; or

(ii) any inferior or cheaper substance has been substituted wholly or in part for the article so as to affect injuriously the nature, substance or quality thereof; or

(iii) any constituent of the article has been wholly or in part abstracted so as to affect injuriously the nature, substances or quality thereof; or

(iv) the article had been prepared, packed or kept under insanitary conditions whereby it has become contaminated or injurious to health; or

where the quality and the purity of it has decreased below the prescribed standards owing to the natural causes which are beyond the control of human agency. Also, where two or more articles of primary food are mixed together and the resultant article of food is stored, sold or distributed under a name, which denotes the ingredients thereof; and is not injurious to health, such an article is not deemed to be adulterated[51].

It is the fundamental principle of criminal jurisprudence that no person should be punished without a guilty mind. Bishop advocates "there can be no crime big or small without an evil mind"[52]. The same view has been reiterated by Salmond and suggested that before the law can justly punish an act, an inquiry must be made into the mental attitude of the wrong doer[53]. Coke, was the protagonist of the maxim '*actus non facit nisi men sit rea*' and this principle was widely applied in the English decisions. Lord Kenyon[54], in 1978 while discussing the importance of this maxim opined that it was a principle of natural justice[55].

Lord Abinger, who followed suit observed "it is a maxim older than the law of England, that no man is guilty unless his mind is guilty"[56]. But over a period of time, due to the steady rise in new kind of offences which could be punished without criminal intent whatsoever, the physical act alone could be punished due to the ensued damage sans mental element, this was a major departure in the deep rooted traditional view of mensrea being an essential part of a crime. A person may be held liable for the sale of adulterated food articles and also for other offences based upon his conduct only without considering his mental element. According to Steven Levitt 'a crime is an act; it

---

(v) the article consists wholly or in part of any filthy, putrid, rotten, decomposed or diseased animal or vegetable substance or is insect infested or is otherwise unfit for human consumption;
(vi) the article is obtained from a diseased animal; or
(vii) the article contains any poisonous or other ingredient which renders it injurious to health; or
(viii) the container of the article is composed, whether wholly or in part, of any poisonous or deleterious substance which renders its contents injurious to health; or
(ix) any colouring matter other than that prescribed in respect thereof is present in the article, or if the amounts of the prescribed colouring ' matter which is present in the article are not within the prescribed limits of variability; or
(x) the article contains any prohibited preservative or permitted preservative in excess of the prescribed limits; or
(xi) the quality or purity of the article falls below the prescribed standard or its constituents are present in quantities not within the prescribed limits of variability, but which renders it injurious to health; or
(xii) the quality or purity of the article falls below the prescribed standard or its constituents are present in quantities not within the prescribed limits of variability but which does not render it injurious to health.

[51] *Ibid.*, sec. 2(ia),(b) to (m).

[52] Bishop Criminal Law, T.H. Flood, (9th edn 1923) available at https://books.google.co.in/books?id= o2 VHAQAAIAAJ&source=gbs_book_other_versions (accessed on 24/2/2018 at 5.07pm)

[53] P J Fitzgerald *Salmond on Jurisprudence*, Universal Law Publishing Co Ltd (12th edn, 2016)

[54] Lord Chief Justice of England

[55] *Fowler* v. *Padjet* 1978 JTR 509,511

[56] *R* v. *Allday*, (1837)8C&P 136

is not an act plus an intent'[57]. The modern judicial trend also seems to suggest that criminality of an act depends upon behaviour, for the court concerns itself more upon behaviour of the individual[58]. In *Pearks, Gunsten Tee Ltd* v *Ward*[59], Channel J, observed " by the general principle of criminal law, if the matter is made criminal offence , it is essential that there should be something in the nature of *mensrea*.....But there are exceptions to this rule and reason for which is that the legislature has thought it so important to prevent the particular act from being committed that it absolutely forbids it to be done; and if it done, the offender is liable to a penalty whether he had any *mensrea* or not".

The departure from mental element as an important element for an offence is that all crimes seem to be the result of the doctrine of *Laissez-faire* which means maximum liberty and freedom from interference by the State, lost its dominancy in favour of social interest in the 20th century. This compelled the individual interest to be relegated to backdrop and became secondary and gave way to social interest. In the light of this, the modern social welfare states have to take care of the health of the people. The State cannot discharge this obligation effectively unless it equips itself with progressive legislations. Eventually, the State to protect the society from the misdeeds of few individuals, wide spectrum of statutory regulations have been passed mandating a set of standards to be maintained in the consumable goods.

The States are basically concerned with purity, minimum quality of food articles, drugs and medicines. Because of this, the 20th century common law lawyers termed the whole group of such offences as "public welfare offences"[60]. Finally, criminologists have also opined that the object underlying the correctional treatment must be changed from mere punishing people to the fruitful one of protecting the social interest. In this 20th century, due to the raising complexities from all corners there is a demand for more and more social regulations to cater to the needs of the society. The courts have also justified the modern legislative stand with convincing emphasis concentrating on the injurious conduct of the defendant rather than on the problem of his degree of guilt. Thus the trend of dispensing with mensrea is not an accident but the

---

[57] Extent and Function of the Doctrine of Mensrea (1923) 171 LL. L Rev

[58] *Sherras* v. *De Rutzen*, (1895) 1 Q. B. 91, *Warner* v. *Metropolitan Police Commissioner*, (1968) 2 All. E R 356

[59] (1902) 2 K.B.I (Kansas Bureau of Investigation)

[60] L M Singh, "Strict Liability under Prevention of Food Adulteration Act 1954:Legislative Consciousness and Judicial Activism", Supreme Court Journal [September-December, 1987] Vol 3, p 121-143

result of changing social conditions. In tune with this trend, the courts have also trodden the new path ignoring the state of mind of the offender.

The Researcher thinks that, the days are not far when the courts might abrade mensrea even in other criminal offences and the physical act may alone be punished in the interest of the society. The Researcher thinks that in heinous crimes, the perpetrators must be absolutely held liable irrespective of his mensrea. This development has caused an apprehension in the minds of the jurists that the mensrea may absolutely be done away with. The Indian Penal Code defines every offence with reference to the specific state of mind of the accused. The wording of the offences are either prefixed or suffixed as knowingly, involuntarily, fraudulently, dishonestly. It is important to understand that offences relating to adulteration have also been defined in the similar way. Section 273 of Indian Penal Code, 1860 provides that whoever knowingly sells the food stuff which is unfit for human consumption, shall be punished. This shows that the much celebrated doctrine of mensrea has not been ignored.

PFA also ignored the doctrine of mensrea and the same is evident from section 19 (1) of the Act which provided that it shall be no defence in a prosecution for an offender pertaining to the sale of the adulterated or misbranded food articles to set up a defence that he was ignorant of the nature, substance or quality of food sold by him. So it is clear that sale of adulterated article is punishable even though the vendor has no knowledge of its being adulterated. 'Knowledge of adulteration constitutes mensrea aggravating the offence"[61]. To keep the PFAA alive and to abreast of the changes in the society, the Act was thoroughly amended in 1976. Section 16 dealing with the Penalties was amended and stringent penalties were provided depending on the gravity of the offence[62]. In order to speed up the proceedings under the Act, Section 16-A was inserted

---

[61]Smt Lajwanti V Ganatra, Advocate, Bombay High Court and V B Ganatra, Advocate Supreme Court, "Government Playing with Public Health' "The Kerala Law Times[1973], p 77-80

[62] Section 16(1A) If any person whether by himself or by any other person on his behalf, imports into India or manufactures for sale, or stores, sells or distributes,—
(i) any article of food which is adulterated within the meaning of any of the sub-clauses (e) to (l) (both inclusive) of clause (i a) of section 2; or
(ii) any adulterant which is injurious to health, he shall, in addition to the penalty to which he may be liable under the provisions of section 6, be punishable with imprisonment for a term which shall not be less than one year but which may extend to six years and with fine which shall not be less than two thousand rupees: Provided that if such article of food or adulterant when consumed by any person is likely to cause his death or is likely to cause such harm on his body as would amount to grievous hurt within the meaning of section 320 of the Indian Penal Code (45 of 1860), he shall be punishable with imprisonment for a term which shall not be less than three years but which may extend to term of life and with fine which shall not be less than five thousand rupees

which ordained Summary trial of cases[63]. It is interesting to note that magistrates were conferred power to award more punishment than they could under the Cr.P.C[64]. A careful reading of the penal provisions indicated that the legislature weighed the cry of millions of people more important than few corrupt persons by prescribing stringent measures. The courts after weighing the individual freedom on the one hand and social interest on the other have stressed on strict liability and the same may be known from *Bhagawan Das Jagdish Chandra* v. *Delhi administration*[65], in which the Supreme Court clearly dealt with absolute liability and observed

*"It is now well established that for establishing an offence under the Prevention of Food Adulteration Act, it is not necessary to establish mensrea either on the part of the manufacturer or distributor or vendor. Even knowledge on the part of all of them that the food was adulterated is not necessary. Ignorance on the part of any one of them that the food was adulterated would not absolve them of liability".*

In *A.P.G&S. Merchants Association* v. *Union of India*[66], the petitioner contended that the Act imposed unreasonable restrictions as Section 16 (1)(a) of the Act created absolute liability and imposed severe penalties for storage and sale or distribution of articles of food found to be adulterated or misbranded or prohibited by law. The Supreme Court rejected this contention and held that for the protection of society, ensuring the purity of articles and preventing malpractices by the trader, the severity of penalties imposed by the Act, is not very disproportionate to the risk involved that it could be deemed to be unreasonable.

---

63 16A. Power of court to try cases summarily.—Notwithstanding anything contained in the Code of Criminal Procedure, 1973 (2 of 1974), all offences under sub-section
(1) of section 16 shall be tried in a summary way by a Judicial Magistrate of the first class specially empowered in this behalf by the State Government or by a Metropolitan Magistrate and the provisions of sections 262 to 265 (both inclusive) of the said Code shall, as far as may be, apply to such trial: Provided that in the case of any conviction in a summary trial under this section, it shall be lawful for the magistrate to pass a sentence of imprisonment for a term not exceeding one year: Provided further that when at the commencement of, or in the course of, a summary trial under this section it appears to the magistrate that the nature of the case is such that a sentence of imprisonment for a term exceeding one year may have to be passed or that it is, for any other reason, undesirable to try the case summarily, the Magistrate shall after hearing the parties, record an order to that effect and thereafter recall any witness who may have been examined and proceed to hear or rehear the case in the manner provided by the said Code.

64 Section 21 Magistrate's power to impose enhanced penalties.—Notwithstanding anything contained in section 29 of the Code of Criminal Procedure, 1973 (2 of 1974), it shall be lawful for any Metropolitan Magistrate or any Judicial Magistrate of the first class to pass any sentence authorised by this Act, except a sentence of imprisonment for life or for a term exceeding six years, in excess of his powers under the said section

65 AIR 1975 SC 1309

66 (1971) 1 SCJ 518: AIR 1971 SC 2346

The Act was amended thrice in 1964, 1976 and in 1986 to keep pace with the new requirements and to make it more suitable to tackle the then emerging issues. These amendments to some extent plugged the loopholes and made punishments more stringent, empowered the consumers and voluntary organisations to play a pivotal role in its effective implementation. In most of the states the implementation of the Act was under the administrative control of the directorate of health sciences. The penalties were set out and appropriate legislative amendments were made considering the gravity of offences and the offences were categorised based on punishment extending upto life imprisonment. The provisions under PFA Rules have been amended nearly 360 times and standards of around 250 articles of food which are of mass consumption have been prescribed [67]. While amending, standards prescribed by codex, social and cultural practices, improvements in the food industry, dietary habits and nutritional status of our people were taken into consideration.

#### 4.5.1.2 Judicial Response under PFAA

In *Northan Mal* v. *State of Rajasthan* [68], the food inspector, purchased chilli powder from the appellant to check adulteration and subjected it for analysis. The public analyst certified that the sample contained ash to the extent of 8.38% by weight. The lower court convicted the appellant. But the apex court on appeal acquitted the appellant and held that it is unsafe to uphold the conviction of the appellant as the 'adulteration found in the sample was marginal' and also did not rule out the possibility of there being an error in the analysis. In *Gauranga Aich* v. *State of Assam*[69],a mere addition of salt to chilly powder was held to make it injurious to health and was held adulterated on the ground that the quality and purity of chilly powder had deteriorated below the prescribed standard. In *State of Rajasthan* v. *Ladu Ram*[70], though the respondent was selling milk without a valid license, was acquitted on the ground that rational reasons were recorded by the magistrate for his acquittal. In *Kailash Chandra* v. *State of U.P*[71], in this case, the quantity of salt found in the sample was marginally excess from what was depicted on the label and the appellant was convicted by the trial court for false and

[67] https://archive.india.gov.in/sectors/health_family/food_prevention.php?pg=2 (accessed on 31/1/2018 at 11.31am)
[68] 1995 CriLJ 2661.
[69] 1990(2)FAC41.
[70] 2002 Cri LJ 426 (Raj).
[71] 2001 All LJ 2753.

misleading statements on label. In appeal, the appellant was acquitted as the court opined that the presence of adulterant was only marginal and not harmful.

In *M. Eswaraiah* v. *State of A.P*[72] the Food Inspector purchased atta and submitted it for analysis for the public analyst. Here also the adulteration being marginal, Revision was allowed and the conviction was set aside. In *Maya Prakash* v. *State of U.P. and Another*[73], after the analysis of ghee sample by the public analyst, it was found that ghee was deficient of prescribed standard of vitamin A and the accused was sentenced and convicted. But on revision, he was acquitted on the ground that the sample was substandard. Similar prosecution lapses were found in *Dilip Singh* v. *State of Rajasthan*[74], *BalKrishan* v. *State of Rajasthan*[75], *Subhash Chander* v. *State of Punjab*[76], *P. Unnikrishan* v. *Food Inspector, Palghat Municipality, Palghat*[77], *Kishan Lai* v. *State of Rajasthan*[78]. In *Motumal* v. *State of M.P*[79], the petitioner was found guilty of selling adulterated peppermint. The Food Inspector got the sample analysed by the public analyst and it was reported that the sample did not conform to the standards without there being any mention of the standard. The trial court convicted the accused and the Sessions court also confirmed the sentence. But on an appeal by the convict, the High Court reversed the judgement and held that there was nothing on record of the public analyst to show that something present in the peppermint injuriously affected the quality or nature of the article or it had any filthy, putrid or rotten or decomposed or diseased animal or was unfit for human consumption.

In *Sawaran Singh* v. *State of Haryana*[80], the Food Inspector purchased milk from the accused and subjected the same for analysis, the report of which showed that the milk was not of prescribed quality. On the request of the accused the second sample was sent for analysis to the Director, central food laboratory. The trial court held the accused the guilty of the offence and convicted him. The accused filed a revision petition in the high court. Though the court held that the public analyst as well as Director, Central Food Laboratory found that sample contained more fat than required, yet it was a marginal deficiency. The court attributed this to the wrongful method of

---

[72] 1999 (1) ALJ 682
[73] 1998 All LJ 116
[74] 1997 RLW (2) 1756.
[75] 1997 RLW (2) 1082
[76] 1997 Cri LJ 563 (Pun).
[77] 1995 Cri LJ 3638.
[78] 2001 Cri LJ 3617 (Raj).
[79] 1999 Cri LJ 4038.
[80] 1998 CriLJ 204 (P&H).

taking the sample of milk, the benefit of doubt was extended to the accused who was ultimately acquitted of the charge.

In *Ranvir Saran* v. *State (Delhi Administration)*[81] a team of Food Inspectors purchased Anardana from the shop of the petitioner and subjected the same to the public analyst for analysis. The analysis report stated that the sample did not conform with the required standards and the magistrate convicted the petitioner after duly satisfying himself that the prosecution had proved the case beyond any doubt. But the session judge reduced the sentence and the high court altogether acquitted the petitioner stating that the entire case was not examined by both the courts in proper perspective. In *M/s Raj Traders* v. *State of H.P*[82], the Food Inspector purchased the mirch kutti for the purpose of analysis and subjected the same for analysis. The article was found containing banned adulterant oil soluble coal tar dye of red colour as an admixture. The shop keeper, his firm, and the petitioner were prosecuted. The main point for consideration in this case was that if the coal tar dye was a non permitted one and if it can be the basis for prosecuting the offenders. The high court ruled that the report does not give any information about the obnoxious nature of the adulterant nor it was shown that the article in question was unfit for human consumption. The accused was acquitted. The interpretation of 'milk', being a primary food was a vexed question to the judiciary.

In *Hariram* v. *State of MP*, the Madhya Pradesh High Court held that milk is not a primary food. But the same court, in the earlier case, *Gopal* v. *State of MP*[83], had taken the contrary view. Court held that the deficiency of content of fat in the milk being a primary food depends on the breed of the cattle, quality of food and the climate, these are factors are beyond the control of human agency. A similar view was expressed in *Food Inspector* v. *State of Kerala*[84], by the Kerala High Court. But the High Court of Punjab and Haryana had held that milk is not a primary food in *Chand* v. *State of Punjab*[85] and *Budha Ram* v. *State of Punjab*[86]. The word 'insect-infestation was also ambiguously interpreted by the courts.

---

[81] 1999 Cri LJ 1082 (Del).
[82] 1995 Cri LJ 1990 (HP).
[83] 1992 Cr. L.J. 2135(2141
[84] 1978 KLJ. 830. *F.I* v. *Narayanan* 1980 KL.J. 454
[85] 1987 E.F.R. 61
[86] 1988 E.F.R 558

In a case popularly known as *Kacheroomal's case*[87] in a revision petition, the High Court ruled that 'presence of living insects is a must for labelling the food article to be insect infested, if the insects are dead, it is not insect-infestation but a case of insect damage only'. On appeal, this view was criticised by Justice Sarkaria, the Supreme court judge and he observed "It would be straining one's common sense to say that an article of food which is infested with living insects and is consequently unwholesome for human consumption ceases to be so and becomes wholesome when these insects die out and infestation turns into an infestation by dead insects"[88]. But the presence of worms in any food article did not make it insect infested as the High Court of Calcutta in *Roller Flour Mill's case*[89] ruled. In *Kacheroomal's* case the word 'insect-infestation' was read along with the expression 'unfit for human consumption'. According to the Court, the former takes the hue from the latter. In such cases, the prosecution had to prove that owing to the presence of filth or insect infestation, the article had become unfit for human consumption.

But this view was later changed by the Supreme Court in *Tek Chand Bhatia's case*[90]. Their Lordships observed: "It seems to us that the last clause 'or is otherwise unfit for human consumption' is residuary provision which would apply to a case not covered by or falling within the clauses preceding it. If the phrase is to be read distinctively, the mere proof of the article of food being filthy, putrid, rotten decomposed ... or insect infested would per se sufficient to bring the case within the purview of the word 'adulterated' as defined in sub clause (f) and it would not be necessary in such cases to prove further that the article of food was unfit for human consumption"[91]. Therefore the prosecution did not have to prove that due to insect-infestation, the article of food has become unfit for human consumption[92]. In a subsequent case *Brahma Das* v. *State of H.P*[93], the accused was prosecuted for selling insect damaged masurdhal. The court ruled that the presence of 36% insect damaged grains comprising of living and dead insects, itself is an indication that the article is unfit for human consumption.

---

87 *Delhi municipality* v. *Kacheroomal* a 1976 SC 394
[88] 1976 SC 396
[89] 1973 FAC 257, *State* v. *Puram Lal A*1985 SC
[90]' *Delhi Municipality* v. *Tek Chand Bhatia* A' 1980 SC 360
[91] 1980 SC 362
[92] '*State of Punjab* v. *Kerwal Krishnan*' 1992 Cr. L.1. 743;' *F.I.* v. *Divakaran*' 1980 KLT 369
[93] A. 1988 SC 1789.

But, over a period of time there seemed to be a deviation from the literal interpretation of the provisions of the Act, like in *Municipal Corporation Delhi* v. *Ved Prakash* [94], where on analysis, the kabuli channa was found to be unfit for human consumption due to 100% infestation but the analysis report did not divulge the kind of infestation. The Delhi high court refused to consider the report because of its ambiguity and held that in as much as the nature of the infestation was not disclosed, it was difficult to see whether and why the article was actually unfit for human consumption. The court further observed that the report ought to disclose the exact nature and the extent of infestation and in the absence of such details, it would be impossible to make out conclusively that the sample was unfit for human consumption within the meaning of Section 2 (i) (f) of the Act. The same Delhi High Court, even in a later case *Municipal Corporation of Delhi* v. *Shanti Prakash*[95], in this case the analysis report only stated that the sample of chilli powder was highly insect-infested and living insects were also found. But, the court held that "unless there is some evidence to show that it was rendered unfit for human consumption on account of such infestation, it would not be held to be adulterated". In the light of these cases and the observations by the court the Researcher strongly opines that this is not the true exposition of law.

The Supreme Court also toed the same line by overlooking the clarion call of the society in *PK Tejani* v. *MR Dange*[96], where the petitioner was a dealer of various items and the scented supari(areca nut was one among them. The Food Inspector purchased a sample of this for analysis and the analysis report revealed the presence of saccharine and cyclamate which were artificial sweeteners. It was pleaded before the court that the dealer believed in good faith that there was no cyclamate in the supari and he also didn't know honestly that saccharine was contraband. The Supreme Court held that in the absence of proof that the addition of saccharine and cyclamate are injurious to health, the food cannot be held to be 'adulterated' as it thought that 'admixture of saccharine and cyclamate is not injurious to health, so it does not amount to adulteration'. Based on this observation, it is implied that if food article which is otherwise adulterated but not injurious to health, it will not attract the penal provision of the Act.

This judicial fiat encouraged the profit maker to indulge in adulteration with impunity on the basis that "*the admixture may not be injurious to health*". In this case

---

[94] (1974) Cr LJ 189 (Delhi) (DB)
[95] (1974) Cr LJ 1086 (Delhi) (DB)
[96] AIR 1974 SC 228

the apex court completely ignored the purpose and spirit of PFAA and though it is appreciable to hold the vendors liable for the violations of law but to require the prosecution to prove that the admixture is injurious to health is deploring. Subsequently in the same year, the Supreme Court in *Ram Labhaya* v. *Municipal Corp of Delhi &others*[97] and *Smt Mani Bai* v. *State of Maharashtra*[98], turned deaf ears to the plea that "it is not injurious to health" or "it does not injuriously affect the quality of the article of food" and opined that the crying need of the society is not only to curb but eradicate the adulteration stock and barrel. It is to everyone's knowledge that the admixture of vanaspati and ghee is not injurious but yet the same was prohibited under the Act[99]. All these developments prompted the legislature to remove the murky clouds by enacting the amendment Act in 1976 which provided that "an article of food shall be deemed to be adulterated if quality of the food, falls below the prescribed standard or its constituents are present in quantities not within the prescribed limit of variability but which does not render it injurious to health"[100].

### 4.5.1.3 Short Comings of PFAA

"Food Inspectors were not made accountable and there was no adequate check on them. More than fifty percent cases of food adulteration submitted for trial were acquitted for lack of proper and adequate attention on the part of authorities responsible for implementing the Act"[101]. "The dropsy-death episode[102] in the edible oil market in 1998 was an indication that Indian industry has a lot of scope for improvement in agro-

---

[97] AIR 1974 SC 789;(1974) 4 SCC 491

[98] (1974) 1 SCJ 712; AIR 1974 SC 434

[99] Rule 44(f) provided Inter alia, "no person shall sell vanaspati to which ghee has been added"

[100] Section 2(ia)(m) substituted by Act 34 of 1976.

[101] Shriniwas Gupta "Consumer Protection Against Food Adulteration In India", Central India Law Quarterly, [January-March, 1992 Part1 1992] Vol V

[102] *State* v. *Kamal Aggarwal*, The present case is one of the cases which were registered in the year 1998, when several instances of persons being affected by disease "dropsy", which results from consumption of adulterated mustard oil, were reported. The accused Kamal Aggarwal, a manufacturer of mustard oil, is alleged to be the source from which the contaminated mustard oil began its journey. As per case of the prosecution, accused Kamal Aggarwal had sold 11850 kg of mustard oil, adulterated with argemone oil, to Rampat Oil Mills on 19.07.98 and from Rampat Oil Mills the said oil was sold to the retailer named Shyam Bihari Gupta,at Pahar Ganj from where complainant Dal Bahadur purchased about two liters of mustard oil. This adulterated mustard oil was consumed by complainant Dal Bahadur and his family members as a result of which Dal Bahadur, his wife Swastika @ Susti and his children Sonia and Sunil got afflicted with "dropsy" disease. Child Sonia, aged about 6 years, was most affected and expired on 06.09.98. Matter was reported to police on 06.09.98 from D.D.U. hospital where Sonia was admitted for treatment. Statement of Dal Bahadur was recorded and a case was registered. During the course of investigations sample of mustard oil was seized from the house of complainant and was sent for chemical examination and resulted positive for argemone oil. But the prosecution has failed in proving its case against the accused beyond reasonable doubt and hence accused is acquitted of offences. Available at https://indiankanoon.org/doc/93497033 (accessed on 5/8/2018 at 4.10pm)

processing and food quality"[103]. "The Act suffered from a number of loopholes. For instance, there was no requirement for training to food inspectors, did not provide for mandatory standardisation of food products, inspectors to the population ratio was missing in the Act" [104]. The Act was not applicable to the imports made by the government of India.

Despite PFAA, the central government imported milo containing poisonous dhatura seeds and fungus and wheat containing poisonous ergot from America in June 1973. "The central government and the state government and the Food Corporation of India are violating the law of the land in not furnishing to the purchasers the compulsory statutory warranty under Section 14[105] of the PFAA, 1954"[106]. "The study discloses that there is hardly any improvement in the enforcement of this social legislation since 1972 when the Law Commission submitted its report. Most of the acquittals under the Act are the result of either the failure to detect or investigate the offence, or the inefficiency in the actual conduct of the case by the prosecuting authorities or the delay in the investigation"[107]. "A perusal of Section 13 of the PFAA will show that only a formal procedure has been prescribed but not a fair and reasonable one"[108].

At the outset, the Act did not prescribe the mandatory standardisation of food products. Though punishment under the Act varied from minimum penalties to life imprisonments, death penalty was conspicuous by its absence however, dreadful the offence of adulteration was. Law must be nurtured with sufficient biting teeth and genuine apprehension should be created in the mind of every person in the trade that any deviation from the set standards shall be visited with exemplary punishment. The Act failed to distinguish between different categories of adulteration, based on their

---

[103] Sathish Y Deodhar, "WTO pacts and Food Quality Issues", Economic and Political Weekly,[July 28-August3, 2001], Vol XXXVI No 30, p 2813-3816

[104] Dhulia, Anubha, "Laws on Food Adulteration: A Critical Study with Special Reference to the Food Safety and Standards Act, 2006", ILI Law Review, (February, 2010) Vol1.

[105] Sec14 PFAA, Manufacturers, distributors and dealers to give warranty.—No manufacturer or distributor of, or dealer in any article of food shall sell such article to any vendor unless he also gives a warranty in writing in the prescribed form about the nature and quality of such article to the vendor: [59] [Provided that a bill, cash memorandum or invoice in respect of the sale of any article of food given by a manufacturer or distributor of, or dealer in, such article to the vendor thereof shall be deemed to be a warranty given by such manufacturer, distributor or dealer under this section. Explanation.—In this section, in sub-section (2) of section 19 and in section 20A, the expression "distributor" shall include a commission agent.

[106] Supra note 51

[107] Vachaspati Magotra, "The Prevention of Food Adulteration Act: Consumer at the mercy of Indolent and Impetuous Enforcement", Indian Socio-Legal Journal [1978] Vol IV, No1, p 43-55

[108] P.A.S Rao, "A Critique on the Prevention of Food Adulteration Act 1954", Chartered Secretary, The Journal of Corporate Executives,[Jan-December 1983]Vol XII, p 822-828

severity of the damage that was caused. The medical officers were to function even as food inspectors which in essence meant they were only part time food inspectors. Food inspectors should have had the necessary background of food science, food technology than the medical background which altogether is a different field.

The rules framed under this Act were very technical and fixed standards for various articles of food and other allied matters without fixing the accountability in case of negligence of the inspectors or any officers who were in charge of the implementation of the Act. Lack of adequate number of food inspectors, laboratories for testing the samples and absence of time frame for testing the samples defeated the purpose of the Act. The prescribed punishments under the Act were too petty and small fines which failed to deter the offenders. Under the Act, no prosecution could be instituted except under the written consent of the central or state government. Where cognisance was taken on the basis of the FIR lodged by the police, but if there was nothing on record to establish that police was authorised by the central or state government to institute the prosecution, the court was held to have no power to hear the complaint thereby rendering the consequent proceedings liable to be quashed[109].

In 2003, the Centre for Science and Environment had disclosed that coke and Pepsi soft drinks were containing pesticides and requested for ban but to no avail. Again after 3 years CSE conducted a detailed survey and released its findings depicting that coke and pesticides contained a cocktail of 3 to 6 pesticides and the average pesticide residues was 24times higher than the standard prescribed by the Bureau of Indian Standards. These soft drink companies are continuously committing crimes under PFAA and the Bureau of Indian Standards Act. Due to all these loopholes it was realised that PFA was not able to tackle adulteration even after five decades of its operation which ultimately paved way for another effective legislation to curb the menace of adulteration in the society.

#### 4.5.2. Food Safety and Standards Act, 2006

Multiplicity of laws relating to food, standard setting and enforcement agencies spread through the different sectors of food, due to which ambiguities and confusions galored in the mind of investors, manufacturers, traders and consumers. "However nothing can be farther from truth. Individual food products are not homogenous across countries, different countries and firms adopt different performance standards and safely

---

[109] *Yamuna Sah* v. *State of Bihar* 1990 (2) FAC 16 (Pat).

and quality norms and, moreover, buyers cannot ascertain the quality of food products merely by physical inspection. To complicate matters further, India had too many archaic food laws and too many ministries implementing these laws. This impedes the healthy growth of the Indian food industry in a liberalised world"[110].

Requirements under various laws regarding admissibility, contaminants, preservatives, levels of food additives, food colours and other allied parameters had varied standards. These standards were often rigid, contradictory and non-responsive to modernisation and scientific improvements. In the light of this scenario, it was realised that multiplicity of laws and their enforcement, specification of different standards under different laws, different enforcement agencies under different laws were unfavourable for the survival and growth of burgeoning food sector/processing industry. It was also realised that the situation was unfavourable for the effective fixation of standards for food and also for enforcement agencies to work efficiently. Foods for special dietary purposes or neutraceuticals[111] or functional foods or health supplements which had flooded the food market were omitted from the purview of all laws. There was no specific law applicable to them.

There was no liability or the concept of self regulation on the food business operator to deal with the safe food under PFAA, but it was the duty of the food inspectors to ensure the availability of safe food to the people. Indiscreet flooding of the market with unscrupulous imported goods was rampant. Food safety[112] became an important health issue. As countries witnessed a growth in International food trade, development of complex food types, processes and handling, there was a parallel rise in the level of awareness on different food and water borne diseases. Food safety became a growing global concern.

In the year 1998, the Prime Minister's Council on Trade and Industry appointed a subject group on Food and Agro Industries, which recommended for a single comprehensive legislation on food under which one food regulatory authority would be established investing it with the power of monitoring the food sector in the entire country. In 2004, the Joint Parliamentary Committee on Pesticides Residues had stressed on the need to merge all prevailing food laws and have a single regulatory body

---

[110] Supra note 103

[111] Nutraceutical is a new word, invented by Dr. Stephen DeFelice in 1989. It is two words put together: nutritional and pharmaceutical. Nutraceuticals are dietary supplements that are also called functional foods. Oxford Dictionary Thesaurus 2001, Indian edition, Oxford University Press

[112] FSSA Sec 3 (q) 'Food safety' means assurance that food is acceptable for human consumption according to its intended use

for all matters concerning food. The committee expressed its concern for public health and food safety in the country. In April 2005, the Standing Committee of Parliament on Agriculture in its $12^{th}$ report expressed its ambition to have a legislation, integrating all the food laws and also urged for expediting the formalities and process of formulating one. In pursuance of these suggestions, the then Member Secretary, Law Commission of India was required to conduct a review of food laws prevailing in various developing and developed countries and the international instruments pertaining to it. After making the required survey at the international scenario, the then Member Secretary, recommended for the promulgation of new food law, for promoting emerging food processing industries taking its income, employment generation and export potential into consideration.

It was also recommended that all existing food laws and orders be subsumed within the impending Act as the trend at the international level was towards modernisation and assimilation of regulations relating to standards of food with the removal of multi level and multi departmental control. It was also suggested to lay a special emphasis on responsibility with the producers/manufacturers of food, safety of food good manufacturing practices, and process control namely hazard analysis and critical point, functional foods, genetically modified food, food recall, risk analysis and emergency controls. All these recommendations and suggestions with appropriate modifications found expression in the way of an integrated food law-Food Safety and Standards Bill, 2005.

As expressed in the Bill, the main objective was to bring about a single Act pertaining to food and to provide for a systematic and scientific development of food processing industries, to establish a single food authority to fix standards for food regulate the production/manufacturing, importing processing and distribution /sale of food. The Bill hoped to achieve the availability of only the safe and wholesome food even to last man in the society. The food authority would be supported by Scientific Committees and Panels by prescribing the standards and by a Central Advisory Committee in prioritising the work. The enforcement mechanism under the Act would include State Commissioner for Food Safety, his officers and Municipal bodies or Panchayati raj.

Among other things, the Bill incorporated the salient features of the erstwhile PFAA. The prescribed standards for food were based on international legislations and

Codex Alimentarius Commission[113]. To concise, the Bill considered the international practices and policy framework regarding providing single window authority to guide and to monitor persons engaged in manufacture, marketing, processing, handling, transportation, importing and sale of food. This Bill was posted for public comments. The President of India signed the Bill on August 23, 2006 and was finally enacted as The Food Safety and Standards Act, 2006 subsuming the PFA and eight other different legislations[114].

## 4.6 Food Safety and Standards Act, 2006 (FSSA)

The FSSA has 12 Chapters containing 101 Sections and Two Schedules. This Act has three tier structure, an Apex Food Safety and Standards Authority of India (FSSAI) at the top assisted by a Central Advisory Committee and various Scientific Panels and Committees.

### 4.6.1 Objective of the Act

The Act has gone all out to fix the science based standards for all kind of articles of food irrespective of whether they originate here or imported. The object of the Act is to have a single reference point for all matters concerning food safety and standards, regulations and enforcement. The primary objective is to have all-inclusive/ comprehensive legislation aimed at ensuring the consumer safety through appropriate food safety management system. It was held *'International Spirits and Wine Association of India* v. *Union of India*[115]' that "for ensuring better consumer safety, the aim appears to be of setting higher standards of quality based on the science".

FSSA has been enacted primarily to guarantee better consumer safety, where every consumer is rest assured of the better quality of the food that he is consuming. For the first time, a novel concept of food safety management system was introduced.

---

[113] "In 1962, the Codex Alimentarius Commission (Codex Alimentarius or Codex) was formed under the joint sponsorship of two United Nations (UN) organizations: the World Health Organization (WHO) and the Food and Agriculture Organization (FAO). Today the 165 countries worldwide that comprise Codex hold two major food-related goals: (1) to protect the health of consumers and (2) to assure fair practices in food trade. These goals are accomplished by development of food standards, food guidelines, codes of hygienic practices, and other actions". Eddie Kimbrell, "WHAT IS CODEX ALIMENTARIUS?" AgBioForum, [2000] Vol 3, Number 4, Pages 197-202 Available at http://www.agbioforum. org/ v3n4/ v3n4a03-kimbrell.pdf(accessed on 12/7/2018 at 1.26pm)

[114] Prevention of Food Adulteration Act of 1954, Fruit Products Order of 1955, Meat Food Products Order of 1973, Vegetable Oil Products (Control) Order of 1947, Edible Oils Packaging (Regulation) Order of 1988, Solvent Extracted Oil, De- Oiled Meal and Edible Flour (Control) Order of 1967, Milk and Milk Products Order of 1992 and also any order issued under the Essential Commodities Act, 1955 relating to food.

[115] AIR 2013 Bom 178 p 183.

Considering the awe-inspiring and vast public interest that this legislation deals with, this cannot be a case where any interim reliefs can be granted pending the final disposal of Writ Petitions. The purpose of FSSA is to consolidate all laws pertaining to food and to launch FSSAI for laying down science based standards for articles of food. FSSA also intended to regulate and monitor the manufacturer, storing, distribution, sale and import of food/articles of food to warrant the availability of wholesome and safe food for human consumption.

### 4.6.2 Salient Features of the Act

Though FSSA incorporated the basic idea of food safety and other related matters from the PFA, yet new measures were introduced for ensuring the basic safety of food were introduced. Many customer friendly measures have been initiated for smooth working of the Act and the following are some of the salient features of the Act,

- Paradigm shift from multi level and multi departmental control to an integrated line of Command i.e. establishment of Food Safety and Standards Authority of India ( FSSAI)[116]
- Cumulative response to important issues like primary/novel/genetically modified food
- Shift from mere regulatory system/regime to self compliance through food safety management systems
- For the first time, exhaustive definition was given to 'food business'[117]
- No person can commence any food business unless he registered or licensed as per FSSA
- Responsibility on food business operators[118] to guarantee that the food is processed, manufactured, imported or distributed is in compliance with FSSA
- General provisions for articles of food[119] in terms of additives[120] or processing aids

---

[116] FSSA, Sec 4.

[117] *Ibid* Sec 3(n) Any undertaking whether for profit or not and whether public or private carrying out any of the activities related to any stage of manufacture, processing, packaging, storage, transportation, distribution, import and includes food services, catering services, sale of food or food ingredients.

[118] *Ibid* , Sec 26.

[119] *Ibid* Sec 19

[120] *Ibid* Section 3 (k) 'Food additive' means any substance not normally consumed as a food by itself or used as atypical ingredient of the food, whether or not it has nutritive value, the intentional addition of which to food for a technological purpose in the manufacture, processing preparation, treatment, packing packaging……..but does not include contaminants or substances added to food for maintaining or improving nutritional qualities.

- Restrictions on advertisements and prohibition on unfair trade practices which are misleading or deceiving or contravenes the provisions of this Act[121]
- Provision for graded penalties based on the severity of offence including civil penalties for minor offences and punishment for serious violations[122]
- Imported articles of food should conform with the standards stipulated by FSSA[123]
- Novel Food Recall Procedure[124] has been introduced to withdraw any unsafe food from the market
- Many customer friendly steps have been taken to implement food safety laws like an online licensing portal and online food import clearance system
- Harmonisation of Indian food standards with that of Codex Standards
- Notified referral labs[125] for the exclusive purpose of testing of food have been established

**4.6.3 Definitions under FSSA**

The Researcher has dealt with few important definitions given under the Act for the purpose of study.

**4.6.3.1 Food[126]**

A simple reading of the definition of food given in the Act indicates that any substance irrespective of whether it is processed, unprocessed or partially processed that is meant for human consumption is considered as food. It was held in *Joshy K V* v. *State of Kerala*[127], that "going by the definition, it only indicates that any substance whether processed, partially processed or unprocessed which is intended for human consumption is meant as food". Here it is evident that the Act is not concerned with the nutritional aspect of the thing meant for human consumption but only monitors the safety aspect of

[121] *Ibid* , Sec 24
[122] *Ibid* , sec 48 to 67
[123] *Ibid* , Sec 25
[124] *Ibid* , Sec 28
[125] *Ibid* Sec 43
[126] *Ibid* Sec 3(1)(j) (j) "Food" means any substance, whether processed, partially processed or unprocessed, which is intended for human consumption and includes primary food to the extent defined in clause (zk), genetically modified or engineered food or food containing such ingredients, infant food, packaged drinking water, alcoholic drink, chewing gum, and any substance, including water used into the food during its manufacture, preparation or treatment but does not include any animal feed, live animals unless they are prepared or processed for placing on the market for human consumption, plants, prior to harvesting, drugs and medicinal products, cosmetics, narcotic or psychotropic substances. Provided that the Central Government may declare, by notification in the Official Gazette, any other article as food for the purposes of this Act having regards to its use, nature, substance or quality.
[127] 2013 Cri L.J 2789at 2795(Ker)

it and the same was demonstrated in *Dhariwal Industries* v. *State of Maharashtra*[128], that "definition of 'food' in the Food Safety Act is wide enough to include gutka and pan masala, it is obvious that the regulations also apply to gutka and pan masala". The definition given in the Act for 'food' includes even liquor which is by nature intoxicating. FSSAI has notified 'The Alcoholic Beverages Regulations 2018'[129], which implies that even liquor has been treated as food by the FSSAI by notifying the standards for it.

Though, tobacco is consumed by humans, thankfully it/its products are kept outside the definition of food. Normally, nobody consumes tobacco as it is, but they chew it in the processed form. According the definition the word food includes, primary food, genetically modified food, infant food, packaged drinking water, alcoholic drinks, chewing gum and including water. But it was held in, *Pyarali K. Tejani* v. *Mahadeo Ramchandra Dange and Others*[130], that " that by itself mean that a product not named in the inclusive definition which is not a food will become a food product unless it is shown that it is eaten with relish by men for taste or for nourishment". In the light of this observation tobacco is included in the definition of food though it satisfies all criteria given in the Act for definition of food. But tobacco in any form need to be avoided may be the principle behind this exclusion.

Therefore, it was held in *M/S Omkar Agency* v. *Food Safety and Standards Authority of India*[131], that "chewing tobacco and its products are covered by the provisions of Cigarettes and other tobacco Products (Prohibition of Advertisement and Regulation of Trade and Commerce, Production, Supply and Distribution) Act, 2003(COTPA) and Food Safety and Standards Act has no application to such products". The fact that tobacco/its products are not included in food is further reinforced by the fact that Food Safety and Standards (Food products Standards and Food Additives) Regulations, 2011, does not define tobacco as no standards could be specified for tobacco. It was held in *Joshy K V* v. *State of Kerala*[132] that "it is found that COTPA is exclusive law, which deals with tobacco and tobacco products; whereas the Food Act is exclusive law, which deals with foods other than tobacco".

---

[128] 2013(1)FAC 26 at p.43(Bom)

[129] These regulations are specifying the standards for the following Alcoholic beverages Distilled Alcoholic Beverage (Brandy, Country Liquor, Gin, Rum, Vodka and Whisky, Liquors or Alcoholic cordial), Wines, Beer

[130] AIR 1974 S.C. 228

[131] AIR 2016 Pat.160 at p173

[132] 2013 Cri L.J 2789 at p.2795(Ker)

#### 4.6.3.2 Food Additives[133]

According to the Act, Food Additives are those which are not consumables as it is but they are added to food either to enhance the taste or appearance of food or to prolong the shelf life of food. They may not have a nutritive value but they play a major role in enhancing the desirability of the food either in terms of taste or appearance or being organoleptic. These food additives may be added at the stage of manufacture/preparation or packing or treatment or storage. It was held in *M/S Chromachemie Laboratory Private Limited, Bangalore* v. *Authorized Officer, Chennai Seaport and Airport, Food Safety and Standards authority of India, Ministry of Health and Family Welfare, Chennai*[134], that Erythritol used in the manufacture of artificial sweetner is a food additive. It was held in *Danisco(India) Private Ltd* v. *Union of India*[135], that food additives are "goods are admittedly not meant for direct consumption but are intended for industrial use in manufacture of yoghurt and fermented milk products".

#### 4.6.3.3 Food Business[136]

The definition given under the Act includes any business/enterprise may be with profit motive or for charity or no profit no loss basis, may be public enterprise or private enterprise, engaged in any activity related to food in the form of manufacture, processing, handling, storage, transportation, distribution and import of food. The food business also includes catering services, sale of food or ingredients for food. The definition under Act encompasses all activities related to food irrespective of the purpose, size and ownership of the enterprise. It was held in *Vishnu Pouch packaging Pvt Ltd* v. *State of Maharashtra*[137] , that even a transportation of food is also a food business as he has to obtain license under Section 31(1) of the FSSA read with Food safety and Standards (Licensing and Registration of Food Business) Regulations.2011.

---

133 FSSA sec3(1)(k) "food additive" means any substance not normally consumed as a food by itself or used as a typical ingredient of the food, whether or not it has nutritive value, the intentional addition of which to food for a technological (including organoleptic) purpose in the manufacture, processing, preparation, treatment, packing, packaging, transport or holding of such food results, or may be reasonably expected to result (directly or indirectly), in it or its by-products becoming a component of or otherwise affecting the characteristics of such food but does not include "contaminants" or substances added to food for maintaining or improving nutritional qualities;

134 AIR 2016(NOC)238 (Mad)

135 AIR 2015(NOC) 219(Del)

136 FSSA Sec3(1) (n) "Food business" means any undertaking, whether for profit or not and whether public or private, carrying out any of the activities related to any stage of manufacture, processing, packaging, storage, transportation, distribution of food, import and includes food services, catering services, sale of food or food ingredients

137 2013(1) FAC 172 (Bom)at p. 177

#### 4.6.3.4 Food Business Operator (FBO)[138]

The definition given for food business operator includes any person who deals with food. He may be a manufacturer/producer, processor, trader, stockist, whole seller, retailer or a transporter, as long as they are all dealing or transacting with food. Each one of them comes under the purview of the Act and they are to comply with the Act.

#### 4.6.3.5 Food Safety[139]

The Act does not detail anything about the term Food Safety and it has only stated that it is an assurance that it is a safe food and the same can be safely consumed by human beings as per their wishes and preferences. Food Safety also means an assurance that the food is acceptable for human consumption according to their intended use and the word 'Standard' under the Act means in relation to any article of food, the standards notified by the FSSAI. It is of vital importance to all consumers and food business operators who are engaged in production, processing distribution and sale. These standards provide consumers, the confidence that the food they consume does no harm to them and they are protected against food adulteration.

#### 4.6.3.6 Food Safety Management System[140]

This means the adoption of good manufacturing practices, hygienic practices, analysing the probable hazards and the critical point where it becomes necessary to take actions to minimise the same. It also includes any other safety measures stipulated by the Act. But the Act failed to specify as to what are good manufacturing practices, what are good hygienic practices and the probable hazards that a food business operator may encounter in the course of his business dealing with food. The Act has left too many loose ends that will have to be tied. The Act should have specified what the good manufacturing practices are, as the word good is abstract and relative. Whatever is good according to one food business operator may not be good according to the other food business operator .Or may be the Act has kept all these terminologies open, so that the same may be interpreted taking a particular situation into consideration. Good

---

[138] FSSA Sec 3(1)(o) "FBO" in relation to food business means a person by whom the business is carried on or owned and is responsible for ensuring the compliance of this Act, rules and regulations made there under;

[139] *Ibid* Sec 3(1)(q) "food safety" means assurance that food is acceptable for human consumption according to its intended use;

[140] *Ibid* Sec 3(1)(s) Food Safety Management System" means the adoption Good Manufacturing Practices, Good Hygienic Practices, Hazard Analysis and Critical Control Point and such other practices as may be specified by regulation, for the food business;

manufacturing practice may differ from case to case basis. Or the FSSA should have borrowed the explanations for these terminologies from Codex Standards.

### 4.6.3.7 Hazard[141]

Hazard means anything in the form of biological, physical or chemical agent in the food making it unsafe for human consumption or the condition of food is such that on consumption it can pose a threat to one's health. The presence of these hazards in food has a direct link with the health of the consumer causing adverse effects.

### 4.6.3.8 Misbranded Food[142]

Misbranded foods are those, 'they are not what they claim to be'. It can partake different forms like when food is offered for sale with claims which may be false, misleading and deceptive. These claims may be made in the form of advertisements or by printing on the packing or the label. The food is also misbranded when someone sells it under a different person's/concern's name, or sells it misrepresenting it to be his products.

The counterfeit articles of food are also misbranded when goods are depicted to belong to one but in reality it is not so, when the ingredients as stated on the package or on the container or on the label is not true, when the contents are depicted to originate from a particular place but in reality, no so, when the articles of food contain any food additive or preservative, but the content list on the label or package does not reflect the

---

[141] *Ibid* Sec 3(1)(u) "hazard" means a biological, chemical or physical agent in, or condition of, food with the potential to cause an adverse health effect

[142] *Ibid* Sec 3(1)(zf) misbranded food" means an article of food –

(A) if it is purported, or is represented to be, or is being – (i) offered or promoted for sale with false, misleading or deceptive claims either; (a) upon the label of the package, or 5 (b) through advertisement, or (ii) sold by a name which belongs to another article of food; or (iii) offered or promoted for sale under the name of a fictitious individual or company as the manufacturer or producer of the article as borne on the package or containing the article or the label on such package; or

(B) if the article is sold in packages which have been sealed or prepared by or at the instance of the manufacturer or producer bearing his name and address but - (i) the article is an imitation of, or is a substitute for, or resembles in a manner likely to deceive, another article of food under the name of which it is sold, and is not plainly and conspicuously labelled so as to indicate its true character; or (ii) the package containing the article or the label on the package bears any statement, design or device regarding the ingredients or the substances contained therein, which is false or misleading in any material particular, or if the package is otherwise deceptive with respect to its contents; or (iii) the article is offered for sale as the product of any place or country which is false; or

(C) if the article contained in the package – (i) contains any artificial flavouring, colouring or chemical preservative and the package is without a declaratory label stating that fact or is not labelled in accordance with the requirements of this Act or regulations made thereunder or is in contravention thereof; or (ii) is offered for sale for special dietary uses, unless its label bears such information as may be specified by regulation, concerning its vitamins, minerals or other dietary properties in order sufficiently to inform its purchaser as to its value for such use; or (iii) is not conspicuously or correctly stated on the outside thereof within the limits of variability laid down under this Act.

same, when the food is meant for special dietary purposes but the label does not mention – under all these circumstances also the food articles become misbranded food.

**4.6.3.9 Standard**[143]

Standards are nothing but the quality of the articles of food as specified by the food authority. For every article of food, different standards are stipulated under the Act and for a safe food, the same will have to conform with these standards.

**4.6.3.10 Unsafe Food** [144]

Under the erstwhile PFAA, 1954 whatever was defined as adulterated food has been rechristened as unsafe food under FSSA. Unsafe food is something which is unfit for human consumption, owing to his obnoxious nature, whose quality is so affected that on consumption, it might injure the health of the consumer. The article of food may be partially or completely composed or have become filthy, putrid, rotten, decomposed or diseased animal substance or vegetable, the article of food may be a substitute of any inferior quality of food for superior quality of food, if the food is processed un-hygienically, if one of the ingredient of a food is a prohibited one, if the banned food additive has been used in the manufacture/preparation of food or if the food is found to have contained residues of pesticides or contaminants or preservatives or food additives beyond the permissible limits- under these circumstances the articles of food become unsafe food for human consumption.

**4.6.3.11 Meaning of Adulteration**

Anything which is not pure, substandard or spurious is deemed to be adulterated[145]. All violations of statutorily prescribed standards of quality and violations

---

[143] *Ibid* Sec3(1) (zu) "standard", in relation to any article of food, means the standards notified by the Food Authority

[144] *Ibid* sec3(1) (zz) "unsafe food" means an article of food whose nature, substance or quality is so affected as to render it injurious to health :— (i) by the article itself, or its package thereof, which is composed, whether wholly or in part, of poisonous or deleterious substance; or (ii) by the article consisting, wholly or in part, of any filthy, putrid, rotten, decomposed or diseased animal substance or vegetable substance; or (iii) by virtue of its unhygienic processing or the presence in that article of any harmful substance; or (iv) by the substitution of any inferior or cheaper substance whether wholly or in part; or (v) by addition of a substance directly or as an ingredient which is not permitted; or (vi) by the abstraction, wholly or in part, of any of its constituents; or (vii) by the article being so coloured, flavoured or coated, powdered or polished, as to damage or conceal the article or to make it appear better or of greater value than it really is; or 8 (viii) by the presence of any colouring matter or preservatives other than that specified in respect thereof; or (ix) by the article having been infected or infested with worms, weevils, or insects; or (x) by virtue of its being prepared, packed or kept under insanitary conditions; or (xi) by virtue of its being mis-branded or sub-standard or food containing extraneous matter; or (xii) by virtue of containing pesticides and other contaminants in excess of quantities specified by regulations.

[145] PFAA, Sec 2

of orders regarding limitation of colouring or preservatives amount to adulteration[146]. In certain cases adulteration renders the food only sub standard but not unsafe or dangerous. But at times adulteration makes the food articles positively injurious and harmful to the health of the consumers. The Researcher thinks that the latter is a case of 'slow and subtle murder'. Adulteration of food can happen at different stages beginning from production, manufacture, distribution and sale. In *Ganeshmal Jeshraj* v. *Government of Gujarat* [147], the Supreme Court observed, "*The small tradesmen who eke out a precarious existence living almost from hand to mouth are sent to jail for selling food stuff which is often not adulterated by them and the wholesalers and manufacturers who really adulterate the foodstuff and fatten themselves on the misery of others escape the arm of the law*".

### 4.6.4 Establishment of FSSAI

Section 4 of FSSA[148] provides for the establishment of FSSAI with its head office at New Delhi. It is a body corporate with a perpetual succession and a common seal with power to acquire, hold and dispose of the property. The food authority can also establish offices at different parts of the country. It was held in *Centre for Public Interest Litigation* v. *Union of India*[149], that "the food authority is also authorised to constitute a Central Advisory Committee, so also Scientific Panels".

#### 4.6.4.1 Composition of the FSSAI

Section 5 of FSSA deals with the composition of the FSSAI. The authority consists of twenty three members; Chairperson and other twenty two members out of which one third shall be women. Out of these twenty two members seven members not below the rank of a joint secretary to the Government of India to represent the ministries or Departments of Central government dealing with agriculture, small scale industries, Commerce, consumer affairs, food Processing, Health, Legislative Affairs, who shall all

---

[146] M C Valson, "Prevention of food adulteration- some basic problems", Cochin University Law Review, Annual Index [1983], p330-334

[147] (1980) 1 SCC, 363

[148] FSSA Sec 4. Establishment of Food Safety and Standards Authority of India. –

1. The Central Government shall, by notification, establish a body to be known as the Food Safety and Standards Authority of India to exercise the powers conferred on, and to perform the functions assigned to, it under this Act.
2. The Food Authority shall be a body corporate by the name aforesaid, having perpetual succession and a common seal with power to acquire, hold and dispose of property, both movable and immovable, and to contract and shall, by the said name, sue or be sued.
3. The head office of the Food Authority shall be at Delhi.
4. The Food Authority may establish its offices at any other place in India.

[149] 2013(2)FAC 135 at p. 139(sc)

be ex officio members. Two representatives from food industry of which one shall be from small scale industries, two representations from consumer organisations, three eminent scientists or food technologists, five members appointed by rotation every three years, two persons from farmer's organisation and one person from retailers' organisation. The Chairperson and members other than ex officio shall hold office for a term of three years from the date on which they assumed office and shall be eligible for re appointment.

**4.6.4.2 Central Advisory Committee**

The FSSAI shall by notification establish Central Advisory Committee [150] consisting of two members each to represent food industry, agriculture, consumers, relevant research bodies and food laboratories and all Commissioners of Food safety and the Chairperson of the Scientific Committee shall be the ex officio member and it shall meet at least three times a year. According to Section 12 of FSSA it is the duty of Central advisory Committee to ensure cooperation between the Food authority and the enforcement agencies and to assist and advice the FSSAI to draw up a proposal for the Authority's work programme, to prioritise the work, in identifying the potential risks, in pooling of knowledge and such other tasks as may be specified.

**4.6.4.3 Scientific Panels**

The FSSAI shall constitute Scientific Panels [151] with independent scientific experts. The authority may establish as many Scientific Panels as it considers necessary in addition to the Scientific Panels on food additives, flavourings, processing aids and materials; pesticides and antibiotic residues; genetically modified organisms and foods; functional foods, nutraceuticals, dietetic products; biological hazards; contaminants in the food chain; labelling; and method of sampling and analysis .

The FSSAI shall constitute Scientific Committee which will have chairpersons of the Scientific Panels and six independent scientific experts who are not belonging to any Scientific Panels. It is the duty of these Scientific Committees to provide the scientific opinions to the Food Authority and have the power to organise and conduct public hearings. The Scientific Committee should also advise on multi sectoral issues which might fall within the competence of more than one Scientific panel. It was held in

[150] FSSA Sec 11
[151] *Ibid* Sec 13

*Centre for Public Litigation* v. *Union of India*[152] , that "the Scientific Committee shall set up working groups on issues which does not fall under scientific panels".

#### 4.6.4.4 Duties and Functions of FSSAI[153]

---

[152] 2013(2) FAC 135 at 139(SC)

[153] Sec 16 FSSA16. Duties and Functions of Food Authority. –

1. It shall be the duty of the Food Authority to regulate and monitor the manufacture, processing, distribution, sale and import of food so as to ensure safe and wholesome food.
2. Without prejudice to the provisions of sub-section (1), the Food Authority may by regulations specify-

a. the standards and guidelines in relation to articles of food and specifying an appropriate system for enforcing various standards notified under this Act;
b. the limits for use of food additives, crop contaminants, pesticide residues, residues of veterinary drugs, heavy metals, processing aids, myco-toxins, antibiotics and pharmacological active substances and irradiation of food;
c. the mechanisms and guidelines for accreditation of certification bodies engaged in certification of food safety management systems for food businesses;
d. the procedure and the enforcement of quality control in relation to any article of food imported into India;
e. the procedure and guidelines for accreditation of laboratories and notification of the accredited laboratories;
f. the method of sampling, analysis and exchange of information among enforcement authorities;
g. conduct survey of enforcement and administration of this Act in the country;
h. food labelling standards including claims on health, nutrition, special dietary uses and food category systems for foods; and
i. the manner in which and the procedure subject to which risk analysis, risk assessment, risk communication and risk management shall be undertaken.

3. The Food Authority shall also-

a. provide scientific advice and technical support to the Central Government and the State Governments in matters of framing the policy and rules in areas which have a direct or indirect bearing on food safety and nutrition;
b. search, collect, collate, analyse and summarise relevant scientific and technical data particularly relating to-
   i. food consumption and the exposure of individuals to risks related to the consumption of food;
   ii. incidence and prevalence of biological risk;
   iii. contaminants in food;
   iv. residues of various contaminants;
   v. identification of emerging risks; and
   vi. introduction of rapid alert system;
c. promote, co-ordinate and issue guidelines for the development of risk assessment methodologies and monitor and conduct and forward messages on the health and nutritional risks of food to the Central Government, State Governments and Commissioners of Food Safety;
d. provide scientific and technical advice and assistance to the Central Government and the State Governments in implementation of crisis management procedures with regard to food safety and to draw up a general plan for crisis management and work in close co-operation with the crisis unit set up by the Central Government in this regard;
e. establish a system of network of organisations with the aim to facilitate a scientific co-operation framework by the co-ordination of activities, the exchange of information, the development and implementation of joint projects, the exchange of expertise and best practices in the fields within the Food Authority's responsibility;
f. provide scientific and technical assistance to the Central Government and the State Governments for improving co-operation with international organisations;
g. take all such steps to ensure that the public, consumers, interested parties and all levels of panchayats receive rapid, reliable, objective and comprehensive information through appropriate methods and means;
h. provide, whether within or outside their area, training programmes in food safety and standards for persons who are or intend to become involved in food businesses, whether as FBOs or employees or otherwise;

The duties and functions of FSSAI have been elaborately dealt with in Section 16 of FSSA according to which it is the duty of the FSSAI to regulate and monitor the manufacture, processing, distribution, sale and import of food. It shall also regulate the standards and guidelines for the accreditation of certification bodies engaged in certification of food safety management systems for food business and notify the accredited laboratories. The duties and functions mentioned in Section 16(2) is in addition to the duties mentioned under Section16 (1). It was held in *Dhariwal Industries* v. *State of Maharashtra*[154], that "Chapter III of the Act contains general principles of food safety and it requires the central government, the state governments and the FSSAI and other agencies implementing the Act to identify the possibility of harmful effects of the food additives on health on the basis of assessment of the available information".

The FSSAI has no unbridled/unfettered power to regulate and monitor various activities in food production. It was held in *Vital Neutraceuticals Pvt Ltd* v. *Union of India*[155], that "it is obvious that the authority has to be within the power and procedure which is prescribed by the various provisions of the Act". The provisions of Section 16(5) also do not give any unrestrained power to FSSAI to give any directions it wants to the Commissioner or other authorities under the Act. The directions will have to be

---

i. undertake any other task assigned to it by the Central Government to carry out the objects of this Act;
j. contribute to the development of international technical standards for food, sanitary and phyto-sanitary standards;
k. contribute, where relevant and appropriate, to the development of agreement on recognition of the equivalence of specific food related measures;
l. promote co-ordination of work on food standards undertaken by international governmental and non-governmental organisations;
m. promote consistency between international technical standards and domestic food standards while ensuring that the level of protection adopted in the country is not reduced; and
n. promote general awareness as to food safety and food standards.
4. The Food Authority shall make it public without undue delay-
a. the opinions of the Scientific Committee and the Scientific Panel immediately after adoption;
b. the annual declarations of interest made by members of the Food Authority, the Chief Executive Officer, members of the Advisory Committee and members of the Scientific Committee and Scientific Panel, as well as the declarations of interest if any, made in relation to items on the agendas of meetings;
c. the results of its scientific studies; and
d. the annual report of its activities.
5. The Food Authority may, from time to time give such directions, on matters relating to food safety and standards, to the Commissioner of Food Safety, who shall be bound by such directions while exercising his powers under this Act;
6. The Food Authority shall not disclose or cause to be disclosed to third parties confidential information that it receives for which confidential treatment has been requested and has been acceded, except for information which must be made public if circumstances so require, in order to protect public health.

[154] 2013(1)FAC26 at p.41 (Bom)

[155] 2014(1) FAC 1 at p.24 (Bom)

within the purview of powers granted by FSSA and should follow the other rules of natural justice.

In *M/S Nestle India Ltd., New Delhi* v. *Food safety and Standards Authority of India, Mumbai*[156], FSSAI had banned the Nestle Maggi Noodles on the ground that it was unsafe for human consumption as it contained lead and monosodium glutamate more than the permissible limit. The Commissioner issued the ban orders as per Section 34 of FSSA dealing with Emergency Prohibition orders. In this case it was contended that at the outset Show cause notice should have been given and the principles of natural justice should have been followed. Nestle was carrying on business for more than 30years in the market and no contamination was reported in the past. Out of 70 samples 50% of the samples were found to be within the permissible limits. Hence, under these circumstances proper opportunity of explaining the case should have been given to the manufacturer. Ban order was set aside based on the ground of violation of principles of natural justice.

Taking a serious view about the menace of adulteration in the society, especially in the case of adulteration of milk which is largely consumed by infants in India, the court in *Swami Achyuthanand Tirth* v. *Union of India*[157], held that "it will be in order, if the union of India considers making suitable amendments in the penal provisions at par with the provisions contained in the State amendments to the Indian Penal Code. It is also desirable that Union of India revisits the FSSA, to revise the punishment for adulteration making it more deterrent in cases where the adulterant can have an adverse impact on health". In this case, the Supreme Court ordered for the setting up of State level Committee headed by the Chief Secretary or the Secretary of Dairy department and District level Committee headed by the District Collector to review the work done to curb the milk adulteration in the district and in the state.

### 4.6.5 General Principles of Food Safety[158]

---

[156] AIR 2016 (NOC) 225(Bom)

[157] AIR 2016 S.C.3626 at 3634

[158] Sec 18 FSSA. General principles to be followed in administration of Act.–The Central Government, the State Governments, the Food Authority and other agencies, as the case may be, while implementing the provisions of this Act shall be guided by the following principles namely:–
(1) (a) endeavour to achieve an appropriate level of protection of human life and health and the protection of consumer's interests, including fair practices in all kinds of food trade with reference to food safety standards and practices; (b) carry out risk management which shall include taking into account the results of risk assessment and other factors which in the opinion of the Food Authority are relevant to the matter under consideration and where the conditions are relevant, in order to achieve the general objectives of regulations; 17 (c) where in any specific circumstances, on the basis of assessment of available

Section 18 of FSSA deals with General Principles as to Food Safety. While implementing the provisions of FSSA, the Central Government and state Governments, FSSAI and other agencies, shall be guided by some of very important principles. They should make an effort to achieve an appropriate level of protection of human life as well as the protection of consumer's interests. They should carry out risk management by considering the results of the risk assessment and other relevant matters. When the possibility of harmful effects on the health of the people identified but scientific uncertainty persists, necessary and proportionate risk management measures will have to be adopted, but nothing more than whatever required in the situation. Such measures will be reviewed within a reasonable period of time. When the FSSAI and the Commissioner are satisfied that the food may present a risk for human health, then

---

information, the possibility of harmful effects on health is identified but scientific uncertainty persists, provisional risk management measures necessary to ensure appropriate level of health protection may be adopted, pending further scientific information for a more comprehensive risk assessment; (d) the measures adopted on the basis of clause (c) shall be proportionate and no more restrictive of trade than is required to achieve appropriate level of health protection, regard being had to technical and economic feasibility and other factors regarded as reasonable and proper in the matter under consideration; (e) the measures adopted shall be reviewed within a reasonable period of time, depending on the nature of the risk to life or health being identified and the type of scientific information needed to clarify the scientific uncertainty and to conduct a more comprehensive risk assessment; (f) in cases where there are reasonable grounds to suspect that a food may present a risk for human health, then, depending on the nature, seriousness and extent of that risk, the Food Authority and the Commissioner of Food Safety shall take appropriate steps to inform the general public of the nature of the risk to health, identifying to the fullest extent possible the food or type of food, the risk that it may present, and the measures which are taken or about to be taken to prevent, reduce or eliminate that risk; and (g) where any food which fails to comply with food safety requirements is part of a batch, lot or consignment of food of the same class or description, it shall be presumed until the contrary is proved, that all of the food in that batch, lot or consignment fails to comply with those requirements.

(2) The Food Authority shall, while framing regulations or specifying standards under this Act– (a) take into account– (i) prevalent practices and conditions in the country including agricultural practices and handling, storage and transport conditions; and (ii) international standards and practices, where international standards or practices exist or are in the process of being formulated, unless it is of opinion that taking into account of such prevalent practices and conditions or international standards or practices or any particular part thereof would not be an effective or appropriate means for securing the objectives of such regulations or where there is a scientific justification or where they would result in a different level of protection from the one determined as appropriate in the country; (b) determine food standards on the basis of risk analysis except where it is of opinion that such analysis is not appropriate to the circumstances or the nature of the case; (c) undertake risk assessment based on the available scientific evidence and in an independent, objective and transparent manner; (d) ensure that there is open and transparent public consultation, directly or through representative bodies including all levels of panchayats, during the preparation, evaluation and revision of regulations, except where it is of opinion that there is an urgency concerning food safety or public health to make or amend the regulations in which case such consultation may be dispensed with: Provided that such regulations shall be in force for not more than six months; (e) ensure protection of the interests of consumers and shall provide a basis for consumers to make informed choices in relation to the foods they consume; (f) ensure prevention of– (i) fraudulent, deceptive or unfair trade practices which may mislead or harm the consumer; and (ii) unsafe or contaminated or sub-standard food.

18 (3) The provisions of this Act shall not apply to any farmer or fisherman or farming operations or crops or livestock or aquaculture, and supplies used or produced in farming or products of crops produced by a farmer at farm level or a fisherman in his operations.

depending on the nature, seriousness and extent of the risk, will initiate appropriate steps to inform the general public about the same and the measures that are undertaken to prevent, reduce and curb the impending risk. If any food fails to comply with food safety requirements is a part of a particular batch, lot or consignment of the food, till the contrary is proved, it will be presumed that all the food in that batch, lot or consignment fails to comply with those requirements and the food will be treated as unsafe.

The FSSAI, while framing Regulations or specifying the Standards, shall take into account the locally prevailing practices and conditions including agricultcural practices, food handling, storage and transport conditions. International standards and practices will also be considered if they are in practice. Food standards will be determined based on the risk analysis. Open and transparent public consultations to be held during the preparation, evaluation and revision of regulations. But if there is an urgency concerning food safety or public health such consultations may be dispensed with, but, such regulations shall be in force for not more than six months. The interests of the consumers will also be considered. Efforts will be made to prevent unfair/ fraudulent/deceptive trade practices and curb the unsafe or contaminated or sub-standard food. The provisions of this Act shall not apply to any farmer or fishermen or farming operations or standing crops or crops of livestock or aquaculture and also to supplies used or produced in farming by a farmer or a fisherman in his operations.

The Researcher submits that, the authorities apart from implementing the Act, are required to play a proactive role of ensuring the availability of safe and wholesome food and to prevent the risk that may be caused to health by the consumption of adulterated food. Therefore, it was held in *Dhariwal Industries* v. *State of Maharashtra*[159], that "it is, therefore, clear that FSSA, 2006 is the comprehensive single special legislation for all food products on the subject of safety and standards". It was held in *Uma exports Ltd* v. *Union of India*[160], that "in the instant case, the goods have been lifted from somewhere be it farm level or some other place and thereafter it has been imported and brought in a different country mixed with minerals. Therefore in the opinion of the High Court, such an item loses its characteristic of it being brought within the definition of crop".

---

[159] 2013(1) FAC 29 at pp41-42 (Bom)
[160] 2013(1) FAC 422 at p 424 (Cal)

### 4.6.6 General Provisions as to Articles of Food

These General provisions deal with the presence of food additive or processing aid, contaminants, naturally occurring toxic substances, heavy metals, pesticides, residues of veterinary drugs and antibiotics and micro biological counts. It also deals with genetically modified foods, organic foods, functional foods, packaging and labelling of foods.

#### 4.6.6.1 Use of Food Additives or Processing Aid

Section 19[161], ordains that no article of food should contain any food additive or processing aid unless it is in accordance with the provisions of FSSA. This Section has also defined processing aid as any substance or material not including apparatus or utensils, and consumed as a food ingredient by itself, used in the processing of raw materials, foods or its ingredients to fulfil a certain technological purpose during treatment or processing and which may result in the non intentional but unavoidable presence of residues in the final product.

The Food safety and Standards (Food Products Standards and Food Additives) Regulations, 2011, have dealt exhaustively with all most all kind of food both natural and manmade. Chapter 2 of these Regulations deal with Food Product Standards classifying the food products under different categories like dairy Products and Analogues; Fats, oils and fat emulsions; Fruit and Vegetable Products; Cereal and Cereal Products; Meat and Meat Products; Fish and Fish Products; Sweets and Confectionary; Sweetening Agents including Honey;  Salt, spices, Condiments and Related products; Beverages(other than Dairy and Fruits and Vegetables based); Other Food Products and Ingredients; Proprietary Food and Irradiation of Food. Chapter 3 of these Regulations deal with 'Substances Added to Food'.

For the purpose of this Regulations Good manufacturing practices mean, that the quantity of the additives added to food shall be limited to the lowest possible level necessary to accomplish its desired effect and the food additive must be prepared and handled in the same way as a food ingredient. The Substances that can be added may in

---

[161] FSSA Sec 19 Use of food additive or processing aid, No article of food shall contain any food additive or processing aid unless it is in accordance with the provisions of this Act and regulations made thereunder.

*Explanation.*– For the purposes of this section, "processing aid" means any substance or material, not including apparatus or utensils, and not consumed as a food ingredient by itself, used in the processing of raw materials, foods or its ingredients to fulfil a certain technological purpose during treatment or processing and which may result in the non-intentional but unavoidable presence of residues or derivatives in the final product.

the form of Food Additives, Colouring Matter, Artificial Sweetners, Preservatives, Antioxidants, Emulsifying and Stabilising agents, Anticaking agents, Anti foaming agents in edible oils and fats, Use of release agents in Confectionary, flavouring agents, Use of Flavour enhancers, Use of Sucrose Aceate Isobutyrate. This chapter also specifies the standards of Additives that may be used with respect to different articles of food. *United Distribution Incorporation* v. *Union of India*[162], was a case concerning the goods which were found to be non complaint with Regulation 2.7.4 of the Food and Additives Regulations. This was a case about filled chocolates. The vegetable fat was found in the filling and because of this FSSAI found it non complaint with regulation of the Food and Additives but this decision of FSSAI was not sustainable as Regulation 2.7.4. of the said Regulations clearly indicates that in case of filled chocolates, the coating shall be of chocolates that meets the requirement of one or more chocolates variants. It is thus clear that the specification with regard to vegetable fat has to be confined only to the chocolate shell and not to the filling. The standards prescribed for chocolate cannot be applied for the filled in chocolates as the filling is distinct from the outer shell. Non clearance of chocolate on the ground that vegetable fat was found in the fillings was held not proper.

#### 4.6.6.2 Contaminants, Naturally Occurring Toxic Substances, Heavy Metals

Section 20[163], of the Act provides that no article of food shall contain any contaminants, naturally occurring toxic substances and heavy metals beyond the quantities as mentioned in the Act. Food safety and Standards (Contaminants, Toxins and Residues) Regulations, 2011, have been framed to regulate the different types of Contaminants, Toxins and Residues that may be present in different kinds of food. Chapter 2 of these Regulations deal with Metal contaminants, Crop Contaminants and Naturally occurring toxic substances, Residues in the form of insecticides, Antibiotic and other Pharmacologically Active Substances. According to these Regulations, Metal Contaminants include Lead, Copper, Arsenic, Tin Zinc, Cadmium, Mercury, Methyl Mercury, Chromium and Nickel. In the long list of Insecticide residues, the prominent are DDT, DDE, Endosulfan A, Endosulfan B, Dicofol and Hydrogen Cyanide. The

[162] AIR 2015 Delhi 31 at p36

[163] FSSA Sec 20, Contaminants, naturally occurring toxic substances, heavy metals, etc. No article of food shall contain any contaminant, naturally occurring toxic substances or toxins or hormone or heavy metals in excess of such quantities as may be specified by regulations

amount of insecticide in the foods shall not exceed the tolerance limit as prescribed in the Regulations.

#### 4.6.6.3 Pesticides, Residues of Veterinary Drugs and Antibiotics and Micro-Biological Counts

Section 21 provides that no article of food shall contain insecticides, pesticides, residues of veterinary drugs, antibiotic, solvent; pharmacological active substances and micro biological counts in excess of such tolerance limit as may be specified by the Regulations. The Regulations provide that no insecticide shall be used directly on articles of food except fumigants registered and approved under the Insecticide Act, 1968.

#### 4.6.6.4 Genetically Modified Foods, Functional Foods, Proprietary Foods

Section 22 indicates that no person shall manufacture, distribute, sell or import any novel food, Genetically Modified Foods, Organic Foods, Functional Foods, Foods for Special Dietary purposes, Nutraceuticals, health supplements, Proprietary Foods and such other articles of food which the Central government may notify, this is to make these foods safe for consumption as these foods are not ordinary foods but meant for a particular/special purpose.

#### 4.6.6.5 Packaging and Labelling of Foods

It was held in *Lochamesh B hugr* v. *Union of India*[164], that Section 23 "mandates that no person shall manufacture, distribute, sell or expose for sale or dispatch or delivery to any agent or broker for the purpose of sale, any packaged food products which are not marked and labelled as per the, The Food safety and Standards (Packaging and Labelling) Regulations, 2011". These Regulations stipulate General Requirements for Packaging [165], labelling. Containers which are rusty, enamelled containers which are chipped, copper and brass containers which are not properly tinned, containers made of aluminium not conforming in chemical composition to IS 20 specification for cast Aluminium and Aluminium alloy for utensils and containers made of plastic materials which do not conform with standards stipulated by this Regulations should be used neither in preparation nor in storing nor in packaging.

---

[164] 2013(1)FAC 239 at p.262 (Kant)

[165] The Food Safety and Standards(Packaging and Labelling)Regulations, 2011, Regulation 2.1.1

This Regulation also specifies the different kinds packaging materials that ought to be used for different kinds of pre packaged food[166]. The FSSAI has regulated food packaging norms after it found contamination of food products through its packaging material. FSSAI has directed all the FBOs not to wrap food items using recycled plastics, news paper and paper sheets. This move was initiated after the results of two studies conducted by FSSAI revealed the presence of chemical contamination and heavy metals from the packing materials into food. New Regulations for packing will come into effect from July1, 2019.

The Regulations contain the General requirements for labelling[167] and directs that every pre packaged food shall carry a label containing information either in English or Hindi and in addition to this any other language, the label shall not contain any statement, claim, design or device which can mislead the people that food products contained in the package have a got nutritive value implying medicinal or therapeutic claims. The label shall be applied in such a manner that it should not be detached from the container. If the container is covered by a wrapper, the wrapper shall contain all necessary information about the product as per the requirements of this Regulation.

The purpose of labelling the packaged products is to ensure that the relevant information pertaining to the product is available on the package for the benefit of the consumers. In addition to the general requirements, the pre packaged foods shall contain information on the label[168]regarding the name of the food item, the ingredients in the descending order of weights and its percentage, nutritional information, declaration regarding the nature of the product veg or Non-veg, declaration regarding food additives, name and complete address of the manufacturer, net quantity, lot/code/batch identification number the date of manufacture and expiry, best before use date, if it is an imported food product-the name of the country of its origin and the name of the importer. It is ordained by this Regulation that the labels should not contain any false or misleading statements.

#### 4.6.6.6 Restrictions of Advertisement and Prohibition as to Unfair Trade

Section 24 lays restriction on advertisement of any food which is misleading and contravening the provisions of FSSA and Regulations made under this Act. It was held

---

[166] *Ibid* 2.1.2
[167] *Ibid* 2.2.1
[168] *Ibid* 2.2.2

in '*Centre for Public Interest Litigation* v. *Union of India*'[169], that "advertisements for carbonated beverages are being monitored by the Advertisement Standards Council of India Regulations, hence do not come under FSSA". This provision makes it clear that no person can engage himself in any unfair trade practice for the purpose of promoting sale, consumption of a particular article of food by making any statement, which falsely represents that the foods are of a particular standard, quality and quantity. Last year, researchers from the AFMC, Pune scrutinised 1200 Indian food advertisements and found that almost 60% of them are misleading in nature, they neither follow FSSAI Rules nor Advertising Standards Council of India codes[170].

No person can make a statement regarding the need for or the usefulness or guarantee of a particular food article. FSSAI's new Rules published on November 19th 2018 on food advertisement has allowed food and beverage companies to use words like natural, fresh, pure and real but with a disclaimer. Erstwhile, the food and beverage companies were not allowed to use words like, fresh, original, traditional, premium, finest, best, authentic, genuine and real on the food labels. Food could be called fresh only if they are not processed in any manner except being washed, peeled, chilled, trimmed or cut. The word natural could be used only for food derived from a plant, animal or microorganism or mineral and to which nothing was been added. FSSAI could not initiate any action against these companies as they had legally valid trademarks because of which now they are required to use the disclaimer to caution the consumers.

However, new Rules have done away with all these strict conditions and the companies are free to use those terms only a disclaimer stating that it is only a brand name or trademark and does not represent the true nature. The FSSAI (Advertising and Claims) Regulations, 2018 provides details as to how the food firms ought to advertise their products avoiding misleading claims and without undermining the importance of healthy and balanced diet. These Rules will be effective from July 1st, 2019.according to these Rules, claims made in the advertisement should be consistent with the information on the label. The labels are ought not to have words implying that the food is recommended, prescribed or approved by medical practitioners or have been approved for medical purposes. When a nutrient comparative claim is made, the food needs to

---

[169] 2013(2) FAC 135 at p.144(SC)

[170] Kalyan Ray, *FSSAI Brings Out New Norms on Advt to Stop False Claims*, Deccan Herald dated 28th November 2018 Bengaluru Edition

have at least 25% higher energy value or nutrient content as against the same food without the nutrient.

### 4.6.7 Provisions Relating to Import of Articles of Food

After the passing of FSSA, Section 25 dictates that no person is allowed to import into India any unsafe /misbranded/substandard food or any food containing extraneous matter. If the Act and the Regulations mandate a licence for the import of a particular food, without the required licence no person can import it. Any article of food which contravenes any of the provisions of the Act also cannot be imported. This provision requires the Central government to follow the standards laid down by the FSSAI and the Rules and Regulations Made there under, while prohibiting, restricting or regulating the import of articles of food under the Foreign Trade (Development and Regulation) Act 1992.

In *Danisco (India) Private Ltd,* v. *Union of India*[171], admittedly, the goods in question were not meant for direct consumption but were intended for industrial use as ingredients in the manufacture of Yoghurt and fermented milk products. No objection certificate for the import of said goods was denied on the ground that the goods in question did not conform with the labelling Regulations under FSSA though the packages substantially complied with the labelling requirements as per the Regulations. Hence rejection of no objection certificate was held not proper. On 21st June 2017, FSSAI banned the import of milk and milk products from China[172]. The Government of India (Directorate General of Foreign Trade, Ministry of Commerce and Industry) on 24/9/2008 vide its Notification No 46 had imposed a ban on import of milk and milk products from China for 3 months.

The ban as a temporary measure wanted other departments to put in place suitable measures to restrict the import of contaminated milk and milk products in the country. The ban was further extended by another 6 months i.e upto 23/6/2009 vide notification No 67 dated 1/12/2008. The ban was extended from time to time through notifications, latest one being Notification No 12/2015-2020 dated 24th June 2016 which extended the ban upto 23rd June 2017. FSSAI, in a meeting on 12/6/2017, with the concerned Departments/Ministries of the Government of India, it was recommended to extend the ban on import of milk and milk products from China for a further period of

---

[171] AIR, 2015 (NOC)219(Del)

[172] Notification No 1-13/FSSA/DP/2008 available at https://www.fssai.gov.in/dam/jcr.../ Advisory_ Milk _Ban_China_22_06_2017.pdf (accessed on 1/8/2018 at 9.00pm)

one year. Accordingly, "Ban on import of milk and milk products including chocolates and chocolate products and candies/confectionary/food preparations with milk and milk solids as ingredients from China may be extended for a period of one year i.e. upto 23rd June 2018 or until their safety is established on the basis of credible reports and supporting data, whichever is earlier"[173].

Again the ban has been further extended by six months i.e till December 23, 2018. "The ban has been extended keeping in mind the food and milk safety. Representatives from the National Dairy Research Institute (NDRI), the animal husbandry department, the Department of Customs and the commerce ministry were present during the meeting. They were of the opinion that the milk that came from China was contaminated with melamine content."[174].

### 4.6.8 Special Responsibilities as to Food Safety

Under the erstwhile PFA the task of ensuring the food safety was cast on the food inspectors. But FSSA endeavours to make the food safe from farm to the fork, hence the primary liability of making the food safe lies at the outset on FBOs.

#### 4.6.8.1 Responsibilities of the Food Business Operator

Section 26(1), FSSA requires all the FBOs to ensure that the articles of food that he deals with satisfy the requirements of this Act all stages of production, processing, import, distribution and sale. A duty is cast on the FBOs not to sell, manufacture, distribute or store any article of food which is unsafe, misbranded or substandard or which contains extraneous material. Without the license, the FBOs also cannot deal with such kind of food which requires the licence from the authorities under the Act. No FBO can deal with any article of food which is for the time being prohibited by the FSSAI or by the government in the interest of the public health. FBOs must ensure that any person suffering from contagious disease is not employed in any activities of the food production/manufacture.

---

173 https://www.fssai.gov.in/dam/jcr.../Advisory_Milk_Ban_China_22_06_2017.pdf (accessed on 1/8/2018 at 10.05pm)

174 Available at https://www.fssai.gov.in/dam/jcr.../FSSAI_News_Milk_Ban_FNB_22_06_2018.pdf (accessed on 1/8/2018 at 10.10pm)

The FBO can sell the articles of food to the vendor only if the FBO guarantees in writing about the nature and quality of such articles. In pursuance of the same, the cash memo or a bill or an invoice issued by the FBO in respect of the sale shall be construed to be a guarantee for the purpose of this Section. Where any food is found to be unsafe, all the food in that batch, consignment or lot will also be presumed to be unsafe unless detailed assessment is carried out within a specified period of time and it is found that there is no evidence of the rest of the food being unsafe. Even if a particular article of food is in any conformity with the standards fixed under the Act, yet the competent authorities can lay restrictions on the food being placed in the market and he can also order for the withdrawal of such food from the market if he thinks that the food is unsafe, after duly recording the reasons in writing for doing the same.

**4.6.8.2 Liability of Manufacturers, Packers, Wholesalers, Distributors and Sellers**

Section 27 (1) relates to the liability of manufacturers, packers, wholesalers, distributors and sellers. Under the Act, both the manufacturer and the packer of any article of food will be liable when any article of food does not meet the requirements under the Act. Section 27(2) provides that the wholesaler or distributors shall be liable for nay article of food, which is, supplied after the expiry date; stored or supplied in violation of the safety instructions by the manufacturer or as stipulated by the Rules and Regulations of the Act; is misbranded or unsafe; un identifiable as far as the manufacturer is concerned; received despite the knowledge that the article of food is unsafe.

Section 27(3) makes the seller liable when he sells any article of food, after the date of its expiry; has kept the food in unhygienic condition; misbranded; unidentifiable of the manufacturer or the distributor and if he has received the articles of food with knowledge that they are unsafe. In *Yash* v. *State of Punjab*[175], the petitioner and his co-accused were carrying on the business of manufacturing adulterated desi ghee and tomato sauce and they were selling the same under the name of Neeraj Premium Quality. There was nothing on record to suggest that they had obtained the license from FSSAI. It was held that "manufacturing of adulterated food articles meant for human consumption amounts to playing with the health of people. There is no extraordinary circumstances in favour of the petitioner which entitle him to pre-arrest bail, a concession to be allowed by the court".

---

[175] 2014(1) FAC 445 at p.446

#### 4.6.8.3 Food Recall Procedures

Section 28 deals with Food Recall procedures which is a novel feature of this Act. Section 28(1) provides that if a FBO thinks that the food which he has processed, manufactured, or distributed or is already there in the market does not conform with the standards set by the Act or the Rules and Regulations made there under, he shall immediately initiate measures to withdraw the same from the market and consumers as well giving reasons for its withdrawal and keep the concerned authorities informed. Section 28(2) requires the FBO to immediately inform the authorities and Section 28(3) requires the FBO to inform the authorities as to the action taken by him to prevent the risks to the consumers and he should also not prevent or discourage any person who wants to help the authorities to prevent or reduce the risk arising from such a food. Section 28(4) mandates the FBO to follow the conditions and guidelines relating to food recall procedures as stipulated by the authorities.

### 4.6.9 Enforcement of the Act

Section 29(1) of the Act has invested the FSSAI and the State Authorities with the power of enforcement of the Act. It empowers these authorities to monitor and ensure that the relevant requirements are fulfilled by the FBOs. The authorities are to maintain a system of control, public communication on food safety and risk, food safety surveillance and other required monitoring activities covering all stages of food business.

#### 4.6.9.1 Food Safety Officers

Section 29(4) has made the Food Safety Officers responsible for the enforcement and execution of the Act within their area. Section 29(5) demarcates the duties of the Food Safety Officers in terms of who should execute the Act where and what kind of cases. Section 29(6) has given the same powers to the Commissioner of Food safety and Designated Officers, as are conferred on the Food Safety Officers.

#### 4.6.9.2 Commissioner of Food Safety of the State (CFS)

The State government shall appoint the 'Commissioner of Food safety' for the State for the effective implementation of the Act. Section 30(2) of the Act has armed the CFS to prohibit the manufacture, storage, distribution or sale of any article of food in the respective state for such period but not exceeding one year. CFS is required to carry out the survey of the industrial units engaged in the manufacture of food to ensure the compliance of the standards notified by FSSAI for articles of food and conduct training programmes both for the persons at the CFS office and also for participants at different segments of food chain for creating awareness about food safety. It is the duty of CFS to ensure an efficient and uniform implementation of the standards and other requirements and also to ensure a high standard of objectivity, accountability transparency and credibility. He can also sanction prosecution for offences which is punishable with imprisonment under this Act.

The CFS can delegate his powers to any officer subordinate to him, as he may deem necessary. It was held in *M/S Omkar Agency* v. *Food Safety and Standards Authority of India*[176], that "gutka is not article of food within the scope of FSSA and its Regulations. Since there has been no standardisation of gutka, the CFS is not competent to issue any prohibitory orders with respect to gutka". It was also held that the prohibitory orders of CFS regarding the use of tobacco and nicotine is not only arbitrary but is also beyond the scope powers of conferred by FSSA. It was held in *Ganesh Pandurang Jadhao* v. *State of Maharashtra*[177], that this provision makes it clear that contravention of the prohibitory orders issued by the CFS would amount to 'failure to comply with directions is a subject matter for adjudication under the provisions of Chapter IX of FSSA. It is not made punishable with imprisonment and cannot be referred to Court, such cases should necessarily go before the adjudicating officer.

**4.6.9.3 License for Food Business**

Section 31 and Food Safety and Standards (Licensing and Registration of Food Businesses) Regulations, 2011, deal with Licensing and Registration of food business. After the FSSA has been promulgated, no person can begin or carry on any food business unless he obtains a license issued by the authorities under the FSSA. But, a Petty Food Business Operator[178] need not obtain license but shall register himself with

[176] AIR 2016 Patna 160 at pp180,181

[177]2016 Crl L.J 2401 at p.2407(Bom)

[178] "Petty Food Manufacturer" means any food manufacturer, who manufactures or sells any article of food himself or a petty retailer, hawker, itinerant vendor or temporary stall holder; or distributes foods including in any religious or social gathering except a caterer; or such other food businesses including small scale or cottage or such other industries relating to food business or tiny food businesses with an annual turnover not exceeding Rs 12 lakhs and/or whose production capacity of food (other than milk and milk products and meat and meat products) does not exceed 100 kg/ltr per day or procurement or handling and collection of milk is up to 500 litres of milk per day or slaughtering capacity is 2 large animals or 10 small animals or 50 poultry birds per day or less. Food Safety and Standards (Licensing and Registration of Food Businesses) Regulations, 2011, 1.2.1;4

the Registering Authority [179] by submitting an application as specified by the Regulations on payment of fees of Rs 100/annum. These petty food manufacturers are required to maintain the basic hygiene and safety requirements specified in Part I of the Schedule 4. He has to declare that he would adhere to these requirements.

In pursuance of the same, the Designated Officer may either grant the license or refuse to grant the license in the interest of the public, only after recording the reasons in writing for doing so or issue order for inspection within 7 days of receipt of an application for registration. In pursuant of the inspection being ordered, the registration shall be granted within 30 days, only after being satisfied with the safety, hygiene and sanitary conditions of the premises. In case the registration is neither granted nor denied nor inspection is ordered within 7 days from the date of application and no decision is communicated to him within 30 days, the petty food manufacturer can begin his food business as if he has got the registration. But it will be incumbent on the FBO to comply with any improvement as required by the Registering authority[180]. The Registering authority shall issue a registration certificate along with a photo identity card which will have to be displayed at a prominent place in the premises or cart or vehicle or any other place where the person carries on sale/ manufacture of food. At least once a year, the Registering Authority or any other authorised officer shall carry out food safety inspection of the registered establishments.

Any FBO or any person carrying on food business at the time of promulgation of FSSA, shall get their existing license converted to license/registration under these Regulations by making an application to the licensing/Registering authority under FSSA within one year of notification of these Regulations. Non compliance of this Rule will attract penalty under Section 55 of FSSA. License for commencing food business, which falls under Schedule 1, shall be granted by the Central Licensing authority and for the rest of the food business, the license will be given by concerned state or union territory's licensing authority. The licensed FBO shall make sure that all conditions of license and safety, sanitary and hygienic requirements specified by the Schedule 4 of the

---

[179] Registering Authority" means Designated Officer/ Food Safety Officer or any official in Panchayat, Municipal Corporation or any other local body or Panchayat in an area, notified as such by the State Food Safety Commissioner for the purpose of registration as specified in these Regulations. Food Safety and Standards (Licensing and Registration of Food Businesses) Regulations, 20111.2.1;5

[180] "State Licensing Authority" means the Designated Officer appointed under section 36 (i) of the Act by the Commissioner of Food Safety of the state or by the Chief Executive Officer of the Food Safety and Standards Authority of India in his capacity of Food Safety Commissioner" Food Safety and Standards (Licensing and Registration of Food Businesses) Regulations, 2011. 1.2.1;6

Regulations are complied with. Licensing authority will undertake regular periodical food safety audit and inspection of the licensed establishment. No person shall manufacture, sell, stock, exhibit or sell any article of food which is subjected to the treatment of irradiation except under a license obtained from Department of Atomic Energy under the Atomic energy (Control of Irradiation of food) Regulations 1996.

An application for grant of license should be made to the concerned licensing Authority as specified in Regulations 2.1.2(3) and 2.1.2(4) accompanied by relevant documents and if the authorities require additional information, FBOs make the same available within the next 30 days. On failure of the FBOs to furnish the sought information within 30 days, the application for license stands rejected. On compliance of these requirements, application ID number will be assigned. Then, the Food Safety officer or any other designated person will inspect the premises as prescribed by FSSAI. On inspection, the officer may issue a notice guiding the FBO on necessary steps to be taken or alterations/modifications to be made to ensure proper sanitary and hygienic conditions as specified by the Regulations.

The applicant shall comply with the same within the next 30 days. Within the next 30 days from the receipt of an inspection report, the Licensing Authority will consider the application and may either grant the license or reject the same after giving an opportunity of being heard and recording the reasons for rejection in writing. In a nutshell, the license will be issued by the concerned Licensing Authority within 60days from the date of the application. If an applicant does not receive any intimation from the Licensing authority of inadequacy or inspection report reflecting defects, the FBO can start business after 60 days as if he has obtained license. In case the application is rejected, an appeal will lie to the CFS. A true copy of the license will have to be displayed in a prominent place in the premises where the FBO will carry on his business.

A single license may be issued for one or more articles of food and for different establishments or premises which are situated in the same area. But, if the articles of food are manufactured, stored, sold or exhibited for sale at different locations in more than one area, separate applications shall be made and separate license may be issued. FSSAI may appoint Designated officer or Food Safety Officer for central government establishments like railways and defence which have many food establishments, to ensure food safety in these establishments and also to carry out food safety audit once a year.

A license or registration once granted shall be in force for a period of 1 to 5 years as chosen by the FBOs and subject to the compliance of conditions if any. Application for renewal of license or registration shall be made not later than 30 days prior to the expiry of the registration or license, the failure of which will attract a fine of Rs 100 for each day of delay. In case the renewal application is not filed as stated above, the registration or the license expires and the FBO will have to stop all business activities. In case, the FBO wants to continue the business, he will have to apply for fresh registration or license all over again by following the due process.

The licensing or the registering authority after giving the FBO a reasonable chance of being heard can suspend the registration or the license in respect of all or any activity for which the registration or the license was obtained after recording the reasons for such suspension if he has complied with requirements mentioned in the improvement notice issued under Section 32 of the Act. Within fourteen days from the date of order of suspension, the registering/licensing authority may order for inspection of the premises. If the registering/licensing authority is of the opinion that even after the inspection report, the FBO has failed to comply with the observations and suggestions made by the authority, may cancel the license/registration after giving the FBO an opportunity.

Not withstanding anything the authorities can suspend or cancel the license or registration in the interest of public health but only after recording the reasons for doing so. Suspension/cancellation of the license/registration will not entitle the FBO for any compensation or refund of fees paid in respect of the registration/license. The aggrieved FBO can make fresh application for registration/license to the authorities only after the expiry of 3months, only if all observations made in the improvement notice have been complied with.

After the grant of license/registration, the FBO is obligated to inform the registering/licensing authorities, about the up to date information of their establishments, shall inform any modification, or additions, changes in product category, layout expansion or closure or any other relevant information, based on which the license was granted and such information should be conveyed before the changes occur. But if the changes alter the information contained in the license shall require the approval of the license before starting of business with such changes and for which the FBO will have to submit the original license to the licensing authorities along with the fees equivalent to one year license fees for effecting the change.

In the event of the death of the holder of the license/registration certificate, the same shall devolve on the legal representative or any family member of the deceased or until the expiry of a period of 90 days from the date of death of the holder or for such long period of time as the Designated Officer may allow. The legal representative or a family member of the deceased may apply to the authorities for transfer of such certificate or license in his favour and the license shall continue till the unexpired term. Every manufacturer, distributor, dealer selling an article of food shall give either separately or in cash memo or label a warranty regarding the article of food.

#### 4.6.9.4 Improvement Notices

Section 32 provides for issuing improvement notices to the FBO who fails to comply with the requirements of the Act. The Designated officer may serve an improvement notice if he has reasonable grounds for believing that the FBO has failed to conform to the requirements of the Regulations. The notice shall contain the grounds that the FBO has not complied with, specify the measures that will have to be undertaken by the FBO within a reasonable time to secure compliance. The failure of the FBO to comply with the improvement notice, the Designated Officer shall suspend the license. In continuation of this, if the FBO does not comply with the Improvement notice, the Designated officer after giving the licensee an opportunity to show cause, cancel the license granted to him. Any person aggrieved by the improvement notice or refusal to issue a certificate as to improvement or cancellation/suspension/revocation of license, an appeal may lie to the CFS whose decision may be final. The appeal may be filed within 15 days from the date of notice of the decision was served.

#### 4.6.9.5 Prohibition Orders

Section 33 empowers the court to issue Prohibition Orders after giving an opportunity of being heard to FBO where the FBO is convicted of an offence under the Act and the court which convicted him is satisfied that his food business is posing health risk to the public. The court can prohibit the use of the process, premises, equipment or treatment for the purpose of food business. The court shall serve a copy of the Prohibition Order on the guilty FBO. Any person, who knowingly contravenes the prohibitory orders of the court shall be guilty of an offence and he will be punishable with a fine which may extend to Rs 3 lakhs. A Prohibition order shall come to end when the court is satisfied on an application by the FBO that he has taken sufficient measures to justify the cancellation of Prohibition order. The application by the FBO should be made within six months from the date of the prohibition order.

#### 4.6.9.6 Emergency Prohibition Notices and Orders

Section 34 provides that after the Designated Officer is convinced that the food business of a particular FBO is posing danger to the public health, he may after serving the notice on FBO, may apply to the CFS for imposing the prohibition. If the CFS is also convinced about the situation, he shall by an order, lay a prohibition, the copy of which will be served on the FBO by the Food safety officer. Any person knowingly contravening such an order shall be guilty of an offence punishable with imprisonment for a term that may extend to two years and with fine which may extend to two lakhs. The Emergency prohibition orders will be withdrawn when the Designated Officer produces a certificate that he is satisfied that the FBO has taken sufficient measures to justify the withdrawal of the prohibition order.

Maagi noodles incidence in 2015 is a classic example for this provision where Nestle, the FBO was issued the emergency prohibition orders on the ground that the noodles manufactured by Nestle contained the food additives i.e. Monosodium glutamate and lead beyond the permissible limit and was also required to recall the Maggi noodles from the market. On the compliance of the requirement under the Act and the certificate issued by the Designated Officer, the prohibition orders were withdrawn and allowed to continue its food business. On 3/1/2019, the Supreme Court revived the government case in the National Consumer Disputes Redressal Commission against the FBO *'Nestle India'*, demanding damages of 640 crore alleging false labelling, unfair trade practices and misleading advertisements of its product Maggi noodles. The apex court on December 16, 2015 had stayed the proceedings before the National Consumer Redressal Commission and had directed the Central Food Technological Research Institute, Mysuru, to place before it the test reports of Maggi noodles. The counsel for Nestle argued that the lab report showed that the lead content in Maggi is within the permissible limits as stipulated by FSSA, 2006. The judges in fact took a dig at the counsel and said that why at all we should be eating anything that contains lead for which the counsel replied that there was some amount of lead in various other products.

### 4.6.9.7 Notification of Food Poisoning

Section 35 requires the FSSAI to notify registered medical practitioners in any local area as specified in the notification, to report all occurrences of food poisoning coming to their notice to such officer as may be specified.

### 4.6.9.8 Designated Officer

The CFS shall appoint the Designated Officer, to be the in charge of food safety administration in such area as may be specified by the regulations. There is one Designated Officer for each district. The Designated Officer shall issue or cancel license of food business operators; prohibit the sale of any article of food which contravenes the provisions of the Act; to receive samples and report of articles of foods from Food safety Officer and get them analysed; to recommend to the CFS for sanction to launch prosecutions; to maintain all records of all inspections made by Food Safety Officers and action taken by them, to get any complaint investigated even against Food Safety Officer.

### 4.6.9.9 Food Safety Officer and His Powers

According to Section 37, the CFS in consultation with the State government can appoint Food Safety Officers for such local areas as it may assign to them for the purpose of performing its functions under the Act. Section 38 of the Act details the powers of the Food safety officer, which include collection of a sample of any article of food, seizing of any article of food which in the opinion of the Food safety officer to be in contravention of the Act or Regulations, power to enter and inspect any place where the article of food is manufactured or processed or stored for the purpose of sale or stored for the manufacture of another article of food. While exercising the powers of entry and seizure, the Food Safety Officer ought to follow the provisions of Criminal Procedure Code[181] which is contemplated under Section 93 of Criminal Procedure

---

[181] Sec 93 Cr P C. When search warrant may be issued-

(1) (a) Where any Court has reason to believe that a person to whom a summons or order under section 91 or a requisition under sub- section (1) of section 92 has been, or might be, addressed, will not or would not produce the document or thing as required by such summons or requisition, or

(b) where such document or thing is not known to the Court to be the possession of any person, or

(c) where the Court considers that the purposes of any inquiry, trial or other proceeding under this Code will be served by a general search or inspection, it may issue a search- warrant; and the person to whom such warrant is directed, may search or inspect in accordance therewith and the provisions hereinafter contained.

(2) The Court may, if it thinks fit, specify in the warrant the particular place or part thereof to which only the search or inspection shall extend; and the person charged with the execution of such warrant shall then search or inspect only the place or part so specified.

Code, 1973. He is also authorised to seize the relevant books of accounts in relation to the food business. When the Food safety Officer seizes any adulterant, the burden of proving that the adulterant is not meant for the purpose of adulteration lies on the person from whom it is seized.

**4.6.9.10 Liability of the Food Safety Officer in Certain Cases**

Section 39 makes the Food safety officer liable under certain circumstances like, when he without any reasonable ground or vexatiously seizes any article of food or adulterant; does any other act which injures any other person without any reason to believe that such necessity existed for the discharge of his duty, he will be guilty of an offence under FSSA and may be liable for a penalty which may extend upto 1lakh rupees. On the other hand if anybody makes a false complaint against the Food Safety officer and the compliant is proved to be false, the complainant will be punishable with fine which shall not be less than Rs50,000 and can extend upto 1lakh rupees.

**4.6.9.11 Purchaser may have the Food Analysed**

Section 40 enables the purchaser of any article of food to get it analysed from the analyst and get the report. But the purchaser, at the time of purchaser should inform the food business operator of his intention of getting the food sold by him analysed. After the analysis, if the report shows that the food is not compliance with the requirements/standard under the Act or regulations, the purchaser is entitled for the reimbursement of the fees that he is paid towards the analysis of such article and the food analyst shall forward the report to the Designated officer for further step for prosecution as per Section 42.

**4.6.9.12 Procedure for Launching Prosecution**

Section 42 has entrusted the Food safety officer with the responsibility of inspection of food business, drawing the sample and sending it to the Food analyst for analysis. The Food analyst after receiving the sample from the Food safety officer shall analyse it and send the report within 14days to Designated officer with a copy to the Commissioner of food safety. After the receipt of the Report, the Designated officer shall decide whether the contravention is punishable with imprisonment or fine only. In

---

(3) Nothing contained in this section shall authorise any Magistrate other than a District Magistrate or Chief Judicial Magistrate to grant a warrant to search for a document, parcel or other thing in the custody of the postal or telegraph authority

case the contravention of the Act is punishable with imprisonment, within 14days, he should send his recommendations to the commissioner of Food safety for sanctioning the prosecution. Now, the CFS will decide whether the matter be referred to a court of ordinary jurisdiction in case of offences punishable with imprisonment for 3 years or a Special court in case of offences punishable with imprisonment for a term exceeding 3years, where such special courts are established and in case no special court is established such cases will be tried by courts of ordinary jurisdiction. The CFS should communicate his decision to the Designated Officer and the concerned Food safety officer should launch prosecution before the court of ordinary jurisdiction or special court and such communication will also be sent to the purchaser if the sample is collected by the purchaser as per Section 40.

#### 4.6.10 Analysis of Food

Analysis of food is governed by Section 43 to 47 of FSSA and also Food Safety and Standards (Laboratory and sampling analysis) Regulation, 2011.

##### 4.6.10.1 Recognition and Accreditation of Laboratories and Research Institutions

According to Section 43, the FSSAI can notify the food laboratories and research institutions accredited by the National Accreditation Board for testing and Calibration Laboratories for the purpose of analysis of samples by the Food Analysts. The Food Authority may establish or recognise one or more referral food laboratories to carry out the functions, frame regulations specifying the functions of food laboratory and referral food laboratory, the procedure for submission of samples of articles of food for analysis/tests to the laboratories. Section 44 provides for recognition of organisation or agency for food safety audit and also to check the compliance with food safety management systems required under FSSA.

##### 4.6.10.2 Food Analysts and their Functions

Section 45 empowers the CFS of the State to appoint Food Analysts for a specified local area. But, any person who has any financial interest in the manufacture or sale of any article of food will not be appointed as a food analyst. Different Food Analysts may be appointed for different articles of food. In *Ramakant Gupta* v. *State of Chhattisgarh*[182], the sample of cow milk was collected by the Food Safety Officer without mixing the container, in which the milk was stored and from which the sample

[182] 2016Crl.L.J 3386 at p.3390(Chhatt)

was collected. The milk in the container was not stirred properly before collecting the sample. On analysis of the sample it was reported that the sample was not in accordance with the standards of FSSA but it was held that 'sampling of cow milk was not done properly by the Food Safety Officer and it cannot be safely held that sample of milk sent to the Food Analyst truly represented the milk to be tested and the sampling is not done in accordance with the Act and therefore the prosecution has failed to bring home the offence under Section 26(2) (i) of the FSSA and consequently penalty imposed was cancelled.

Section 46 enumerates the Functions of food analyst. On receipt of a package containing a sample for analysis from a food safety officer, the food analyst should ensure that the container is properly sealed, in case the container is broken or is unfit for analysis, he should within 7 days from the date of receipt of sample inform the Designated Officer about the same and request him to send the second part of the sample. The Food analyst, within a period of fourteen days from the date of receipt of the sample should send the report to the Designated Officer indicating the method of sampling, in case the analysis is requested by the purchaser of article of food as per Scetion40, to him with a copy to the Designated Officer. In case the Food Analyst cannot analyse the sample within fourteen days, he shall inform the Designated Officer and the CFS explaining the reasons for it and shall also specify the time to be taken for analysis. An appeal against the report of the Food Analyst will lie to the Designated officer who in turn may refer the matter to the Referral food laboratory for its opinion.

In *Cargill India Private Ltd* v. *State of Uttarakhand*[183], the High Court directed the Food Analyst and Analyst of Referral Lab to be summoned for cross examination . It was held that Reports of public analyst and Referral labs have no evidentiary value unless is proved by the experts. In *Dharampal Satyapal Ltd* v. *State of Kerala*[184], the Designated Officer had not heard the appellant or considered the grounds raised by him in the appeal on the merits but rejected the appeal on the ground that the appeal is not supported by report from an accredited laboratory. In the absence of a stipulation in the Act that an appellant challenging the accuracy of the analysis of the report should produce the report of an accredited laboratory, the impugned order cannot be sustained. Therefore, the petitioner is entitled to succeed. The Designated Officer should have heard the petitioner and decided on merits.

---

[183] 2013 Cr. L. J (NOC) 140 (Uttar)
[184] AIR 2014 Ker 51 at p.53

In *Narayani Flavours and Chemicals Pvt Ltd* v. *The Auth. Officer, Food Safety Standard Authority of India*[185], as per the Regulations under the Act, the Designated officer shall dispose of the appeal within 30 days of its filing. In this case, the statutory period had expired and the department had recommunicated their stand which they had taken earlier against which an appeal was pending and the said subsequent stand was brought to the notice of the High court by the petition. The authorities cannot keep the appeal in suspended animation for a long time. Therefore, the High Court directed the Designated officer before whom the appeal was pending to dispose of the appeal within a fortnight from the date of communication of the order. In *M/S Nestle India Ltd* v. *Food safety and Standards Authority of India, Mumbai*[186],writ petition was filed challenging the imposition of ban on Nestle product Maggi Noodles. It was alleged that there was clear violation of principles of natural justice. Writ petition was held maintainable in spite of availability of alternate remedy of appeal under the FSSA.

**4.6.10.3 Mode of Sampling and Analysis**

Section 47 deals with the procedure for collection of sample and analysis of the same. When a Food safety Officer collects a sample of food for analysis, he must give a notice in writing to the person, from whom the sample is collected, of his intention to get it analysed. Divide the sample into four parts and mark and seal each part separately and take the signature or thumb impression of the person from whom the sample is collected. But if the person, from whom the sample is collected, refuses to sign or affix his thumb impression, the Food safety Officer shall call upon one or more witnesses and take the signature or thumb impression of that person.

"The PFA requires that the food inspector find at least two witnesses in the presence of whom this sampling is done. Now to find two witnesses is the second part of the harrowing stage simply because no one is willing to come forward and subject himself to endless visits to the courts for the next several years. Usually, the food inspectors beg, plead, cajole and request whoever they can to become witnesses for the sampling! But of course Indian ingenuity being what it is, most of them get away by giving false names and addresses and thereby it is impossible to find them at the time of the court's hearing"[187]. The Researcher knows that the situation is no different today.

185 2014(1) FAC 440at pp441, 442(Cal)
186 AIR 2016(N.O.C)225(Bom)
187 Srivatsa Krishna, "PFA: Dead in Letter, Dying in Spirit", The Administrator, [April-June, 1996] Vol XLI p 161-162

Out of the four parts of the sample, the Food Safety Officer shall send one part to the food analyst after intimating the Designated Officer of the same, handover two parts to the Designated Officer for safe custody and send the remaining part to the accredited laboratory for analysis if the FBO has requested for doing so. If the test reports of the Food analyst and the accredited laboratory vary from each other then the Designated officer shall send one part of the sample which is kept under his safe custody to referral laboratory for analysis, whose decision will be final.

The Food Safety Officer collects the sample, by the immediate succeeding working day, send the sample to the concerned Food Analyst for analysis and report. But, in case the sample sent to the Food Analyst is lost or damaged or unfit for analysis, the Designated Officer shall, on requisition made to him by the Food Analyst or the Food safety officer, despatch one of the Parts sent to him to the Food Analyst for analysis. In case of imported articles of food, authorised officer of FSSAI shall take its sample and send it to the Food Analyst who shall send the report within a period of 5 days to the authorised officer.

### 4.6.11 Offences and Penalties

Lord Hewart, C.J, in *Cotterill* v *Pern*[188], held, "the element of mensrea is not required beyond the point that the facts must show an intention on the part of the person accused to do the act forbidden". In *Kat* v. *Dimenz Lord Goodard*[189], C.J. observed, 'where a statute forbids the doing of a certain act, the doing of it in itself implies mensrea.... By throwing an onus on the defendant of proving that he acted without intent to defraud, it seems to me that the statute assumes that the doing of the act alone implies an intent to defraud'

By the latter half of 19$^{th}$ century it was accepted that law can't afford to inquire into an individual's actual mental state before punishing him. Thus the doctrine of 'strict liability' was applied where neither knowledge nor intention nor negligence was required for conviction[190]. Hence, though there must always be an actus reus, there need not always be a mental element in relation to each part of the actus reus. Now strict liability has been replaced by 'Absolute liability' which in effect means the accused is guilty without any mental element at all and that he has no defence either at common law or under statute. Difficulties of proof can be overcome if the prosecution does not

[188] (1936) 1K.B. 53 (T.A.C).
[189] (1952) 1K.B.34.
[190] Hart, 'The Aims of Criminal Law' (1958) 23 Law and contemporary problems 401.

have to prove mens rea. It is easier to enforce the law when mensrea is irrelevant than when the prosecution have to prove it [191] . According to Fitzerald, P.J. 'strict responsibility' has become a necessary evil[192]. Hence as on today, absolute liability is the principle for any offence under the FSSA.

#### 4.6.11.1General Provisions Relating to Offences

According to section 48, a person may make any article of food injurious to health by means of adding any article or substance to the food; using any article as an ingredient in the preparation of food; abstracting any constituents from the food or subjecting the food to any other process or treatment which might render it injurious, with the knowledge that the article of food may be sold or offered for sale or distributed for human consumption. To determine whether any food is unsafe or injurious to health, regard shall be had to the normal conditions of use of the food by the consumer and its handling at each stage of production, processing and distribution, information provided to the consumer - on labels or other information regarding the avoidance of specific adverse health effects from a particular food or category of foods not only the immediate or short term or long term effects of that food on the health of a consumer but also on subsequent generations.

The probable cumulative toxic effects should also be considered. In determining the safety of food regard must also be had to the fact where the quality or purity of the food article being a primary food, has fallen below the specified standard or its constituents are present in quantities beyond the acceptable limits, may be due to natural causes or beyond the control of human agency, then such primary food shall not be deemed to be unsafe or substandard or food containing extraneous matter as no standard is specified for a primary food under FSSA.

#### 4.6.11.2 General Provisions Relating to Penalty

Section 49 relates to general provisions relating to penalty. It provides that while adjudging the quantum of penalty, the adjudicating officer or the Tribunal shall have due regard to the amount of gain or unfair advantage derived by the FBO by the contravention of the Act, the amount of loss caused to any person as a result of contravention, the repetitive nature of the contravention, whether the contravention was

---

[191] Mensrea In Statutory Offences available at http://shodhganga.inflibnet.ac.in/bitstream/ 10603/ 132472/ 7/07_chapter%202.pdf (accessed on 11/3/2018 at 10.52 pm)
[192] *Ibid*

wilful and any other relevant factor in the instance. Section 50 deals with penalty for selling food not of the nature or substance or quality demanded by the customer. It provides that any person, who sells to the purchaser's prejudice any food which is not in compliance with the provisions with the Act in terms of its nature of substance or quality demanded by the purchaser, shall be liable to a penalty not exceeding 5 lakh rupees. But, petty FBOs for such non compliance shall be liable for a penalty not exceeding 25000 Rs.

Section 51deals with a situation when any person either by himself or through somebody manufactures for sale or stores or sells or distributes or imports any article of food for human consumption which is substandard, shall be liable to a penalty which may extend to 5 lakh rupees. In *Anil Kumar Srivastava* v.*State of Bihar*[193] , the petitioner was posted as Assistant Godown Manager of State Food Corporation. The State Development Officer took the sample of the wheat from the godown and the wheat was found to be contravening the standards of the Act. It was submitted by the petitioner that he was not responsible for maintaining the standard of the grains as they were supplied by the Food Corporation of India and this fact that they were supplied by Food Corporation of India was not disputed. Considering the aforesaid arguments the court ordered the petitioner to be released on anticipatory bail.

Section 52 deals with penalty for misbranded food. Any person who whether by himself or through somebody on his behalf, manufactures for sale or stores or sells or distributes or imports any article of food meant for human consumption which is misbranded, shall be liable for a penalty which may extend to Rs 3 lakh. The adjudicating Officer may issue direction to the person found guilty under this Section to take corrective action to rectify the mistake or such article of food shall be destroyed. Section 53 deals with penalty for misleading advertisement. Any person who publishes or is a party to the publication of an advertisement which falsely describes any food or it is likely to mislead as to the nature or substance or quality of any food or gives false guarantee will be liable for a penalty which may extend to Rs 10 lakh.

According to Section 54, the penalty for food containing extraneous matter including the imported food may extend to Rs1lakh. Section 55 provides for a penalty for failure to comply with the directions of Food safety Officer, which may extend upto Rs 2 lakhs. But it was held in *State of Rajasthan* v. *Hat Singh* that 'where an act or an

---

[193] 2013(2) FAC 173 at p 174(Pat)

omission constitutes an offence under two enactments, the offender may be prosecuted and punished under either or both enactments but shall not be liable to be punished twice for the same offence'. In this case the bench said that prosecution under two different Acts is permissible if the ingredients of the provisions are satisfied on the same facts. The bench interpreted the scope of Section 188 of IPC and observed: "Section 188 of the IPC does not only cover breach of law and order, the disobedience of which is punishable. Section 188 is attracted even in cases where the act complained of causes or tends to cause danger to human life, health or safety as well".

Penalty for unhygienic or unsanitary processing or manufacturing of food is prescribed by Section 56, a penalty which may go upto Rs 1 lakh. Section 57 mandates that if any person whether by himself or through somebody else, imports or manufactures for sale or stores or sells or distributes any adulterant shall be liable for a penalty not exceeding Rs 2 lakhs in case the adulterant is not injurious, but if the adulterant is injurious- the penalty may extend upto Rs 10 lakhs. In this case the accused cannot set up a defence that he was holding such adulterant on behalf of somebody. Section 58 prescribes the penalty for contravention of the Act or any Rules and Regulations framed under the Act for which no specific penalty is provided but, which may extend to Rs 2 lakhs.

**4.6.11.3 Provisions for Punishment**

Section 59 prescribes punishment for unsafe food. Where any person who whether by himself or through somebody else, manufactures, sells, stores, distributes or imports any article of food which is unsafe shall be punished with an imprisonment for a term which may extend to 6 months and also fine which may extend to Rs 1 lakh, in case such contravention does not result in any injury. If such unsafe food results in a non grievous injury to the consumer, the imprisonment may extend to 1year and also fine which may extend to Rs 3 lakh. If contravention results in a grievous injury, the punishment would be imprisonment which may extend to 6 years and also fine which may extend to Rs 5 lakh. But where such contravention results in death, the imprisonment would be not less than 7 years but which may extend to imprisonment for life and also with fine which shall not be less than Rs 10 lakhs.

Section 60 mandates that without the permission of the Food safety officer, if any person retains, removes or tampers with any food, vehicle, equipment, package or labelling or advertising material or any other thing that has been seized, he shall be punished with imprisonment for a term which may extend to 6months and also with fine which may go upto Rs 2 lakhs. If any person provides any false information or produces any false or misleading document wilfully, will be punishable with imprisonment which may extend to 3months and also with fine which may extend to Rs 2 lakhs. If any person resists, obstructs or attempts to obstruct, impersonate, threaten, intimidate or assault a Food Safety Officer in exercising his functions he will be punishable with imprisonment for a term extending to 3 months and also with fine which may extend to Rs 1 lakh.

Section 63 provides that, if any person or FBO excepting the petty FBO, himself or through any person who is required to obtain license- manufactures, sells, distributes, stores or imports any article of food without license, shall be punishable with imprisonment for a term which may extend to 6months and also with a fine that may extend to Rs 5 lakhs. Section 64 prescribes punishment for subsequent offences. If any person after having been previously convicted for an offence under this Act, subsequently commits and is convicted of the same offence, he will be liable to twice the punishment, which might have been imposed on a first conviction, subject o the condition that the punishment being maximum provided for the same offence and a further fine on daily basis which may extend upto Rs 1 lakh, where the offence is a continuing one and his license stands cancelled. The court may also publish the offender's name and the place of residence, the offence and the penalty imposed at the offender's expense in such newspapers or in such other manner as it thinks fit. The cost of all this shall be treated to be a part of the cost attending the conviction and shall be recoverable as a fine.

#### 4.6.11.4 Compensation in Case of Injury or Death of Consumer

Section 65 deals with compensation in case of injury or death of consumer. When any FBO manufactures, sells, imports, distributes any article of food causing injury to the consumer or his death, the court may direct him to pay a compensation to the victim or to the legal representative of the victim, a sum not less than Rs 5 lakh in case of death, not exceeding Rs 3 lakh in case of grievous injury and not exceeding Rs 1 lakh in all other cases of injury. The compensation shall be paid within 6 months from the date of occurrence of the incident and in case of the death of the victim, an interim relief shall be paid to the next of kin within 30 days of the incident.

Where any person is guilty of causing grievous injury or death, the Adjudicating officer or the Court may cause the name and place of residence of the guilty, the offence and the penalty imposed to be published at the offender's expense in such newspapers or in such manner as the Adjudicating Officer or the Court may deem fit. The expenses of such publication will be treated as part of the cost attending the conviction and shall be recoverable in the same manner as fine. The Adjudicating Officer or the Court may also order for cancellation of the license and also for recall of food from market, forfeiture of establishment and property in case of grievous injury or death of the consumer and prohibition orders may also be issued.

**4.6.11.5 Offences by Companies**

Section 66 deals with offences by companies. Generally, when any company is guilty of committing any offence under any law, every person who was in charge of and responsible to the company for conduct of the business as well as the company shall be deemed to be guilty of the offence and shall be liable to be proceeded against and punished accordingly. Where a company has different establishments or branches or different units in any branch, the concerned head who is nominated by the company to be responsible for food safety will be liable. But such person will not be held liable, if he proves that the offence was committed without his knowledge or that he exercised all due diligence to prevent the commission of such offence.

But if it is proved that the offence has been committed with the consent or connivance of or is attributable to negligence on the part of any director, manager, secretary or other officer of the company, such a person will also liable to be proceeded against and punished as per FSSA. But, the Researcher thinks that the doctrine of strict liability has been given a go by here. If the Board of directors of a company can be made accountable for other offences under different laws, why should be exempted under FSSA. It is human nature that if one is allowed to scot free after proving that the offence was committed without his knowledge, than everybody proves the same and escape from the clutches of law.

**4.6.11.6 Penalty for Contravention of the Act in Case of Import of Articles of Food to be In Addition to Penalties**

Section 67 provides for penalty for contravention of the Act in case of import of articles of food to be in addition to penalties provided under any other Act. This provision makes the importer of any article of food which is in contravention of the Act, liable not only under FSSA but also under Foreign Trade (Development and Regulation) Act, 1992 and the Customs Act, 1962. Such article of food which contravenes the standards of FSSA shall be destroyed or returned to the importer only if permitted by the competent authority under the Foreign Trade (Development and Regulation) Act, 1992 or the Customs Act, 1962.

### 4.6.12 Adjudication and Food Safety Appellate Tribunal

Chapter X of the Act deals with adjudicatory process and the setting up of Food Safety Appellate Tribunal.

#### 4.6.12.1 Adjudication

Section 68 of the Act stipulates that an Adjudicating Officer under this Act shall not be below the rank of an Additional District Magistrate and shall follow the procedure as prescribed by the central government from time to time. His powers are that of a Civil Court and the proceedings will be judicial proceedings as per Section 193 and 228 of the Indian Penal Code 1860. He shall follow the principles of natural justice and only on being satisfied that the accused has contravened the Act or the Rules or the Regulations under the Act, shall impose such penalty as he deems appropriate in tune with the provisions pertaining to the offence. But while judging the quantum of penalty, he shall weigh the guidelines specified in Section 49 of FSSA.

It was held in *Ramakant Gupta* v. *State of Chhattisgarh*[194], that "by virtue of section 68 of the Act holding of an enquiry by the Adjudicating Officer is sine qua non for arriving at the conclusion that such a person has committed the contravention of the provisions of FSSA or the Rules and Regulations made there under and satisfaction can be arrived into after making an enquiry into all relevant facts". In *PradeepKumar Gupta* v. *State of UP*[195], on the analysis of Sections 68,70 and 71 of FSSA, it was held that an appeal against the order of Adjudicating Officer and Tribunal cannot be filed as criminal appeal and would lie only as a civil miscellaneous appeal. Similarly the appeal against any order of the Tribunal would lie to the High Court not as criminal appeal but as an appeal from order on the civil side.

---

[194] Crl.L.J. 3386 at p3389(Chhatt)
[195] 2016 Crl.L.J 122 at p128(All)

#### 4.6.12.2 Power to Compound Offences

Under Section 69, the Commissioner of Food Safety can permit the Designated officer to accept from petty food manufacturers who themselves manufacture and sell article of food, hawkers, itinerant vendors, retailers and temporary stall holders against whom a reasonable belief exists that the offence under FSSA has been committed by them, payment of money which may not be more than 1lakh rupees after the composition of the offence which such person is suspected to have committed. On payment of such money by the suspect, he will be discharged if he is in custody and no extra proceedings will be initiated against such a person, but this provision cannot be used to compound an offence for which punishment of imprisonment has already been prescribed under the Act.

#### 4.6.12.3 Establishment of Food Safety Appellate Tribunal.

Section 70 of FSSA enables the Central government or the state government as the case may be to establish one or more Food Safety Appellate Tribunal to try appeals from the judgements of the Adjudicating Officer passed under Section 68FSSA. The Central or the State government shall prescribe the areas and matters over which the Tribunal may exercise its jurisdiction. The Presiding Officer of the Tribunal who shall not be below the rank of a District Judge will be appointed by the Central or the State government as the case may be. It was held in *Narayan Malviyas* v. *State of MP*[196], that "in view of availability of remedy of appeal against the impugned order a petition under Art 226 and 227 of the Constitution of India could not be entertained by the High Court".

In this case the counsel withdrew the petition reserving the right to file an appeal before the appellate tribunal. In *Pourushottam Khemuka* v. *The State of Madhya Pradesh*[197], the counsel for petitioners could withdraw the petition with a liberty to avail the remedy before the appellate Tribunal. But it was, however, submitted that the interim direction granted on 19.9.2013 may be continued for a certain period till petitioners avail such remedy before the tribunal. "It was further ordered that till 21.2.2014 the interim order dated 19.9.2013 shall remain in operation". It was held in

[196] 2013(2)FAC 463 at p 464 (MP)
[197] 2014(1)FAC 433 at p 434 (MP)

*Amrinder Chopra* v. *State of Punjab*[198], that "the Tribunal may consider the condonation of delay by the petitioner".

#### 4.6.12.4 Procedure and Powers of Tribunal

Section 71of FSSA dictates that the Tribunal shall not be bound by the procedure as prescribed by Civil Procedure Code but shall follow the principles of natural justice and shall be subject to the provisions of FSSA. But it shall have the same powers as that of a civil court for the purpose of discharging its duties under the Act. The Tribunal is endowed with powers to regulate its own procedure and can also decide the place at which it can have its set up. Proceedings of the Tribunal shall be deemed to be judicial proceedings. The appellant may appear either party in person or through legal practitioners.

Limitation Act is applicable to file an appeal before the tribunal. Anybody aggrieved by the decision or order of the Tribunal may prefer an appeal to the High Court within 60 days from the date of communication of the decision of the Tribunal. If there is a delay in filing an appeal before the High court, the High Court only on satisfying itself that the appellant was prevented by sufficient cause from filing an appeal within 60 days, can allow the same to be filed within the next 60 days and thus condoning the delay and may entertain the appeal.

#### 4.6.12.5 Civil Court Not to Have Jurisdiction

Section 72 of FSSA bars the Civil Court from trying any suit or proceeding pertaining to matters which an Adjudicating Officer or the Food Safety Tribunal is empowered under the Act. No injunction shall be granted by any authority or the court with respect to any action initiated or to be initiated in continuation of the power conferred by the Act.

#### 4.6.12.6 Power of Courts to Try Cases Summarily

Section 73 mandates that the court should try the cases summarily and provides that the offences not triable either by an adjudicating officer or by the Tribunal, shall be adjudicated either by a Judicial Magistrate of the First Class or by a Metropolitan Magistrate and in case of conviction, a sentence of imprisonment for a term not exceeding one year may be passed. But, if it appears to the Magistrate at the beginning of the summary proceeding that the case is such that a sentence of imprisonment

---

[198] 2014(1)FAC 444 at p 445 (P&H)

exceeding one year may have to be passed or it is not appropriate to try the case summarily, he shall recall the witness who may have to be examined and rehear the case. At this juncture, the Researcher thinks that this procedure leads to multiplicity of suits, the FSSA should have conferred both civil and criminal jurisdiction on one judge than haggling between the Tribunal and the Magistrates, like in the case of Family Courts, the presiding officer will have concurrent jurisdiction to award maintenance under Section 125 of Cr.P.C, 1973, this would have been a welcome move.

#### 4.6.12.7 Special Courts and Public Prosecutor

Section 74 empowers the central government or the state government to establish as many Special courts as may be necessary to try cases under the Act relating to grievous injury or death of the consumer for which the FSSA has stipulated the punishment of imprisonment for more than 3 years. The trial in Special Courts will be prioritised over the trial of any other case under FSSA against the accused in any other court and the trial in Special courts will be concluded first and the trial in other courts will be kept in abeyance.

But, here the Researcher emphatically notes that the FSSA does not have a provision which enables the owner for seeking the release of the food item and the vehicle seized, from the authorities. Likewise, when food items and vehicles are seized, applications are filed for seeking the release of the same. But, the authorities under the Act expressed their inability to release the vehicle on the ground that they lack such powers under FSSA. When similar applications are filed before the Magistrates, they also express their inability in this regard. The aggrieved persons are left in the lurch without their being remedy to their problems.

#### 4.6.12.8 Appeal

Aggrieved by the decision of the Special Court, under Section 76 of FSSA, an appeal may be filed at the respective High court within forty five days from the date on which the order was served. The High Court shall dispose the appeal by a bench consisting of not less than two judges. Section 77, FSSA mandates that no court can take the cognisance of an offence after the expiry of one year from the date of commission of an offence. But this will not prevent the Food Safety Commissioner from approving the prosecution extending the period upto three years. It was held in '

*M/S Garima Milk and food Products Limited* v. *State of Rajasthan*[199], that the Commissioner of Food Safety after recording the reasons, can further extend the period of limitation upto three years using his discretionary power. During the trial of any offence under the Act, if the court is satisfied on evidence that any importer, manufacturer, distributor or dealer has colluded, then the court under Section 78 can implead such importer, manufacturer or dealer as the case may be.

#### 4.6.12.9. Magistrate's Power to Impose Enhanced Punishment

For a term of imprisonment exceeding six years, Section 79 authorises the Magistrate to enhance the punishment that is to pass any sentence authorised by this Act. On 20th of September 2018 in *State of Maharashtra and Anr.* v. *Sayyed Hassan Sayyed Subhan and Ors*[200],that food safety violations can be prosecuted under IPC as well. Supreme Court ruled that "where an act or an omission constitutes an offence under two enactments, the offender may be prosecuted and punished under either or both enactments but shall not be liable to be punished twice for the same offence".

It also held that the Food Safety Officers can lodge complaints alleging offences under the provisions of the Indian Penal Code against persons, transporting, stocking and or selling the prohibited goods. In this it was observed that there is no bar for prosecution under the Indian Penal code just because the provisions in FSSA prescribe penalties. The opposite view adopted by the High Court was set aside and the Supreme Court further observed that referring to *State of Rajasthan* v. *Hat Singh*[201], that prosecution under two different Acts is permissible only if the ingredients of the provisions are satisfied on the same facts but the offender shall not be punished twice for the same offence.

#### 4.6.12.10 Defences Which May or May Not be Allowed in Prosecution

Section 80 of the Act enumerates different defences that may or may not be allowed in prosecution. Those defences have been categorised as the following-defence relating to publication of advertisements, defence of due diligence, defence of mistaken and reasonable belief not available and defence in respect of handling food, defences of significance of the nature, substance or quality of food. In any defence of due diligence, the defence that all reasonable precautions due care has been taken to prevent the

[199] 2017(1)FAC 65 p66 (Raj)
[200] Criminal Appeal No.1195 of 2018 arising out of Special Leave Petition (Criminal) No.4475 of 2016
[201] (2003) 2 SCC 152

commission of an offence. This one defence can vitiate the whole gamut of offences under FSSA. This defence hits the concept of absolute liability below the belt, which means that even after committing a grave mistake under FSSA if a person is allowed to set up this defence and prove the same, the very purpose of bringing in FSSA is defeated. Chapter XI, Sections 81to Section 84 deals with Finance, Accounts. Audit and Reports.

### 4.6.13 Miscellaneous

Miscellaneous provisions are contained in Chapter XII, Section 85 to 101. Section 89 mentions that FSSA has an overriding effect over all other food related laws. Section 91empowers the Central Government to make rules regarding the administration of this Act and for carrying out the provisions of the Act. Section 92 empowers the Food Authority to make regulations and the important ones are the power to limit the levels of additives, limits of quantities of contaminants, toxic substances and heavy metals, limits of residues of pesticides and veterinary drugs, the ways of marking and labelling of food articles, guidelines and conditions with regard to food recall procedures, procedure for getting the food analysed and the procedure to be followed by food laboratories for the analysis of food and to give guidelines for participation in Codex Meetings and to prepare response to matters pertaining to Codex.

As per Section 94, the State government, subject to the powers of the Central government and FSSAI can make rules and regulations after the approval by FSSAI. Section 96 enunciates that a penalty imposed under this Act, if is not paid, shall be recovered as an arrear of land revenue and the defaulter's license will be suspended till the penalty is paid. Section 101 confers power on the central government to make such provisions not consistent with the provisions of this Act, as may appear to be necessary, for removing any difficulty in giving effect to the provisions of this Act.

### 4.6.14 True Incidents that Depict the Implementational Collapse of FSSA

Food regulation anywhere in the world is a difficult and challenging task. Food regulation is expected to be a facilitator for the industry, but very often they both are at loggerheads. The FSSA is very impressive and ambitious in its object but the implementation though may not be impossible yet is a colossal task and fraught with many challenges. The following quoted incidents demonstrate the failure of the Act.

- Few countries rejected milk produced from Kolar district from Karnataka due to fluoride content above the permissible limit[202] but we were oblivious of this contamination and consumed it.
- Karnataka State Human Rights Commission issued a notice on January 30th to Labour Commissioner seeking his response to the complaint that 60 employees in one of the garment factories in Bangalore fell ill after consuming contaminated water[203]. The Commission took notice of the media reports and issued a notice suo-moto.
- In a crackdown on the food adulteration, the south zone police at Charminar at Hyderabad raided two milk adulteration units and apprehended two persons. They mixed skim milk powder with milk powder of inferior quality[204].
- On March 8th 2017, three children aged between 14 and 15yrs died hours after consuming the food for dinner at a hostel run by a residential school[205]. The students died, hours after they complained that the sambar served during dinner was bitter.
- As many as 14 students took ill after consuming lunch at the SC/ST student's hostel at Yadgiri, a place at Karnataka on 10th March 2017[206]. These students complained that the contaminated water used to cook the food had resulted in the incident.
- A news paper reported that stale food was served in Rajdhani Express, one of the prestigious trains run by our railways[207]. The passengers complained that the food served for dinner was stale and that bad smell emanated from food packets. If this is the scenario in state owned enterprise, one can imagine the scenario at hotels, motels and other private owned enterprise.
- A total of 60 students took ill after having the mid-day meals provided to them at government higher primary school at Benakatti in Bagalkot District[208]. One of the students complained that she found a lizard in the sambar, after rice and sambar were served to them and soon the students started vomiting.

---

[202] *'High fluoride content: Milk exported from Kolar rejected'*. Deccan Herald, Bengaluru Edition dated July 18, 2017,.
[203] *Ibid 'KSHRC issues notice',* January 31st, 2014,
[204] m.timesofindia.com/city/Hyderabad (accessed on January 4th at 10.00pm)
[205] Deccan Herald, Bengaluru Edition dated March 11, 2017
[206] *Ibid*
[207] *Ibid* 29th March 2017
[208] *Ibid* 15th March 2017

- The National Human Rights Commission suo-moto issued a notice to food regulator FSSAI over reports of pesticides being found in food items more than the prescribed limit[209]. In its notice, NHRC directed FSSAI to submit its reply within eight weeks from the date of receipt of notice. Maximum Residual Levels of pesticides in fruits and vegetables and other food products have been fixed under the Food Safety and Standards (Contamination, Toxins and Residues) Regulation, 2011. Presence of pesticides residues beyond these levels in food products including food, vegetables and meat is treated as a violation of the said regulations, which attracts penal action under the Act.
- A pregnant woman discovered a dead lizard in her French fries at a fast food outlet in Kolkata[210].
- In one of the biggest hauls ever, FSSAI officials seized around 34,000kg of tea leaves adulterated with synthetic colours from a godown in Arumbakkam. They also seized 300kg of tartrazine, carmoisine and sunset yellow, artificial colours that are permitted by the FSSAI in certain types of food that are not consumed on daily basis. For every kilo of tea leaves, they had added more than 10 grams of these colours. In this case the proprietor confessed that he sold at least 2000kg of adulterated tea daily. FSSAI permits only eight synthetic colours to be added to specific foods but not to tea leaves. The studies have shown that these permitted clours are consumed on a prolonged basis, they could cause cancer, nervous disorders, toxicity or heart diseases as they stay in the body for more than a month. According to the Food Safety and Standards Regulations, 2011, the quantity of colour should be a milligram or less for a kg of food[211].
- Ninety five students fell ill after consuming midday meal, in which a lizard was found, at Gaddikeri Government High School on December 12th 2018[212].
- On 14th December 2018, 12 devotees were killed and another 70 devotees fell ill and were admitted to hospitals after consuming prasada at the temple in Chamarajanagar[213]. The death toll rose to 15.

---

209 http://economictimes.indiatimes.com/industry/cons-products/food/nhrc-notice-to-fssai-over-reports-on-pesticides-in-food-items (accessed on 3rd April 2017 at 8.20pm)

210 '*When health is at stake*'. Deccan Herald, Bangalore Edition dated 5th April 2017,

211 "34,000 kg adulterated tea worth Rs 70 lakh seized", Times of India, Tamil Nadu Edition, dated 21st June 2018

212 '*Lizard in midday meal, 95 students fall ill*', Deccan Herald, Bengaluru Edition, dated 13th December 2018

213 *Ibid 'Poison-laced 'prasada' kills 12 at temple'*, dated 15th December 2018

- On 20/12/2018, over 100 children fell ill after consuming their midday meal in separate incidents in Ballari and Bagalkot districts[214].
- On 31/12/2018, over 100 kids became ill after consuming food in which they had spotted a lizard[215].
- This is in continuation of the above mentioned incident, where the children living at the government run Balamandira, Siddapura complained that they are used to seeing the worse stuff in their food. Stones and worms are routinely served along with rice and sambar[216].
- On January 26th 2019, two women died after consuming temple prasada in Chintamani in Chikkaballapur district[217].
- On February 13th 2019, a total of 54 children fell ill in two separate incidents after consuming food cooked in hostel and a school in Chikkaballapur districts.[218]

These instances are not the only incidents of fall out of FSSA, but only a few which saw the light of the day; many such instances go unreported owing to so many reasons. All these instances go unheeded and also raise a question if state owned enterprises are exempt from the purview of the Act. In spite of the Act being in place for more than a decade, we are not able to prevent deaths and injuries caused due to unsafe food. At this juncture the Researcher is sceptical about the effective implementation of the Act and has thus undertaken to study the difficulties involved in the implementation of the Act.

### 4.6.15 Conclusion

The Act is very impressive in its object and ambition, but implementation of it in its true spirit is a mammoth task. *"Improve quality, you automatically improve productivity"*- W. Edwards Deming[219]. Indian food industry comprises enterprises of different sizes such as organised sector, small scale and unorganised sectors. The domestic market in itself is quite big. The requisites of standards in each sector are different. The food chain has different stake holders ranging from a small farmer to

---

[214] *Ibid 'Over 100 kids ill after eating midday* meal', dated 21st December 2018
[215] *Ibid 'Lizard in dinner:Over 100 kids ill'*, dated 1st January 2019
[216] *Ibid 'What is on the menu? Anna sambar, worms and stones'*, dated 2nd January 2019.
[217] *Ibid '2 women die after eating prasada, 10 in hospital'* dated 27th January 2019.
[218] *Ibid' 54 Children Fall Ill After Consuming Food'*, dated 14th February 2019.
[219] William Edwards Deming (1900-1993) is widely acknowledged as the leading management thinker in the field of quality. He was a statistician and business consultant whose methods helped hasten Japan's recovery after the Second World War and beyond. Available at https://www.google.co.in/ search? ei=6nl FW-TYOoiBvwTYnJSoBw&q=who+is+w+edwards+deming&oq=who+is+W+Ed(accessed on 11/7/2018 at 9.18am)

street vendors to retailers to the big industrialists. The protocols for standardisation of food products should keep in mind the actual users of these standards, the culture, the environment and the infrastructure of the country where they are to be implemented. Lack of infrastructure is significantly contributing to the implementation challenge of the Act.

Man power training should be prioritised as the implementation of the Act mainly depends on the experts who man the various committees. There are some premier institutes like, Central Food Technology Research Institute, International Food Technology Training centre, Indira Gandhi National University, Indian Council of Agricultural Research, industrial associations like Confederation of Indian Industry, The Federation of Indian Chambers of Commerce and Industry offer graduate, post graduate, diploma and certificate courses in various aspects of food sector like food analysis, quality assurance, food safety. These courses must be made compulsory for those working in the food sector. Training the food handlers on the various aspects of the Act is the need of the hour.

Despite the stringent provisions of FSSA against unsafe food and the enforcement mechanism armed with vast powers and heavy punishment for the violators, the fact remains that the consumer is still getting unsafe/adulterated and misbranded articles of food to the peril of his health. We can still encounter dhabas lined up through the length and breadth of the country on all national and state high ways which are oblivious of FSSA, we can still see the small eateries selling chats in unhygienic conditions, we will still continue to buy food articles from the bakeries next door unsure of its safety and standards, we still consume food cooked in some other place and brought to another place to be served, we will still relish paani poori, all time favourite snack of everyone despite watching video on the social media as to how the paani is prepared by mixing toilet cleaner and the dough for poori is kneaded by stamping with one's foot, pooris are prepared in slums in unhygienic condition.

It is not enough if the State alone ventures out to stop the availability of unsafe food in the market but all of us owe a responsibility to be cautious. None of us even bother to know if the person who is selling the food has registered himself or if he is the license holder issued from FSSAI. The citizens need to be pro active in nation building. It is always easy to blame the system without realising that all of us are part of the system and we also owe a duty to the society. We the general public should co-operate by joining our hands realising that it is for our own good/interest that the state has brought this legislation. Whenever we are buying food/articles of food, the least that we can do is if to clarify and confirm ourselves by asking the food vendor if he has registered himself or has obtained the licence as the case may be and if he has obtained the permission from FSSAI, to display the registration certificate or the license in a prominent place at the place of business and to look at the label if the expiry period/best before period is not over.

A report in one of the leading news papers[220], has reported that the central government has asked Karnataka to ensure foolproof monitoring of the midday meal scheme even as 156 children across the state took ill last year due to unhygienic food or negligence while serving it. A news paper report[221], reported that a study was conducted by 2 member team of the Indian Institute of Technology-Bombay which found the presence of micro plastic[222] in several table salt brands in the country. The study found that 63.76 microgram of micro plastic was found per kilogram of salt tested. The incidents like these are an evidence of the fact that FSSA is not working to its full capacity. As the government itself has admitted, last year also children suffered after consuming mid deals, which means that it is not a stray case and keep happening at regular intervals.

Every one of us would have suffered from food borne disease but fail to attribute it to the consumption of contaminated food, owing to our faith that the food that we consume is safe. Right to life includes right to have nutritious food as remarked by the apex court in an arena of cases. But the miraculous mocking of the food borne illnesses is constantly witnessed not as sporadic events but as a common and casual incident. The studies have shown that recurrent attacks of food borne diseases, if not treated in time could result in long lasting effects such as joint inflammation or kidney failure. In extreme cases the person may die of food borne illness. Food safety should be seriously viewed as a shared responsibility among the producers, consumers, governments, industry and all food business operators. Food safety is not a superficial process but it involves not only to curb adulteration and presence of toxins but to safeguard it against the presence of pathogens like microbes, viruses and parasites which can cause food borne diseases. Food safety should become a way of life and not a statutory obligation.

---

[220] "Govt working to make midday meals safer", Deccan herald, dated 15/7/2018 Bengaluru edition.

[221] *Ibid* "IIT-B study finds microplastic in table salt brands" dated 4/9/2018

[222] Microplastics are small pieces of plastic usually measuring less than 5mm in diameter

# CHAPTER V

# FOOD SAFETY AND STANDARDS AND THE CONSUMER LAWS

## 5.1 Introduction

The modern day man has inherited a technical tradition which has enabled him to manufacture things which are far beyond the wildest fancy of the people living in the last century. Today, nobody can escape from the attraction of modern living. The innate urge of the people to lead a supposedly better life have introduced a myriad of canned and cooked food stuffs, soft drinks, bottle juices that it is impossible for a common man to know if what he is consuming is safe and good. Taking the advantage of the ignorance of the consumers, the unscrupulous manufacturers, distributors, and other food business operators have resorted to various undesirable and dangerous means of adulteration of food articles. In a developing country like ours, the life of a common man has been rendered worse by the adulteration of the basic need-food articles. In the present day increasing tempo of industrialization and urbanisation, the need to supply the unadulterated food-stuffs to the citizens in this socialistic welfare state has naturally assumed great importance[1]. It is an accepted truth that the consumers of all goods need to be protected in India but the protection of those consuming food-stuff is all the more of larger importance. It is apparently so because public health even in the words of Roscoe Pound, is the most important social interest[2] as the food products are to make or mar the health of the citizens and has been recognised as a fundamental right of all persons in Indian society.

## 5.2 Development of Consumer Protection Laws in India

Consumer protection is not a novel phenomenon in the Indian society which has been a prominent socio economic problem. Speaking about its origin, the problem of consumer is deep rooted in ancient Indian jurisprudence than an eye can see. Consumer protection is deep rooted in the rich Indian civilisation dating back to the vedic age i.e., 5000 BC to 2500BC. Consumer was a victim at the hands of the trader and he has been exploited since ancient periods. In ancient India human values were held in high esteem and ethical practices were considered inviolable. The welfare of the subjects was of utmost importance for the rulers for which they regulated not only the social conditions but

[1] Mrs Vijaya Sharma, "Adulteration of Food and Drinks & Protective Legislation in Consumer's Interest", The Indian Journal of Legal Studies, [1987],Vol 9, p 152-161

[2] Roscoe Pound while classifying social interests under six heads, placed public health at the top of it.

also economic life of the mass, which in turn gave rise to various trade restrictions to guard the interests of the consumers/buyers.

### 5.2.1 Consumer Protection in Ancient India

The Vedic age in India has been considered as the first literary source of evolution of civilisation. Vedas are the repository of knowledge and culture and gave clear ideas about the legal concepts in a developed civilisation. Matters pertaining to criminal wrongs and civil rights are discussed in detail in Vedas. Throughout the ancient period, one comes across four broad types of relevant criminal offences-adulteration of food stuff, charging of excessive prices, fabrication of weights and measures, and selling of forbidden articles for which statutory measures and punishments have been recommended, from time to time by the leading texts of the time. Prominent among them are: the Manusmriti (800 BC to 600 BC), Kautalya's Arthasastra (400 BC to 300 BC), Yajnavalkyasmrti (300 BC to 100 BC), Naradasmrti (100 AD to 200 AD), Brihaspatismrti (200 Ad to 400 AD), and Katyayanasmrti (300 AD to 600 AD)[3]. In ancient India, society as a whole practiced and followed Dharma[4] which was a guiding principle governing all human relations. The philosophy of Dharma owed its origin to Vedas which were the revelations of god himself.

Through *Srutis*[5] and *Smritis*[6], many writers and philosophers of ancient India have prescribed to the king the codes about the method of ruling the state and its subjects and consumer protection was also dealt with in their writings. An average Indian consumer is known for his patience and tolerance because he considers the receipt of defective goods or services as an act of fate or unfavourable planetary position in his horoscope. When a new product purchased by him turns out to be defective from day one, he takes it silently, blaming it on his fate or as the consequences of the wrongs committed by him in his previous birth[7]. Manu, the ancient law giver, wrote about ethical trade practices and prescribed a code of conduct to traders and specified punishments to those who committed crimes against the buyers. For example, he referred to the problem of adulteration and said "one commodity mixed with another must not be sold (as pure), nor a bad one (as good) not less (than the proper quantity or weight) nor anything that is at hand or that is

[3] Singh Gurjeet, *The Law of Consumer Protection in India Justice Within reach*, Deep & Deep Publications, New Delhi, (1996) at 44

[4] Codes of morals. They also deal with the rules of conduct, law and custom

[5] Revelation

[6] Tradition

[7] K Shrinivasan. "*A New Era in Consumerism*" The Consumer, the Hindu Folio dated 31/10/1990 available at https://www.thehindu.com/folio/fo9910/99100060.htm (accessed on 1/7/2018 at 7.00pm)

concealed"[8]. The punishment "for adulterating unadulterated commodities and for breaking gems or for improperly boring (them)" was the least harsh[9]. "Man who behaves dishonestly to honest customers or cheats in his prices shall be fined in the first or in the middle most amercement[10]. There was a process to inspect all weights and measures every six months, and the results of these inspections were duly noted[11].

Manu smriti and Yajnavalkya smriti recommended that a person who deals in false gold and one who sells unclean meat ought to be maimed and he should also be made to pay the highest amercement. All these measures depict how efficiently Manu dealt with consumer matters and many of which are still relevant today. Use of defective and false balance and tampering the weights and measures was also another criminal acts that were indulged by the trading community in the ancient period. Manu advocated that once in every six months, the weights and measurements should be checked and duly marked by the king. The trading community also used to exploit the innocent consumers by charging excessive prices. To save the consumers from this arbitrary practice. Manu advocated that it was the duty of the king to fix the rates both for the purchase and sale of the marketable goods after considering the original cost of the goods, transport charges, incidental charges and also the profit margin. Manu recommended for this kind of fixation of prices publicly either once in five nights or fortnightly. Yajnavalkyasmriti stated that the sale and purchase should be conducted daily according to the value fixed by the king. Demanding higher price was punishable[12].

Kautilya's Arthasastra comes next in line to deal with the role played by the state in monitoring the trade and its duty to thwart crimes against consumers. In Kautilya's Arthashastra, there are references to the concept of consumers' protection against exploitation by the trade and industry, short weights and measures, adulteration and punishment for these offences[13]. Chapter II, in book IV, "The Removal of Thorn" of the

---

[8]Manu, "*The Laws of Manu*", trans, George Buhler, Clarendon Press, London(1886) VolXXV available at https://archive.org/stream/lawsofmanu00bh#page/n7/mode/2up/search/food (accessed on 5/7/2018 at 10.45pm)

[9]The Sacred Books of the East 'The *Laws of Manu*' trans, G.Buhler, Atlantic Publishers & Distributors, (2002) Vol XXV available at https://books. google.co.in/books?id =MdnDmbEqwY4C&pg =PA393&lpg =PA393&dq=The+punishment+"for+adulterating (accessed on 3/7/2018 at 12.30pm)

[10] *ibid*

[11] Dr. A. Rajendra Prasad "*Historical Evolution of Consumer Protection and Law in India:* A Bird's Eye View" available at http://www. jtexconsumerlaw.com/v11n3/jccl_india.pdf (accessed on 11/1/2017 at 3.30pm)

[12] Singh Gurjeet, *The Law of Consumer Protection in India Justice Within reach*, Deep & Deep Publications, New Delhi, (1996) at 46

[13] K Shrinivasan. "*A New Era in Consumerism*" The Consumer, the Hindu Folio dated 31st October, 1990 available at https://www.thehindu.com/folio/fo9910/99100060.htm (accessed on 1/7/2018 at 7.00pm)

Arthashastra of Kautilya deals with Consumers Protection against Merchants[14]. Kautilya observed, "the superintendent of standardization should cause factories to be established for the manufacture of standard weights and measures." he further said "[the superintendent] should cause a stamping [of the weights and measures] to be made every four months. The penalty for unstamped [weights] is twenty seven panas and a quarter. [Traders] shall pay a stamping fee amounting to one kakani every day to the superintendent of standardization"[15]. The chief controller of private trading kept a watch over merchants, inspecting their weights and measurements periodically and ensuring prevention of hoarding, adulteration, or add extra mark-ups[16].

Adulteration of grains, oils, alkalis, salts, scents and medical articles with similar articles of low quality shall be punished with a fine of 12 panas. According to Kautilya, 'the trade guilds were prohibited from taking recourse to black marketing and unfair trade practice'[17]. Several punishments were prescribed for smuggling and adulteration of goods, public health was protected by punishing the adulteration of food products. The state (king) was responsible to cut out the trade of goods that were harmful to the country and that are insignificant. The goods that are highly beneficial and rare were made duty- free as mentioned in the second book. It is clear from the following *sloka* –

*Rastrpeedakrm bhandamuchindyadphalm ch yet.*
*Mahopkarmuchshulkm kuryadvijm ch durlabham*[18].

Which means , along with consumers, the rights of the traders were also considered in mind. They were also well protected. Kautilya said that if after purchasing an article one does not receive it, the fine was twelve *panas*. But this was not in case of defect, a sudden calamity or unsuitability. The policy of consumer protection is apparent in the *Arthashastra*. It describes the role of state in regulating trade practices and its duty to avert crimes against consumers. Kautilya says that

*Prajasukhe sukham rajha prajanam ch hite hitem.*

*Natmpriyam hitm rajha prajanam tu priyam hitem*[19]

[14] Kautila Arthashatra, Shamashastry R (Trans) Mysore: Mysore Printing and publishing House (8th edn 1967), p115

[15] R.P. Kangle, '*The Kautiliya Arthasastra – Part III – A Study*'. Motilal Banarsidass Publishers(2000) p116
[16] V.P. Varma, '*Ancient and Medieval Indian Political Thought*', Varma Publishers, Patna, (1997), p.92.
[17] R.P. Kangle,*The Kautiliya Arthasastra–Part III–A Study* 116 Motilal Banarsidass Publishers (2000) p137
[18] Sunitha Devi, "Kautilya on Consumer Protection", Scholars Journal of Economics, Business and Management[2015] 2(4), p 310-313
[19] Sunitha Devi, "Kautilya on Consumer Protection", Scholars Journal of Economics, Business and Management[2015] 2(4), p 310-313

which means, in the happiness of the subjects lays the happiness of the king and in what is beneficial to the subjects his own benefit. What is dear to him is not beneficial to the king, but what is dear to the subjects beneficial to him. Thus the top most priority was given to the welfare of people. An essential aspect of the welfare of the population was the protection of the interests of the consumers.

A license to sell goods was required for every trader. Foreign traders had required to obtain permission for selling goods. Goods were sold in such a way that consumers had easy access to them. Domestically produced goods were sold in a particular market and imported commodities were sold in many places. But both types of goods were sold by keeping in mind the favour of subjects. Foreign goods were sold in many centers probably to make them easily available. Health of the general public was protected by punishing the adulteration of consumables. All these measures establish the efforts and effectiveness of ancient society in protecting the consumers. We can see the resemblance between the consumer protection during Kautilya's time and the present Consumer Protection Act 1984.

### 5.2.2 Consumer Protection in the Medieval Period

Consumer protection continued to be of prime concern for rulers in medieval India. Consumer protection can be traced back to the Mughal times. India came under the Muslim rule from 712 AD to 1765 AD. During the Muslim rule, various units of measures and weights were used. During the Delhi Sultanate period, especially the period of the Khiljis, the laws elaborately dealt with Consumer protection[20]. Sultan Alauddin Khilji (1296 AD to 1316 AD) had introduced strict price control measures based on production costs[21]. Allauddin khilji was known for his market reforms. He had established separate shopping centres for different consumables like grains, cloth, sugar, dry fruits, herbs, butter and other miscellaneous commodities.

The market controller, the state intelligence officers, and the Sultan's secret agents each submitted independent reports on these shopping centers to the Sultan. Even a minor violation of the rules would not be tolerated Hoarding of food grains was an offence. The grower was required to sell their grains for cash in their fields only at fixed prices and they were not allowed to take the food grains home for private sale. The shopping centre meant for general commodities was under the direct supervision of the commerce ministry. Alla-

---

[20] V Balakrishna Eradi,Judge, Supreme Court of India (Retd) "*Consumer Protection Jurisprudence*" Lexis Nexis (2005) p185

[21] Chapter 3 *Development of consumer protection Act in India* available at http://shodhganga.inflibnet. Ac .in/bitstream/10603/12669/7/07_chapter%203.pdf (accessed on 14/1/2017 at 11/14pm)

ud-din's Minister of Commerce was the Superintendant of weights and Measures and also the Controller of the commercial transactions and was assisted by the Superintendents for each commodity.

### 5.2.3 Consumer Protection in the Modern India

During the British rule, various legislations concerning consumer interests were passed and they were the Indian Contract Act of 1872, the Sale of Goods Act of 1930, the Indian Penal Code of 1860, the Drugs and Cosmetics Act of 1940, the Usurious Loans Act of 1918, and the Agriculture Procedure (Grading and Marketing Act) of 1937 which provided legal protection to consumers. For almost 55 years, the Sale of Goods Act 1930 was the only legislation dealing with Consumer Protection in India which provided for main protection to the buyer against the seller of the defective goods in Section 16 of the Act[22]. This Act provides for exceptions to the principle of Caveat Emptor[23]. We have definitely travelled a long distance from the ancient Roman maxim Caveat Emptor which enjoins upon the purchaser to be cautious. Once the goods were sold, a buyer could not complain about the quality or the quantity of goods as the principle of Caveat Emptor ordained the buyer to be cautious to ensure the quality and the suitability of goods as per his requirement. A duty was cast on the buyer to be cautious in all his dealings and could not hold the seller liable for anything.

This privilege was being misused by the business community and a need was felt for a move to protect the interests of the buyer as well. Over the years, attempts have been made to protect the consumer from unscrupulous sellers by many legislations. The Indian Penal Code 1860 has a number of provisions to tackle the offences against the consumers, dealing with offences related to false weights and measures[24], the sale of adulterated food

[22] Sale of Goods Act, S.16 " Subject to the provisions of this Act and of any other law for the time being in force, there is no implied warranty or condition as to the quality or fitness for any particular purpose of goods supplied under a contract of sale, except as follows: (1) Where the buyer, expressly or by implication, makes known to the seller the particular purpose for which goods are required, so as to show the that the buyer relies on the seller's skill or judgment and the goods are of a description which it is in the course of the seller's business to supply (whether he is the manufacturer or producer or not), there is an implied condition that the goods shall be reasonably fit for such purpose: Provided that, in the case of a contract for the sale of a specified article under its patent or other trade name, there is no implied condition as to its fitness for any particular purpose (2) Where the goods are bought by description from a seller who deals in goods of that description (whether he is the manufacturer or producer or not), there is an implied condition that the goods shall be of merchantable quality: Provided that, if the buyer has examined the goods, there shall be no implied condition as regards defects which such examination ought to have revealed. (3) An implied warranty or condition as to quality or fitness for a particular purpose many be annexed by the usage of trade. (4) An express warranty or condition does not negative a warranty or condition implied by this Act unless inconsistent therewith.

[23] let the buyer be beware

[24] Indian Penal Code 1860 Sections 264-267.

and drinks, the sale of noxious food or drink[25]. "In India after independence, the seed of 'consumerism' was sown by Late C. Rajagopalachari, who setup the first Consumer Centre at Madras before 1950. In the late1950's an attempt was made to usher in a consumer movement in the country on the advice of the Planning Commission, but they failed"[26].

## 5.3 United Nations Guidelines on Consumer Protection[27]

In modern times, protection of consumer rights can be traced back to 1962. On 15/3/1962, the United States president in address to the congress, proclaimed the consumer Bill of Rights. The proclaimed message enunciated the right to choice, the right to information, the right to safety and the right to be heard, which subsequently included The problem and difficulties faced by a common man in day today life are nothing but consumer problems and these problems are omnipresent due to which even the UN thought it fit to issue guidelines about the consumer protection.

The respective states have based their consumer protection legislations on this foundation. The United Nations Guidelines for Consumer Protection (UNGCP) are "a valuable set of principles for setting out the main characteristics of effective consumer protection legislation, enforcement institutions and redress systems and for assisting interested Member States in formulating and enforcing domestic and regional laws, rules and regulations that are suitable to their own economic and social and environmental circumstances, as well as promoting international enforcement cooperation among Member States and encouraging the sharing of experiences in consumer protection."[28].

These guidelines were first adopted by the General assembly on 16th April 1985 and later modified by the Economic and Social Council by Resolution[29]. The objectives of these guidelines are to assist countries in achieving or maintaining adequate protection for their population as consumers[30]and to encourage high levels of ethical conduct for those

---

[25] *Ibid* Sections 270-276

[26] Swami Narayana S, "*Consumer Protection : No Luke warm Attitude*" Vol.9, Consumer Confrontation, Ahmadabad (March — April, 1989) at 12

[27] Available at http://www.un.org/esa/sustdev/publications/consumption_en.pdf accessed on 18/1/2018 at 1.29pm

[28] http :// unctad .org / en/ Pages /DITC /Competition Law / UN- Guidelines –on -Consumer-Protection.aspx (accessed on 18/1/2017 at 12.48pm)

[29] E/1999/INF/2/Add.2 Expansion of the United Nations guidelines on consumer protection to include sustainable consumption available at http://www.un.org/documents/ecosoc/docs/1999/e1999-inf2-add2.pdf (accessed on 23/5/2017 at 1.08pm

[30] Available at http://www.un.org/esa/sustdev/publications/consumption_en.pdf (accessed on 18/1/2018 at 12.59pm)

engaged in the production and distribution of goods and services to consumers[31]. These guidelines for consumer protection stated that "when formulating national policies and plans with regard to food, governments should take into account the need of all consumers for food security and should support and as far as possible, adopt standards from the Food and Agriculture Organisation's and World Health Organisation's Codex Alimentarius"[32] An intergovernmental group of experts on consumer protection law and policy was established to overlook the implementation of the guidelines, to offer a platform for consultations, to provide required technical assistance and to update the guidelines periodically.

## 5.4 Food and Agriculture Organisation Voluntary Guideline 9 on Food Safety and Consumer Protection.

In the World Food Summit 1996: 5years later, it was unanimously agreed to set up an intergovernmental working group to develop voluntary guidelines to achieve the progressive realisation of the Right to food. Voluntary Guideline 9 issued by Food and Agricultural Organisation, UN deals with 'Food safety and Consumer protection'. It is ordained that the state should take measures to ensure that all food which is available in the market is safe and is in conformity with the food safety standards[33]. States are directed to ensure that food business operators are educated about the safe practices in their activities. State should also educate the consumers about the safe storage, food borne diseases and food safety matters[34]. States should initiate measures to protect the consumers from deception and misrepresentation in the packaging and other related matters and help the consumers by providing appropriate information on marketed food. The states are directed to provide recourse for any harm caused by either unsafe or adulterated food. And such

[31] 1(c) Available at http://unctad.org/en/PublicationsLibrary/ditccplpmisc2016d1_en.pdf (accessed on 18/1/208 at 12.55pm)

[32] Sheeba Pillai, " Right to Safe food: Laws and Remedies", The Banaras Law journal, [2012] , Vol 41, p119-135

[33] Voluntary Guideline 9.1 States should take measures to ensure that all food, whether locally produced or imported, freely available or sold on markets, is safe and consistent with national food safety standards available at http://www.fao.org/3/a-y7937e.pdf accessed on 18/1/2018 at around 12.10 pm

[34] *ibid* 9.6 States should ensure that education on safe practices is available for food business operators so that their activities neither lead to harmful residues in food nor cause harm to the environment. States should also take measures to educate consumers about the safe storage, handling and utilization of food within the household. States should collect and disseminate information to the public regarding food-borne diseases and food safety matters, and should cooperate with regional and international organizations addressing food safety issues.

measures should be in agreement with WTO agreements[35]. States are called upon to cooperate with the stake holders in addressing the food safety issues and ensure their participation in framing the relevant policies[36].

## 5.5 Human Rights of Consumers

Every person is consumer in the modern days. During the earlier times when the barter system was in vogue the risks of deception were comparatively less. In the present scenario, where multinational companies are ruling the roost and the indigenous traders will have to vie with them to be in the market, unethical and immoral practices in trade have found in roads in the day today activities, due to which the consumer needs an entire system of legislations and protections at national as also at international level to receive quality and quantity for a fair price paid. It is undeniable that every consumer enjoys certain human rights from which the following may be derived:

1. The right to have information regarding quality, purity and the price of various articles to make proper choices[37]
2. The right to be assured that the weights and measures being used are standard and correct[38]
3. The right to have fair distribution of the essential goods and services at reasonable rates[39]
4. The right to be protected from goods and services which are hazardous to life and property[40]
5. The Right to be protected from the unfair trade practice or from other unscrupulous exploitations[41]
6. The Right to consumer education and awareness[42]

---

[35] *ibid* 9.7 States should adopt measures to protect consumers from deception and misrepresentation in the packaging, labelling, advertising and sale of food and facilitate consumers' choice by ensuring appropriate information on marketed food, and provide recourse for any harm caused by unsafe or adulterated food, including food offered by street sellers. Such measures should not be used as unjustified barriers to trade; they should be in conformity with the WTO agreements (in particular SPS and TBT).

[36] Ibid 9.9 States are encouraged to cooperate with all stakeholders, including regional and international consumer organizations, in addressing food safety issues, and consider their participation in national and international fora where policies with impact on food production, processing, distribution, storage and marketing are discussed.

[37] Derived from '*Freedom of thought and expression*' under Art 18 of ICCPR 1966 and Art 19 of UDHR, Art 13 of American Convention on Human Rights 1969

[38] *Ibid* Art 27(2) '*Right to Protection of Moral and Material Interest*'

[39] *Ibid* Art 7 '*Right to equality, equal protection before the law and non-discrimination*', Art 2,3,14,26 of ICCPR 1966

[40] *Ibid* Art 25'*Right to a standard of living adequate for health and well being*'

[41] *ibid*

[42] *Ibid* Art 26 '*Right to Education*' Art 26 of UDHR and Art 13 and 14 of ICSCER

7. The Right to be heard and to get speedy justice through the enforcement of the basic human rights of the consumers in case of infringement[43]

## 5.6 Indian Constitution and Consumer Rights

It is now widely acknowledged that the true indicator of development of the country and progressiveness of civil society is the level of consumer awareness and consumer protection, the main reason being the rapid increase in the variety of goods and services, thanks to the modern technology. Adding to the woes is the growing size and complexity of production and distribution systems, sophisticated marketing & advertising practices, mass marketing methods and rise in consumers mobility has resulted in fall in the personal interaction between sellers & buyers. All these factors have necessitated for the increased need for consumer protection. The Universal Declaration of Human Rights has influenced Constitutions of various nations. Our Indian Constitution also bears the impact of the Universal Declaration and this has been recognised by the Supreme Court of India[44]. The constitution of India is a social document, it is not made only to provide a machinery of government to maintain law and order and to defend the country, and the founding father of the constitution had a glorious vision of the establishment of a new society in India imbuing with high ideals for guaranteeing the multidimensional welfare of the people[45]. In a welfare state like India, it is the bounden duty of the state to protect the interests of the consumers.

Part IV of the Indian constitution ordains the state to raise the level of nutrition and standard of living of its people and to improve the public health which is its primary responsibility. It is obvious that if the food is adulterated and wholesome nourishment is denied to the society by such adulteration, the state aim will stand thwarted to some extent[46]. The government has to play a prominent role through law in re-defining and reframing the relationship between the producer, seller and Consumer[47]. Though the word 'consumer' is not to be found in the Constitution, the consumer breathes and peeps out through many of the blood vessels of the Constitution[48]. Like in many countries, India also has large segments of people who do not have sufficient resources to live under reasonably

---

[43] *Ibid* Art 8 '*Right to approach justice system*'

[44] K C Daiya, "Consumer Protection Laws in India-step Toward Protection of Human Rights", The Indian Journal of Legal Studies, [1989],Vol IX, p220-231

[45] O Koteshwara Rao, '*Constitution, State and Consumer Welfare*', Eastern Book Company. (1981), p 211

[46] '*Radhakrishna* v *State of Uttar Pradesh*', 1974 FAC 2551 (All)

[47] R N Sharma, "Constitutional Postulates to Maintain Public Health and to prevent Food Adulteration: an Appraisal", the Indian Journal of Legal Studies, [1989 ]Vol IX, p 158-162

[48] Rao Koteswara P, Leelakrishnan P, '*Constitution, State and Welfare*' in Consumer Protection and Legal Control, Eastern Book Co, Lucknow, (1981)

good conditions of health and decency. Therefore the society in which they live has obligations to provide support and that support is not a charity to the citizen but as of a right[49]. Consumer justice implies securing to the consumer, commodities or services equivalent to the payment made by him without violating legally or commercially prescribed or impliedly agreed or understood quantity and standard[50]. In order to save the gullible consumer from suffering from ill health for no fault of his, adulteration is required to be rooted out necessitating the prevention of adulteration of articles of food.

Article 14 of our Constitution guarantees equality before law to all the persons which in essence means, producers, sellers and consumers are all equal before law either for receiving reward or punishment. Hence the state is enjoined constitutionally to protect the consumers. Consumer protection under Constitution implies the passing of required laws and enforcing them to prevent malpractice in trade, commerce and business, defrauding and exploiting the consumer. Article 19(1) (g) guarantees to all citizens a right to carry on any profession, occupation, trade or business, thereby ensuring that the state cannot prevent a citizen from carrying on a business except by a law imposing reasonable restrictions in the interest of general public which implies that no producer can manufacture or sell goods which may be harmful to the consumers. But, Article 19 (2), mandates that no such rights can be enforced where the business is immoral or dangerous. Hence, Consumer Protection Act is in line with reasonable restrictions imposed by law.

There can be reasonable restrictions on 'business on the streets[51]' and 'any harmful trade[52]' or 'dangerous trade[53]' reasonable restrictions can be imposed for public convenience also[54]. While imposing restrictions on the manufacture, import, sale or distribution of any product on the ground of public health, reliance should be placed on its scientific testing by experts[55]. In *Dr. Shivarao Shantaram Wagle and Others v. Union of India and Others*[56], the Supreme Court adopted the same principle when it was required to issue directions to the government to refrain from releasing the Irish butter for human consumption, imported under the *EEC Grant-in-Aid for Operation Food Programme*, on

---

[49] Charles A Reich, "*Individual Rights and Social welfare: the Emerging Legal issues*", [1965] Yle LJ 74

[50] RN Sharma, "Constitutional Postulates to Maintain Public Health and to Prevent Food Adulteration :An Appraisal", The Indian Law Journal of Legal Studies, [1989]Vol IX p158-162

[51] *Pyare Lai* v. *Delhi Municipality*, AIR 1968 SC 133, 138.

[52] *Hari Shankar* v. *Deputy Commissioner*, AIR 1975 SC 112

[53] *Lakhan Lai* v. *State of Orissa*, AIR 1977 SC 722

[54] *Ebrahim* v *Regional Transport Authority* (1953) SCR 290,299

[55] Chapter-2Consumer Protection Laws In India available at http://14.139.60.114:8080/ jspui/bitstream/ 123456789/687/7/Consumer%20Protection%20Laws%20in%20India.pdf(accessed on 13/4/2018 at 2.46pm)

[56] AIR 1988 SC 952.

the ground that the butter may be contaminated by Chernobyl nuclear fallout. In this matter, the Supreme Court appointed a committee to probe and report on 'whether milk and dairy and other food products containing man made radio nuclides within the permissible limits as stipulated by Atomic Energy Regulatory Board on 27th August 1987 are safe and harmless for human consumption'. The Supreme Court refused to issue the restrictive orders regarding the release of butter in question for human consumption, in the light of the committee's report which found the butter to be safe for human consumption.

Article 21 entitles Right to life to every individual. So the state is directed to protect the right of consumers. Article 38 provides that the state shall secure a social order for the promotion of welfare of the people and shall effectively work to achieve a social order in which justice, economic and political shall inform all the institutions of the national life[57]. Article 39(e)[58] directs the state to follow a policy by which the strength of the workers who constitute the major portion of the society is secured and maintained. This is possible only if the workers or consumers receive good food. Hence, adulteration of food materials and creation of artificial scarcity thus increase in price is contrary to this directive.

Article 39(f)[59] imposes a duty on the state to provide opportunities and facilities for the children to develop in a healthy manner for which proper nourishment is crucial. This directive can only be achieved only when unadulterated and nutritious food is available. Article 43[60] imposes an obligation on the state to endeavour to build economic organisation or to make suitable legislation to ensure a decent standard of life to all the workers, who form a majority of population of India and who are in turn consumers. Article 47 [61]imposes

---

[57] Article 38: *State to secure a social order for the promotion of welfare of the people*
*(1) The State shall strive to promote the welfare of the people by securing and protecting as effectively as it may a social order in which justice, social, economic and political, shall inform all the institutions of the national life*
*(2) The State shall, in particular, strive to minimize the inequalities in income, and endeavor to eliminate inequalities in status, facilities and opportunities, not only amongst individuals but also amongst groups of people residing in different areas or engaged in different vocations*

[58] Article 39(e) *that the health and strength of workers, men and women, and the tender age of children are not abused and that citizens are not forced by economic necessity to enter avocations unsuited to their age or strength*

[59] Article 39(f) that *children are given opportunities and facilities to develop in a healthy manner and in conditions of freedom and dignity and that childhood and youth are protected against exploitation and against moral ...*

[60] Article 43: *The State shall endeavour to secure, by suitable legislation or economic organisation or in any other way, to all workers, agricultural, industrial or otherwise, work, a living wage, conditions of work ensuring a decent standard of life and full enjoyment of leisure and social and cultural opportunities and, in particular, the State shall endeavour to promote cottage industries on an individual or co operative basis in rural areas*

[61] Article 47: *Duty of the State to raise the level of nutrition and the standard of living and to improve public health The State shall regard the raising of the level of nutrition and the standard of living of its people and the improvement of public health as among its primary duties and, in particular, the State shall endeavour*

a duty on the state to raise the level of nutrition and the standard of living to improve public health which indirectly says that the state is obligated to ensure food items meant for human consumption must be free of contamination and adulteration and the same is safe. The Constitution has distributed the subjects, relating to product and service regulation, between the centre and states for their better quality and efficiency[62]. Most of the subjects concerning consumer protection related to products have been placed in the concurrent list[63]and the relevant entry is: Adulteration of foodstuffs and other goods[64].

### 5.7 Consumer Protection Act 1986 (CPA)

In the light of United Nations Guidelines on Consumer Protection and specific provisions in Indian Constitution for protecting the interests consumers, a need was felt to craft a specific legislation to protect the interests of the consumers which gave rise to the promulgation of Consumer Protection Act, 1986 whose objective is to provide cheap, simple, and quick justice/remedy to the affected consumers. In the year 1984, Law Commission of India under the chairmanship of Mr KK Mathew submitted its 105th Report on "Quality Control and Introspection of Consumer Goods" to Ministry of Law, Justice and Company Affairs, Government of India. The subject was taken up by the Law Commission suo motto and recommendations were made[65]. A major breakthrough came during the 1986, when the parliament passed the CPA.

The CPA is a boon to the consumers as it is cost effective, consumer friendly, inexpensive procedure and wide jurisdiction which has created a confidence in the minds of people that it is a common man's legislation. The grievances of the affected consumer is filed in the form of a complaint, the consumer need to pay only a nominal fee and another major procedural flexibility is the option of the consumer either to engage an advocate or if the consumer prefers, he can represent himself. These simple and user friendly measures drive consumers to avail themselves of the benefits of the CPA.

---

*to bring about prohibition of the consumption except for medicinal purposes of intoxicating drinks and of drugs which are injurious to health*

[62] Dr Ashok Patil " *Products Safety Standards & Consumer Protection",* KILPAR, Bengaluru [2008]

[63] Constitution of India, Article 246, Seventh Schedule, List III-Concurrent List

[64] *Ibid* entry 18

[65] Law Commission of India, 105th Report on "Quality Control and Introspection of Consumer Goods" October 1984 and the Recommendations were(1) to ensure that the quality of goods is sold according to the standards laid down under the proposed law; (2) for the constitution of advisory councils with reference to particular classes of goods or particular industries;(3) for the appointment of public analysts for such areas as may be assigned with power to examine any notified goods and sold in the market to ensure that they conform to the quality laid down under the Act

The CPA initiated a legal revolution by ushering in the period of consumers and coining new legal culture among the common mass to resort to CPA. The National Commission at the centre, the State Commission at the state level and the District forums at the district level are working together revolutionising the legal system. A consumer forum/commission is required to follow the principles of natural justice and is not bound by the strict rules of procedure enunciated in Code of Civil Procedure 1908. CPA 1986 included the Essential Commodities Act of 1955, the Prevention of Food Adulteration Act of 1954 and the Standard of Weights and Measures Act of 1976. The modern food laws must be more precise in their application, more specific and complete in content and should be able to take account of situations beyond national borders.

Protection of the consumer has been extended to the control of false descriptions of products, nutritional declarations and misleading claims in labelling and advertising. The advantage of these Acts is that they do not require the consumer to prove mens rea as in the case of criminal offence. The offenses are of absolute liability, and not dependent on any intention or knowledge. An attempt is made to look at consumer protection as public interest issue rather than private issue. Law of torts also provided for remedies but Law of Torts imposes heavy responsibility on the plaintiff to prove negligence and standard defences can be set up for the same. Law of Torts is not a codified law in India because of which injured consumers ought to pursue legal remedies under different laws.

The concept of 'consumer protection has three dimensions, namely, the physical protection of the consumer, protection of his economic interests and the most important public interest. The first aspect includes measures to protect consumers against products which may be injurious and unsafe for human consumption. The second aspect protects the consumers against the deceptive and other unfair trade practices providing therewith adequate means to get their grievances redressed. The third aspect refers to measures ensuring adequate supply of goods thus preventing a situation of monopoly in the society at large. Under the Consumer protection Act there is a change in the nature of liability from the negligence liability to the absolute liability and also breaking the shackles of the privity of contract.

It is the responsibility of the state not only to legislate and enforce the measures for protecting the interest of the consumers but also to construct a beneficial framework of relations between the producers, sellers and the consumers, this obligation of the state is the essence of a welfare state[66] and a socialist state as mentioned in the preamble of our Constitution. The state should protect the interests of the consumers and to protect the interests of the consumers of food is all the more crucial for the state as the public health depends on the food that we all consume. 'Public health' in the words of Roscoe Pound is the most important social interest[67] and the same idea[68] has been echoed by the Supreme Court in *Vincent v Union of India*[69]. The proportion of the consumers of food in the society is so large that it includes almost the entire society comprising of mostly illiterates, ignorant and innocents in the Indian scenario, which is therefore a gigantic task[70].

### 5.7.1 Definitions

The Researcher has examined only some of the definitions under CPA for the purpose of the study

#### 5.7.1.1Complaints against Hazardous Goods

Section 2(1) (c) (v) enables a person to file a complaint about any goods which are offered for sale and which may be hazardous to life and safety when used. This section also includes the goods which contravenes any set standards relating to safety and if the trader could have known with his due diligence that the goods so offered are unsafe to the public.

#### 5.7.1.2 Consumer

*"Consumer by definition includes us all. They are the largest group in the economy affecting and affected by almost every public and private economic decisions. Two thirds*

---

[66]Article 38 :State to secure a social order for the promotion of welfare of the people
(1) The State shall strive to promote the welfare of the people by securing and protecting as effectively as it may a social order in which justice, social, economic and political, shall inform all the institutions of the national life

[67]Roscoe Pound classified social interests under six heads and placed public health at the top.

[68]"Maintenance and improvement of public health have to rank high as there are indispensable to the very physical existence of the community and on the betterment of these depends the building up the society which the constitution makers envisaged. Attending to public health in our opinion, therefore, is of high priority perhaps the one at the top.

[69]AIR1987 SC 990

[70]C Gopala Krishnamurthy "Enforcement of Prevention of Food Adulteration Act 1950", in Dr Ramesh V Bhat & B S Narasinga Rao's (ed) book '*National Strategy for Food Quality Control*', National Institute of Nutrition, Hyderabad [1985] expresses *"control of quality of food stuffs by preventing their adulteration is a gigantic problem all over the world. It is much more so in an economically developing country which are technologically lagging behind in the process of production, preservation and packing of food articles. In advanced countries like USA, USSR, Japan, France and Germany, it is possible to exercise rigid controls over the quality of Food articles in view of the high order of general awareness in the people and a sense of realization of their rights, privileges and responsibilities in the task of the cooperation with the law enforcing agency. In our country, the complacent attitude of a common man through illiteracy ignorance and dejection has led to a state of mass inertia, with the result the entire burden of controlling the quality of food stuffs rests with the government and its enforcement machinery".*

*of all spending in the economy is by consumers. But they are the only important group in the economy who are not effectively organised, whose views are often not heard"*- John F Kennedy. 'Consumer is not human subject, but is instead a role that is played by human subjects. This is an undisputed affirmation that every human being qualifies as a consumer and it is integral to his survival'[71]. The dictionary meaning of consumer is 'a person who consumes, especially one who uses a product or a purchaser of product with a relatively long useful life[72]. In simple parlance, a consumer is one who consumes[73]. The definition given in Section 2(1) (d)[74] of CPA is a so comprehensive that it covers not only the consumer of goods but also consumer of services. The definition is wide enough not only to include the person who buys the goods for consideration but also any user of such goods with the approval of the buyer[75]. It also covers any person hires or avails of any services for consideration and includes the beneficiary of such services when the same is availed of with the approval of the hirer.

Consequently, 'it includes anyone who consumes goods or services at the end of chain of production'[76] which in essence means that it includes not only the person who buys any goods for consideration but also any user of such goods when it is with the approval of the buyer. This extended meaning of the definition is crucial as the goods are not used by the buyer alone by also by his family members, relatives and friends. A supplier of goods is not a consumer under the Act[77] Though under the general principles of Law of Contract such users are not entitled to sue the supplier of goods on the ground of

---

[71] JN Barowalia, '*Commentary on the Consumer Protection Act 1986*' Universal Law Publishing Pvt Ltd, (2nd edn, 2002).
[72] Concise Oxford Dictionary, (11th edn, 2008)
[73] Bryan A Garner, 'Black's Law Dictionary' West Group Publication (7th Revised edn, 1999)
[74] Section 2 (1) (d) "consumer" means any person who—
(i) buys any goods for a consideration which has been paid or promised or partly paid and partly promised, or under any system of deferred payment and includes any user of such goods other than the person who buys such goods for consideration paid or promised or partly paid or partly promised, or under any system of deferred payment when such use is made with the approval of such person, but does not include a person who obtains such goods for resale or for any commercial purpose; or
(ii) hires or avails of any services for a consideration which has been paid or promised or partly paid and partly promised, or under any system of deferred payment and includes any beneficiary of such services other than the person who 'hires or avails of the services for consideration paid or promised, or partly paid and partly promised, or under any system of deferred payment, when such services are availed of with the approval of the first mentioned person but does not include a person who avails of such services for any commercial purposes;
[75] In *Dinesh Bhagat* v *Balaji Auto Limited*, (1992) IIICPJ 272 (Delhi CDRC) a scooter purchased by a person was in possession of another person, the complainant from the date of purchase and he had been using and taking it to the respondent for repairs and service. The respondent never objected that he was not entitled to use it or get the repairs done. The Delhi State Commission held that the complainant was using it with the approval of the owner and , therefore he was a consumer under the Act
[76] *Morgan Stanley Mutual Fund* v *Kartick Das & others* (1994) 11 CPJ 7 (SC)
[77] *Glass Studio* v *Collector of Central Excise* (1991)IICPJ 585 (Orissa CDRC)

'privity of contract'[78], yet under the CPA, a complaint may be made by any user of goods though he is not a party to the contract for purchase of those goods.

As per sub clause (i) of Section 2(1) (d) of CPA, any person buying goods for consideration with the purpose of the resale or for a commercial purpose, then he is not a consumer within the meaning of the Act[79]. This beautiful catch is only to protect the hapless individual consumers. If this provision were to be not there, the very purpose of CPA would have been defeated, as only the big business houses would have been benefitted and the individual consumers would have been relegated to the backdrop. But if a person buys something with the intention of making use of the same for a commercial purpose and if that is his self employment or for earning his livelihood, then that purchaser is treated as a consumer under the CPA and can take recourse to CPA. In a nutshell, it is clear that parliament intended to exclude from the definition of consumer not only the persons who obtain goods for resale but also those who purchase goods for carrying on a large scale activity with a view to earn profits[80].

**5.7.1.3 Goods**

Section 2 (1) (i) of CPA provides for the definition of goods. But it has not independently defined the term and has referred to the Sale of Goods Act, 1930[81] and has stated that whatever has been stated under Sale of Goods Act with regard to 'Goods', the same is applicable here also. This definition of goods includes all kinds of movable property, including stocks, shares, growing crops. In continuation of these items when food/articles of food is a subject matter of any contract of sale, food/articles of food can also be construed as goods in this context. Section 16 of Sale of Goods Act, 1930[82] deals

---

[78] The rule of privity of contract means only the parties to the contract can sue and not strangers

[79] *Sri Lakshmi Narayana Rice Mills* v. *The Food Corporation of India* (1995)IIICPR 630(NCDRC)

[80] *Per Majority, Syno Textiles Pvt Ltd* v. *Greaves Cotton & Company Ltd* (1991)1 CPR615(NCDRC)

[81] According to section 2(7) of the Sale of Goods Act, 1930, Goods means every kind of movable property, other than actionable claims and money; and includes stocks, shares, growing crops, grass, and things attached to or forming part of the land which are agreed to be severed before sale or under the contract of sale.

[82] Sec16of Sale of Goods Act, 1930, Implied condition as to quality or fitness.- Subject to the provisions of this Act and of any other law for the time being in force, there is no implied warranty or condition as to the quality or fitness for any particular purpose of goods supplied under a contract of sale, excepts as follows:-
(1) Where the buyer, expressly or by implication, makes known to the seller the particular purpose for which the goods are required, so as to show that the buyer relies on the seller's skill or judgement, and the goods are of a description which it is in the course of the seller's business to supply (whether he is the manufacturer or producer or not), there is an implied condition that the goods shall be reasonably fit for such purpose. Provided that, in the case of a contract for the sale of a specified article under its patent or other trade name, there is no implied conditions to its fitness for any particular purpose.
(2) Where goods are bought by description from a seller who deals in goods of that description (whether he is the manufacturer or producer or not), there is an implied condition that the goods shall be of merchantable

with Quality of goods which may also be the subject matter of the contract of sale. According to this provision, there is no implied warranty or condition as to the quality of goods or fitness for a particular purpose. But, when a consumer specifies the purpose for which he is buying the goods and also stipulates the standard for the quality of goods, if the goods do not fit into the purpose for which it was bought and if the goods are not in conformity with the decided standard, the consumer may rescind the contract of sale. In the context of the Research, if the consumer specifies the purpose for which he is buying the food/articles of food after fixing the standard for quality, if the goods sold by the seller does not conform with the requirement of the standard and the purpose, the same may be rejected by the purchaser/consumer.

Section 17 of Sale of Goods Act, 1930[83] deals with sale of goods by sample, which means that if the seller of goods has already given a sample of goods of a particular standard and quality, when he delivers the bulk order, the bulk should correspond with the sample already accepted. In case these two do not match, the bulk may be rejected by the purchaser, in tune with the contract for sale by sample. In a contract for sale by sample, implied condition is, the quality of the bulk should correspond with the sample already accepted and the goods should be free from defect which may render the goods unmerchantable but which may not be apparent on reasonable examination.

### 5.7.1.4 Manufacturer

Section 2 (1) (j)[84] deals with who is a manufacturer under CPA. According to this provision any person who himself manufactures goods or only assembles different parts manufactured by someone else or who gets the goods from somebody but puts his mark or

---

quality. Provided that, if the buyer has examined the goods, there shall be no implied conditions as regards defects which such examination ought to have revealed.

(3) An implied warranty or condition as to quality or fitness for a particular purpose may be annexed by the usage of trade.

(4) An express warranty or conditions does not negative a warranty or condition implied by this Act unless inconsistent therewith.

[83] Section 17 in The Sale of Goods Act, 1930, Sale by sample.—

(1) A contract of sale is a contract for sale by sample where there is a term in the contract, express or implied, to that effect.

(2) In the case of a contract for sale by sample there is an implied condition—

(a) that the bulk shall correspond with the sample in quality;

(b) that the buyer shall have a reasonable opportunity of comparing the bulk with the sample;

(c) that the goods shall be free from any defect, rendering them un - merchantable, which would not be apparent on reasonable examination of the sample.

[84] Sec 2 (1)(j) Manufacturer means a person who

(i) Makes or manufactures any goods or parts thereof; or

(ii) Does not make or manufacture any goods but assembles parts thereof made or manufactured by other; or

(iii) Puts or causes to be put his own mark on any goods made or manufactured by any other manufacturer

name and sells under his name are all presumed to be manufacturers. Any manufacturer/maker of article of food or any person who only mixes the raw material and comes out with a finished food product/article of food or any person who buys articles of food from someone else but sells under his name and mark are construed to be manufacturer of articles of food in the context of the Research.

### 5.7.1.5 Unfair Trade Practices

There is now a greater awareness that consumers need to be protected not only from the effects of restrictive trade practices but also against unfair practices which are resorted to by the traders to mislead or dupe the consumers. If a consumer is deceitfully induced into buying goods which are not of required quality, it is obvious that the consumer is being tricked to pay for the merchandise on false representation. The consumer protection Act must have a positive and active role to play in the society. The *Sachar Committee* [85] ,recommended that the unfair trade practices like misleading advertisement and false representations, bargain sales, offering of gifts and prizes with an intention of not giving them, conducting promotional contests, bait and switch selling, supplying goods that do not comply with safety standards and hoarding and destruction of goods should be prohibited.

With the steady rise in prices and cost and standard of living, consumers have become cost conscious. Misusing the helplessness of the consumers, the traders have clandestinely resorted to various means of unethical practices having scant respect for the health of fellow human beings. The traders have begun to play truant with the life of the people by tricking them to buy adulterated/substandard/spurious/unsafe articles of food and drinks at a cheaper rate thus making huge profits. It is very shocking to know that poisonous elements are often added to articles of food which may cause innumerable diseases which ultimately may lead to the death of the consumer. Here the Researcher has dealt with limited kind of unfair trade practices that the manufacturers of articles of food might indulge in and the consumer complaints with respect to food and articles of food in the light of Food Safety and Standards Act 2006. Some of the offences against consumers and unfair trade practices by the traders or manufacturers of food can partake the following forms,

#### 5.7.1.5 .1 Deceptive Advertising

[85] Report of the High Powered expert Committee on Companies and MRTP Acts (1978)

Another area of consumer anxiety has been consumer credit and deceptive advertising. The word advertising has been derived from the French word 'adverter' which means to give notice and in business parlance it connotes 'to give information about a new product, its quality and durability'. But in reality it is exactly opposite of what it ought to be. Advertising gives fictitious and hallucinated information which is far from truth and inflicts evil effects on consumers and the society at large. The major problem in our country is illiteracy of the consumers. This conjured information may be seriously taken by gullible innocent consumers who form a major chunk of our population. Just because they are not literate and not informed, the state cannot be absolved of its duty to protect their interest. Sec2 (1) (r) of CPA describes unfair trade practice. Sec 2(1)(r) (vi) and Sec 2(1)(r) (ix) include the acts which can be termed as misleading advertisement False and deceptive advertisement may violate consumers right to information, right to safety, right to choice, harmful to the children, harmful to the health of the common people, mislead the general public by puffery, mislead & destroy the lives of youngsters through false advertisement. Parle's mango drink "Maaza" gave the advertisement of Maaza Mango. The advertisement implied that the soft drink was prepared from original mango while actually preservatives were added to it[86] .

### 5.7.1.5.2 Defective Goods

Sec 2(1) (f)[87] deals with defective goods. A complaint may be filed in respect of goods which suffer from one or more defects. In this context, defect means any imperfection, shortcomings or fault in the purity, standard, potency, quantity or quality which is required to be maintained under any express or implied contract or as claimed by any trader or under the law for the time being in force. The 1993 amendment empowered the consumer to file a complaint not only after he purchased the goods but also even if there is a mere agreement to buy the same.

### 5.7.1.5.3 Misbranded Articles

Misbranding of articles is another unfair trade practice as dealt under CPA. The definition of misbranding is 'to brand (as a food item or drug) falsely or in a misleading

---

[86] Dr Swati Snha and Dr Pradip Kumar Das "False and Misleading Advertisements and protection of the consumers:Indian scenario", International Journal of Law and Legal Jurisprudence Studies, available at http://ijlljs.in/wp-content/uploads/2014/10/false_Advt.pdf (accessed on 04/07/2018 at 10.21pm)

[87] Sec 2(1)(f) "defect" means any fault, imperfection or shortcoming in the quality, quantity, potency, purity or standard which is required to be maintained by or under any law for the time being in force or under any contract, express or implied, or as is claimed by the trader in any manner whatsoever in relation to any goods;

way; specifically: to label in violation of statutory requirements'[88]. In the context, the Researcher has confined her studies to Misbranded food[89]. It may be an imitation of any food article which may deceive the consumer into believing that it belongs to a company under the name of which it is sold. Many a time, it may be falsely stated to be a product of some place but in reality it may not be so. Though the article of food may be genetically modified, this fact may be concealed and depicted to be an organic one. In *V K Srinivasan* v. *Food Inspector*[90], the Madras held that the packets of tea were not labelled as samples and when these packets were exposed for sale, it amounted to misbranding of such packets as there was no other literature to depict that they were part of tea received in the shop for sale.

**5.7.1.5.4 Spurious Goods and Services**

Section 2(1) (oo)[91] of the Act defines Spurious goods and services. For the purpose of the Research topic, the Researcher considers only the spurious goods. Spurious goods mean such goods which are claimed to be genuine but in fact they are not so. Food and articles of food when they don't meet the standards required by law for the time being in force can be referred to as spurious goods. Spurious goods can par take the form of adulteration also. Adulteration in ordinary sense means an act of making an inferior article

---

[88] https://www.merriam-webster.com/dictionary/misbrand (accessed on 3/2/2018 at 6.12pm)

[89] The Food Safety and Standards Act, 2006 Section 3 (1) (*zf*) "misbranded food" means an article of food-
(A) if it is purported, or is represented to be, or is being-
(i) offered or promoted for sale with false, misleading or deceptive claims either; (a) upon the label of the package, or
(ii) sold by a name which belongs to another article of food; or
(iii) offered or promoted for sale under the name of a fictitious individual or company as the manufacturer or producer of the article as borne on the package or containing the article or the label on such package; or
(B) if the article is sold in packages which have been sealed or prepared by or at the instance of the manufacturer or producer bearing his name and address but-
(i) the article is an imitation of, or is a substitute for, or resembles in a manner likely to deceive, another article of food under the name of which it is sold, and is not plainly and conspicuously labelled so as to indicate its true character; or
(ii) the package containing the article or the label on the package bears any statement, design or device regarding the ingredients or the substances contained therein, which is false or misleading in any material particular, or if the package is otherwise deceptive with respect to its contents; or
(iii) the article is offered for sale as the product of any place or country which is false; or
(C) if the article contained in the package-
(i) contains any artificial flavouring, colouring or chemical preservative and the package is without a declaratory label stating that fact or is not labelled in accordance with the requirements of this Act or regulations made there under or is in contravention thereof; or
(ii) is offered for sale for special dietary uses, unless its label bears such information as may be specified by regulation, concerning its vitamins, minerals or other dietary properties in order sufficiently to inform its purchaser as to its value for such use; or
(iii) is not conspicuously or correctly stated on the outside thereof within the limits of variability laid down under this Act.

[90] (1965)1 M L J 58

[91] Sec 2(1)(oo) spurious goods and services" mean such goods and services which are claimed to be genuine but they are actually not so"

for a superior one in order to gain illegitimate profits. It is the primary duty of any state to protect the health of the people. In order to ensure this right, the state at the outset ought to ensure that there is no adulteration of food articles. All violations of prescribed standards of purity or the quality by the manufacturers, with respect to the use of additives or colour or preservatives amount to adulteration. Adulteration of food is often perceived to be subtle murders inflicted on the community; it mostly starts with the manufacturers and travels to the whole sellers and retailers[92].

### 5.7.1.5.5 Violation of Standards of Weights and Measures

A standard means for measurement of quantity or quality of anything is prescribed at the international level which has been accepted the world over. These measuring units are essential not only for maintaining reliability in foreign trade but also for the protection of the consumer. But more than often these standards are violated. Violations of weights and measures are no less harm than the adulteration of food.

The action under this legislation can be invoked by an aggrieved consumer by filing a complaint[93] against defective goods[94], deficient services[95], excess price, hazardous goods when the same is hazardous to life and safety when used, are being offered for sale to the public in contravention of the provisions of law for the time being in force, requiring the traders to display information in regard to the contents, manner and effect of the use of

---

[92] Dr Rifat Jan, *'Consumerism and Legal Protection of Consumers: with a critical and explanatory commentary & latest case Law'*, Deep and Deep Publication Pvt Ltd New Delhi (1st edn, 2007), p 312

[93] Section 2(c) "complaint" means any allegation in writing made by a complainant that –
(i) an unfair trade practice or a restrictive trade practice has been adopted by any trader;
(ii) the goods bought by him or agreed to be bought by him suffer from one or more defects;
(iii) the services hired or availed of or agreed to be hired or availed of by him suffer from deficiency in any respect;
(iv) a trader has charged for the goods mentioned in the complaint price in excess of the price fixed by or under any law for the time being in force or displayed on the goods for any package containing such goods,
(v) goods which will be hazardous to life and safety when used, are being offered for sale to the public in
(a) in contravention of any standards relating to safety of such goods as required to be complied with, by or under any law for the time being in force;
(b) if the trader could have known with due diligence that the goods so offered are unsafe to the public
(vi) services which are hazardous or likely to be hazardous to life and safety of the public when used, are being offered by the service provider which such person could have known with due diligence to be injurious to life and safety
with a view to obtaining any relief provided by or under this Act;

[94] Section 2(f) "defect" means any fault, imperfection or shortcoming in the quality, quantity, potency, purity or standard which is required to be maintained by or under any law for the time being in force or [15] [under any contract, express or implied, or] as is claimed by the trader in any manner whatsoever in relation to any goods

[95] Section 2(g) "deficiency" means any fault, imperfection, shortcoming or inadequacy in the quality, nature and manner of performance which is required to be maintained by or under any law for the time being in force or has been undertaken to be performed by a person in pursuance of a contract or otherwise in relation to any service

such goods where the trader could have known that the goods are unsafe to the public. Under these circumstances the consumer of food/articles of food can file a complaint before the appropriate forum depending on the relief claimed.

### 5.8 Consumer Helpline

Our central government, in order to protect the interests of the consumer a separate Ministry has been dedicated to manage consumer affairs and food &civil distribution. A toll free National Consumer Helpline[96] has been set up in which different sectors are included and food safety is one of them. This helpline also suggests few indicative household tests[97] to detect adulteration in 14 common food items that are used by a common man on a day today basis. The advantage of these tests is, they are not complex either to perform or to understand. They are simple and can be conducted by housewives at home. This helpline also gives details of standard test laboratories which could be approached by a common man to get any goods/product tested for quality.

### 5.9 Jurisdiction of Consumer Courts

Section 11 of CPA provides for the jurisdiction of the District Forum. Sec11(1) deals with pecuniary jurisdiction of the District Forum that it has jurisdiction to entertain complaints where the value of the services or the goods or the compensation claimed does not exceed Rupees Twenty Lakhs, which means complaints involving claims more than twenty lakhs cannot be entertained by the district forum. Later on, it was clarified in '*Farook Haji Ismail* v *Gavabhai Bhesania*' [98] that the pecuniary jurisdiction of the consumer court depends neither on the value of the subject matter nor on the relief allowed by the forum but on the amount of relief claimed by the consumer.

Section 17 deals with the jurisdiction of the State Commission. State Commission has been vested with three types of jurisdiction, namely- original, appellate and revisional jurisdiction. As on today under Sec 17(1)(a) (i) of the Act, the State Commission can entertain complaints where the value of the goods or services and compensation claimed exceeds Rupees twenty lakhs but does not exceed Rupees1crore. But it was in '*Akhil Bharathiya Grahak Panchayat* v *Life Insurance Corporation of India*'[99], confirmed that the

---

[96] TOLL FREE NO. 1800-11-4000 SMS No. 8130009809 (24 Hrs.) available at https://consumeraffairs.nic.in/forms/contentpage.aspx?lid=614 (accessed on 1/7/2018 at 7.30am)

[97] Available at http://nationalconsumerhelpline.in/Datafiles/Adulteration_tests.pdf (accessed on 1/7/2018 at 7.00am)

[98] (1991)II CPJ452 (Guj.CDRC)

[99] (1991)1CPR 112(Maha CDRC)

pecuniary jurisdiction does not depend on the value of the subject matter or on the relief granted by the commission but on the amount of relief claimed which includes even the compensation. Sec 17 (1)(a)(ii) of the Act empowers the state commission to entertain appeals against the order of any district forum within the state. Sec17 (1) (b) of the Act confers Revisional jurisdiction on the state commission. If it appears to the state commission that the district forum has exercised jurisdiction not vested in it by law or has acted illegally or with material irregularity beyond the its jurisdiction, it can call for any records and pass appropriate orders in any consumer dispute which is pending before or has already been decided by the district forum. Sec 21 of the Act deals with the jurisdiction of the National commission. As in the case of the state commission, the national commission also enjoys original, appellate, revisional jurisdiction coupled with the power of review. The original jurisdiction of the National Commission extends to a complaint where the value of the claim in a complaint exceeds Rupees 1 crore. Whenever there is an error apparent on the face of the record may be one of fact or of law, Sec 22(2) of the Act empowers the national commission to review any order made by it.

## 5.10 Section 72 of FSSA 2006 *vis-a-vis* Section 3 of CPA

Section 72 of FSSA, 2006[100] provides that no civil court shall have jurisdiction to entertain any suit or proceeding in respect of any matter which an adjudicating officer or the tribunal is empowered by or under this Act to determine and no injunction would be granted by any court or authority in respect of any action taken or to be taken in pursuance of any power conferred by or under this Act'[101]. This clause prohibits a civil court from entertaining any proceeding or suit in relation to any matter for which an Adjudicating Officer, who has been appointed under this Act or the Tribunal, which is constituted under this Act is empowered to entertain. No court or any other authority can issue injunction in respect of any action initiated or to be initiated in continuation of the power conferred by or under the Act.

But, this provision under FSSA will not bar the Consumer courts from entertaining consumer complaints with respect to food/articles of food. Section 3 of CPA[102], makes it

[100] FSSA, Sec 72 'No civil court shall have jurisdiction to entertain any suit or proceeding in respect of any matter which an Adjudicating Officer or the Tribunal is empowered by or under this Act to determine and no injunction shall be granted by any court or other authority in respect of any action taken or to be taken in pursuance of any power conferred by or under this Act'.

[101] Sumeet Malik, 'Handbook of Food Adulteration &Safety Laws' Eastern Book Company ( 2011) p 143

[102] CPA, Sec 3 'Act not in derogation of any other law.—The provisions of this Act shall be in addition to and not in derogation of the provisions of any other law for the time being in force'.

clear that the provisions of CPA do not have an overriding effect on other legislations but are in addition to the other existing laws and there has to be harmonious application with the provisions of other laws. The provisions of CPA are supplementary in nature and do not encumber the remedies available to the consumers under any other law for the time being in force. The provision of Section 3 CPA, establishes that the forum, reliefs, procedure and means of adjudication as provided under the Act is an additional dispensation and not in derogation of any other law for the time being in force. It is for the consumer to choose a forum which is convenient for him to seek remedy for the loss that he suffered. In *Life Insurance Corporation of India* v *Uma Devi*[103], it was contended that the claim under the insurance policy was repudiated by the corporation in exercise of the powers conferred under Sec 45 of the Insurance Act[104], 1938. This Act provided a complete setup for redressing the grievances and hence the complainant should have approached the city civil court and not a redressal forum under CPA. The National Commission turned down the contention and held:

*"As the provisions of the Act are in addition to, and not in derogation of any other law for the time being in force the State Commission has the jurisdiction to entertain the compliant and to investigate whether the repudiation was justified or not, and to grant such relief as it deems fit, if it is satisfied that there was a deficiency in service"*

But, it was held in *Priti Sinha* v. *Bihar State Housing Board*[105], that when the complainant has already chosen a forum and was unsuccessful, she cannot be allowed to institute fresh proceedings under CPA. It was held in *Sudarshan Chits(India) Ltd* v. *official Liquidator*[106] that the provisions of CPA do not abrogate the provisions of any other law for the time being in force. *'As such, the consumer Forum is an alternative Forum established under the Act to discharge the functions of a civil court'*[107]. In the light of the above stated provisions and the decided cases, the Researcher concludes that the reliefs granted under CPA are in addition to the provisions of FSSA. Therefore, the consumer of food/articles of food can choose to file a complaint either in consumer court to claim damages or before the Adjudicating Officer for levy of fine & for penal punishment.

---

[103] (1991) 3 Comp LJ 171 (NCDRC)
[104] Insurance Act,1938 Sec 45, no life insurance policy can be called into question on grounds of mis-statement or wrong disclosure after two years of the policy coming into force. However, if the insurer is able to prove that the claim was fraudulent, it need not be passed.
[105] (1996) 1 CPR 369 (Bihar CDRC)
[106] (1992) 1 Comp LJ 34 (Mad.HC)
[107] Prof (Dr.) V K Agarwal, *'Bharat's Consumer Protection Law &Practice'*, Bharat Law House, (7th edn 2016), p 342

The most recent case being that of the Maggi Noodles. In April 2015, the food regulator at Uttara Pradesh recalled a batch of of 2lakh packs of Maggi instant noodles after there were reports about the presence of lead and food additives like Monosodium glutamate, in them beyond the permissible limit. For the first time, in the history of CPA, the central government filed a complaint before the National Consumer Redressal Commission under Section 12(1) (d)[108] against Nestle India. On May 29th 2015, the government requested the FSSAI to have a check in this matter. Subsequently FSSAI collected more samples of Maagi noodles from different states for testing. The laboratory results established that the parameters were not within the normal range because of which 2lakh packs of Maggi instant noodles were recalled from the market. But later on fresh batch of Maggi noodles were subjected to laboratory tests where they were proved to be safe human consumption, in pursuance of which they were reintroduced in the market. In the mean while, popular hindi film actors who had endorsed Maggi Noodles were served with legal notice on the claims that they had acknowledged in the advertisement which were thought to be misleading the public.

### 5.11 Procedural Advantages of FSSA Over CPA

The focus of both the Acts are different. FSSA, 2006 fixes the scientific standard for the quality of food/articles of food, nips the problem of adulteration/unsafe food at the bud and at every stage of food production chain, punishes the food business operators in case of any deviation by them. FSSA makes effort to ensure that the food is safe from the grass root level. It makes every food business operator liable in the entire food chain. If the food /articles of food are not in conformity with the standards, it ordains the manufacturer not to release it into the market. In case the food in the market is found to be hazardous, the same may be recalled from the market. FSSA takes all measures to ensure that the available food is safe and not noxious. The persons who are part of adjudicating process are all trained. Adjudicating personnel are more focussed as they are dealing with only matters concerning food. FSSA also regulates the food distributed by any anganawadis to small children, articles of food given to the *pregnant* ladies, mid day meals at the state run schools, akshya patra yojanas or any food issued by the state machinery, which means FSSA will neither consider whether one has paid consideration for the food nor he is the

---

[108] Section 12(1) (d) CPA, the Central or the State Government, as the case may be, either in its individual capacity or as a representative of interests of the consumers in general.

beneficiary of a welfare programme, even the food/food grains issued by the state under NFSA is not spared ahd it attracts the provisions of FSSA. In case of any deviation, the provisions provide for levy of fine and for punishment to trudge to ensure that only safe food is available to the consumers.

If we examine the provisions of CPA, if one has to take recourse to CPA, at the outset he has to be a consumer in the real sense of the term as per CPA. It means that only if he has bought the food for consideration and for his personal use, he will be a consumer, thus entitling him to file a complaint at the consumer courts. But wherever, a person consumes food not bought by him but is a beneficiary of a social legislation or a welfare scheme, he cannot be a consumer which disentitles him to take recourse to consumer court. CPA provides for the payment of damages or the replacement of the defective article along with the cost, to the affected but has no jurisdiction to punish the offender. The reliefs contemplated under CPA are compensatory but not punitive and the Researcher strongly thinks that reliefs under CPA are only damage control measures but is not deterrent in nature. CPA comes to play only when the consumer complains and does not have suo motto powers like in the case of FSSA, but FSSA strives to create a situation where the consumer can be rest assured about the quality and safety of the food/food articles that he is consuming. FSSA creates an ideal situation where CPA can only be damage control tool. In the light of all these facts, the Researcher thinks that it is advisable for a consumer to avail of the adjudicating process under FSSA than under CPA.

## 5.12 Code of Ethics for Food Trade and Industry

The Confederation of Indian Food Trade and Industry is the highest organisation for food trade and industry in India. It has been founded by companies and associations who have resolved to provide adequate, safe, sound and wholesome food to the consumers at large. Considering the need to evolve the principles of vibrant consumer welfare which may supplement and complement the national food legislations like National Food Security Act 2013 and Food safety and Standards Act 2006, CIFTI has drawn up a Code of Ethics for Food Trade and Industry and has strongly urged all the food business operators to adopt them.

## 5.13 Conclusion

The Researcher agrees and acknowledges that the level of consumer awareness and their protection is a true indicator of progressiveness of civil society and development of the country. The primary benefits accruing to the consumers from the statutory measures are, they are protected from the inevitable & clandestine exploitation by the unscrupulous. The organised society is one which is properly regulated by way of the accepted norms of behaviour and only such society has a capacity for growth and development. In the present day status, the accepted norms of behaviour which are instrumental in regulating the social order are the laws, conferring upon the members, the rights and duties and prescribing the sanctions for the violations of such norms. Thus, for the existence, growth and development of the society, the interest of the members should be protected.

It is, therefore in the interest of the existence and growth of a society that consumer protection is extended as a right of the members of such society[109]. In case, it is not protected, the result may be chaos in the society which may result in disintegration and degeneration of the society. In such a scenario, the pre-contractualist situation as the natural law thinkers advocate, the state shall lose its existence. The nation, made to consume sub-standard and adulterated food-stuffs will naturally tend to grow sub-standard and weak. Increased mobility of consumers has resulted in drastic reduction of personal interaction between sellers and buyers which has fuelled the need for increased consumer protection. .

24th December of every year is celebrated as National Consumer Rights Day, in India, as the Consumer Protection Act, 1986 was enacted on that day. Observation of Consumers day alone without creating awareness among the people about their Rights as consumers, does not help much. Law is a means to an end, which fails only sometimes to organise an orderly society. Though the executive and judiciary are playing prominent roles in curbing adulteration, the public participation in the same direction cannot be ignored. Unless there is a concerted effort by the people, by the state and by the authorities, the legislations alone cannot bring about the desired effect. The Researcher is convinced that unless a common man becomes aware of the gravity of the quandary of adulteration of food, his rights and corresponding duties to the society in which he lives, 'the Dracula of adulteration pervades in the society eternally and engulfs it's preys-the healthy consumers'. 'There can be no salvation without cooperation'.

---

[109] Dr Subhash C Sharma, "Consumer protection and the Prevention of Food Adulteration Act", Central India Law Quarterly, Annual Index,[1995] Vol VIII,p 187-194

The legislature and judiciary have endeavoured to protect the innocent and gullible consumers from unscrupulous traders and food business operators but the Researcher strongly feels that it would be impossible to eradicate the deep rooted evil of adulteration unless the incorruptible and dedicated law enforcement agencies are in place. It will not be out of place to note that fault lies not with our laws but with the implementation. This alarming evil development can partly be attributed to the slackness of law enforcement machinery. Though India is not the only country suffering from the vice of adulteration of articles of food, of late it has reached catastrophic heights. Way back in 1983 itself, UN expert had commented "*a whole generation of Indians have grown up not knowing the taste of unadulterated food*"[110].

Consumer helpline number should be displayed everywhere and the awareness should be created among the people to use it whenever they are affected as a consumer. The consumer must be provided with statutory remedy under tort for claiming damages against the adulterators. As Mahatma Gandhi has remarked *"A customer is the most important visitor on our premises. He is not dependent on us. We are dependent on him. He is not an interruption in our work. He is the purpose of it. He is not an outsider in our business. He is part of it. We are not doing him a favour by serving him. He is doing us a favour by giving us an opportunity to do so".* The Researcher thinks that a consumer should be made to feel like a king in the competitive market. Public spirited advocates and consumer organisations should come forward to volunteer their services in initiating law suits without remuneration or may be only a nominal remuneration from the consumers. Consumer courts should be conferred with such powers that the same should enable them to initiate action against the erring producers, suo motto, which to some extent can avoid the delay in bringing the offender to books. As on today only the aggrieved consumer can file a complaint to initiate the action. "In the United States, a series of cases beginning in the 1920's have developed consumer rights against producers and distributors of defective goods so that now even a non-purchaser user can collect in strict liability against a party in the production or distribution chain"[111].

---

110 'Where consumers are easy prey', Indian Express Daily News Paper, 16th September 1983, Delhi Edition

111 Alan M Katz, "The Law against Food Adulteration: A Current Assessment and a Proposal for an Enforcement alternative", Journal of the Indian Law Institute, [1977] Vol 1.p123-129

A drastic measure that the Researcher suggests is that consumer complaints should be filed in the form of public interest litigations, though it may not afford any monetary compensation, yet the same may bring about radical changes to curb adulteration. Adulteration of food should be treated as a matter of public interest as all of us are consumers. The concept of locus-standi or affected party should be done away with. Instead, if anyone is allowed to complain to the authorities under the CPA, the same may act as a means to speed up the procedure. The complaint under CPA should be treated as a public interest litigation as the whole society is a consumer and any member of the society need to be allowed to bell the erring cats.

People are neglectful of their rights which have forced the state to educate the consumers and create awareness among them. People can be educated by various means which may include exhibitions, street plays, display by flash mobs, talks, film shows, demonstrations, advertisements via mass media and publicity materials. People should also be educated about the need to organise themselves into some united body/organisation/association. Whatever cannot be achieved individually, can be achieved by a group having common interest as the Researcher thinks that 'strength lies in unity'. Once the people are united, their voice can get stronger. The consumers association can have their own well equipped laboratory and trained staff which may be free from the clout of the authorities. These consumer associations with the help of the experts can educate the mass about the simple tests that may be conducted by a common man in his house to detect the adulteration of day today food articles. Once the food is found to be contaminated, street plays /exhibitions/shows of adulterated food stuffs should be initiated to educate the people of the adulteration and urge the people to boycott & denounce the use of the same.

Voluntary organisations and Nongovernmental organisations should also be encouraged to establish the laboratories whose findings or reports should be admissible in the judicial process to prosecute the delinquent manufacturer/trader. Consumer Protection Councils need to be established at every district panchayat levels investing it with the power of instituting legal action on behalf of aggrieved consumers against erring traders. In every food processing industry, there is a department for quality control which will subject the product to quality control measures before the same is released into the market.

But, the flip side is, as this department comes under the purview of the producer, the Researcher is highly apprehensive and wonders if high degree of credibility can be placed on them and if the department can be free of the clout that may be exercised by the vested interests. Therefore, it is advisable that the government should establish wings of quality control at every food processing industry independent of the management of the unit which prevents adulteration at the source and the safe food will be ensured in the market. Same manufacturing unit should not be permitted to produce non consumables and consumables meant for human consumption as there may be chances of mix up of both by accident or over sight or by negligence. More laboratories should be established even at the district level which helps in speedy disposal of cases. Consumer groups should involve themselves in the legal aspect of enforcement

# Chapter - VI

# International Conventions and Declarations on Food Safety and Standards: An Analysis

## 6.1 Introduction

'Food is a central activity of mankind and one of the single most significant trademarks of a culture' remarked Mark Kurlansky, food writer and author of 'Choice Cuts'. This is a paradox of plenty. There is plenty and more of it, yet there are those who have none of it. It is there but yet it is not there where it ought to be there and where it is most required. Here the Researcher is referring to food. Not many deny its existence but falter while making this Right available for everyone. Food always lures and evades hungry. In this era of increased state obligation at a multilateral level, no national campaign on the Right to food can be undertaken disregarding the international commitment of a state. By entering into multilateral agreements and treaties, states bind themselves to respect and abide themselves to many legal obligations.

Ensuring the Right to adequate and safe food and the fundamental right to be free from hunger is a matter of international law, enshrined in various international instruments. Here the Researcher emphatically notes that no convention uses the word safe food that's because, to identify something as food, it needs to be safe for human consumption at the outset and only then it can be treated as food. It is widely presumed that safety and standards are inbuilt in the word food. This is the reason we don't find any Convention or Declaration expressly dealing with food safety and standards. But for Codex Alimentarius Commission, we do not have any authority or instrument dealing with safe and standard food.

The right to safe food cuts across the entire spectrum of human rights. Human rights are not territorial in nature. However, the obligations to protect and employ the same in the municipal sphere are primarily domestic, in the relationships between respective states and their own people where the major obligations of national governments are towards the people living under their jurisdictions. As Jean Ziegler[1] remarks '*In a world overflowing with riches, it is an outrageous scandal that almost 900 million people suffer from hunger*

[1] The Special Rapporteur on the Right to Food,

*and malnutrition and that every year over 6 million children die of starvation and related causes. We must take urgent action now. This is my ongoing fight for the Right to food'.*

However, it should be documented that the international community also has obligations to respect basic human rights and one of them is the Right to food. Hungry person of any nationality in any part of the world is also a part of the world community and he has rights claims not only in relation to his own nation but also in relation to the world at large. The human right to adequate food will be hollow and narrow in its application if obligations to honour this right are restricted only to one's own government or one's own state as the Researcher strongly believes that 'children born into poor countries are not born into a poor world'. Universality can be attributed to Human rights in the real sense only when the international community lends a helping hand and steps into the shoes of the national government which has failed to fulfill its obligation to respect this basic right, transcending all national boundaries and does what needs to be done to assure the realization of those rights.

Paradoxically, despite there being a plethora of international instruments to strengthen the 'Right to food', we are not able to achieve a major breakthrough and hunger is still dogging the society. There seems to be no effective mechanism in place and no firm commitment on the part of the world community with regard to the human right to adequate food due to which even today we find more than 900 million people who are starving for food throughout the world, most of them in developing countries[2]. In this context the Researcher legitimately believes that, may be the obligations of the international community need to be well articulated and the same should be implemented in true spirit with lot of fervour.

The human right to live in dignity, free from want, is itself a fundamental right, and is also essential to the realization of all other human rights – rights that are universal, indivisible, interconnected and interdependent[3]. There is increasing recognition worldwide that food and nutrition is a human right, and thus there is a legal obligation to assure that all people are adequately nourished. The articulation of food and nutrition rights in modern

---

[2] The Committee on Economic, Social and Cultural Rights General Comment No. 12: The Right to Adequate Food (Art. 11) para 5

[3] Para 1.4 Government Of India, Law Commission Of India, 'Need for Ameliorating the lot of the Have-nots' – Supreme Court's Judgments Report No. 223 April 2009

international human rights arises in the context of the broader human right to an adequate standard of living[4]. Universal and sustainable food security which in turn ensures a continuous availability of food is a means to achieve the social, economic and human development objectives that members of the world community agreed upon at various world conferences in Rio, Vienna, Cairo, Copenhagen, Beijing and Istanbul. The right to adequate food is also emphasized in the most basic international human rights treaties, including the Universal Declaration of Human Rights, the International Covenant on Economic, Social and Cultural Rights, the Convention on the Elimination of All Forms of Discrimination Against Women, the International Convention on the Elimination of All Forms of Racial Discrimination, and the Convention on the Rights of the Child.

The Right to food has been essentially recognized in a wide range of human rights instruments. In the year 1941, the US President Franklin D Roosevelt's "Four Freedoms" speech outlined one of the freedoms as 'the Freedom from want'[5]. These freedoms were generally refined and became the basis of the *United Nations Charter*[6] and more significantly in the Universal Declaration of Human Rights which encompassed the concern of "freedom from want" through inclusion of economic and social rights, particularly by recognizing this right as a component of an adequate standard of living[7]. Different International Instruments dealt with the Right to food expressly and some of these instruments impliedly supported the Right to Food.

### 6.2 Universal Declaration of Human Rights 1948 (UDHR)

In order to know the factors that led to the promulgation of UDHR, at the outset, it is necessary to know the genesis of United Nations Organisation. The movement for the protection of human rights gained momentum after the World War II. '*The war proved to many, the close relationship between outrageous behavior by a government towards its own citizens and aggression against other nations, between respect for human rights and the*

[4] George Kent University of Hawai'i March 12, 2002 '*THE HUMAN RIGHT TO FOOD IN INDIA*' available at http://www.earthwindow.com/grc2/foodrights/HumanRightToFoodinIndia.pdf (accessed on 27/12/2017 at 3.35pm)
[5] http://www.history.com/this-day-in-history/franklin-d-roosevelt-speaks-of-four-freedoms (accessed on 22/6/2016 at 1.19pm)
[6] Article 1(3) of the UN Charter 1945
[7] Article 25 of the UDHR 1948

*maintenance of peace'*[8]. The necessity for Bill of Rights at the international level was felt at the time of framing of the Charter of the UN. '*At the time of creating the UN, there was not much time to prepare and insert such a Bill of Rights and it was understood that one of the first duties of the organisation would be to frame such a Bill'*[9]. There was a division of opinion under which some wanted it in the form of a 'Declaration' while the others desired for a 'Convention'. As a compromise, it was decided that two documents should be prepared, a 'Declaration' wider in content and more general in expression and a 'Convention' on those matters which "might lend themselves to formulation as binding obligations[10]. Since its inception, the UN has identified access to adequate food as both an individual right and a collective responsibility.

The Universal Declaration of Human Rights[11] is a milestone document in the history of Human Rights and is considered as the International Magna Carta[12]. It was drafted by representatives with different legal and cultural backgrounds from all regions of the world to give it a Universal character which was greeted well by all the nations. The declaration was proclaimed by the United Nations General Assembly in Paris on 10th December 1948[13] as a common standard of achievements for all peoples and nations. This 'path finding' instrument provides a generally acceptable catalogue of man's inalienable rights and has had a remarkable influence on further developments, at both the international and domestic levels[14]. The UDHR is not a formal legal document but is an important norm setting instrument which has immensely influenced the human rights. Being a declaration it does not cast legal obligations on the states to adhere to it mandatorily but it only implores the states to aspire towards moral obligations.

Besides the preamble, the UDHR comprises 30 articles, the first 21 Articles relate to civil and political rights and the next 6 Articles to economic, social and cultural rights. The

---

[8] Gurdip Singh, *International Law*, Eastern Book Company, Lucknow. (3rd edn 2015), p 338

[9] KC Joshi, *International Law & HUMAN RIGHTS*, Eastern Book Company, (3rd edn 2016) p.201

[10] UN Doc. E/CN.4/21 of 1-7-1947

[11] The Universal Declaration of Human Rights (Universal Declaration) is an international document that states basic rights and fundamental freedoms to which all human beings are entitled.

[12] Magna Carta, meaning 'The Great Charter', is one of the most famous documents in the world. Originally issued by King John of England (r.1199-1216) as a practical solution to the political crisis he faced in 1215, Magna Carta established for the first time the principle that everybody, including the king, was subject to the law, Magna Carta remains a cornerstone of the British constitution even today: https://www.bl.uk/magna-carta/articles/magna-carta-an-introduction#sthash.KEjqyJq4.dpuf

[13] General Assembly Resolution 217 A

[14] JG Starke, *Introduction to International Law*,(10th edn 2007), p 364

last three Articles are of general application and have bearing upon the entire declaration. The Preamble of the Charter has reaffirmed the faith in fundamental human rights, in the dignity and worth of the human person and in the equal rights of men and women and has determined to promote social progress and better standards of life .......[15]. As is the case with the Preamble of the UN Charter, the Preamble of the UDHR also bears ample testimony to the fact that it was a normative response to the brutalities of the World War II.

The rights stipulated in the declaration are most comprehensive but it was not designed or proposed as an enforceable treaty obligation and was only a broad classification and recommendation of policy. It recognizes the inherent dignity and equal & inalienable rights of members of the human family. Article 3[16] is perhaps the heart of the substantive provisions in the Declaration. Fundamentally, it guarantees to every human being the basic rights to life, to liberty and affords security to every person. Article 25(1)[17]of the Declaration clearly enunciates the Right of every individual and his family for a better standard of living. Clause 2 of Article 25[18] stresses upon the need to protect the motherhood and childhood as the Researcher also thinks that 'healthy citizenry is the asset of every nation' and it is common knowledge that for better health, safe food is a pre requisite. So Right to food was first established under the non-binding yet universally recognized instrument. Hence UDHR is a common standard of achievement for all peoples and all nations.

But, the Researcher believes that this Declaration was not effective in safeguarding the basic rights as it was not binding on the states and UNO could not enforce it. "*After more than fifty years of Universal Declaration of Human Rights, it is important to reaffirm the fundamental right of everyone to be free from hunger and to take particular note of the needs of people affected by all types of disasters. Universal rights require universal*

---

[15] We The Peoples Of The United Nations Determined to save succeeding generations from the scourge of war, which twice in our lifetime has brought untold sorrow to mankind, and to reaffirm faith in fundamental human rights, in the dignity and worth of the human person, in the equal rights of men and women and of nations large and small, and to establish conditions under which justice and respect for the obligations arising from treaties and other sources of international law can be maintained, and to promote social progress and better standards of life in larger freedom,

[16] Article 3.Everyone has the right to life, liberty and security of person.

[17] Article25. (1) Everyone has the right to a standard of living adequate for the health and well-being of himself and of his family, including food, clothing, housing and medical care and necessary social services, and the right to security in the event of unemployment, sickness, disability, widowhood, old age or other lack of livelihood in circumstances beyond his control

[18] Motherhood and childhood are entitled to special care and assistance. All children, whether born in or out of wedlock, shall enjoy the same social protection.

*actions"*[19]. In the debate on the UDHR, the US delegate *Mrs Roosevelt* stated that the Declaration was neither a treaty nor an international agreement and did not impose any legal obligations and Mr. Cassin of France considered it of a wide moral scope[20]. Most of the heads of countries were not convinced about its multicultural character and deemed it to be unusually broad. Third world countries then and majority of Muslim countries regarded the concept of Human rights as 'western concept' and neither did they evince any interest in its importance nor in its implementation. Many jurists and thinkers thought that the human rights were imaginary. Jeremy Bentham, the founder of modern utilitarianism believed only in the force of positive legislation. For him, human rights were, "nonsense upon stilts" and feared that they were powerful rhetoric in the hands of rulers and a substitute for effective legislation[21]. It has been observed that the Declaration is not binding, although there is some support for the view that it may properly be resorted to for the interpretation of the provisions of the UN Charter in the matters of human rights and fundamental freedoms[22].

Prior to the adoption of the Universal Declaration of Human Rights in 1948 which was a non-legally binding document, broad agreement existed that the rights which were mentioned in the Declaration were to be upgraded into legally binding obligations over a period of time through the negotiation of treaties. In tune with this agreement in 1966, two separate treaties, encompassing almost all the rights enshrined in the Universal Declaration of Human Rights, were adopted after nearly 20 years of negotiations: one for economic, social and cultural rights, the International Covenant on Economic, Social and Cultural Rights and another for civil and political rights, the International Covenant on Civil and Political Rights. Furthermore, the right to food finds a place in the common Article 1 of both the Covenants which provides that 'all people may.........freely dispose of their natural wealth resources' and consequently 'in no case people be deprived of its own means of subsistence'[23].

---

[19] Yogendra Kumar Srivastava, "Right to Food: A Human Right", The PRP Journal of Human Rights[January-March 2001],Vol 5 Issue 1, p11-13

[20] K C Joshi, *International Law & HUMAN RIGHTS*, Eastern Book Company,( 3rd edn 2016) Page No503, Louis B. John,"The Universal Declaration of Human Rights: A Common Standard of Achievement" in L.M. Singhvi(Ed), Horizons of Freedom I (1968) p 8.

[21] The Impact Of The Universal Declaration Of Human Rights On The Study Of History Antoon De Ba History and Theory 48 (February 2009), 20-43 Wesleyan University 2009 ISSN: 0018-2656 available at http://www.inth.ugent.be/wp-content/uploads/2012/05/Impact-UDHR.pdf accessed on 12/8/2017 at 7.00pm

[22] Oppenheim's '*International Law*', Vol. I (1992) p.1002

[23] "*The right to food"*, Report of the Special Rapporteur, E/CN.4/2001/53, 7th February 2001, Page 14, Para 42,

## 6.3 International Covenant on Economic, Social and Cultural Rights, 1966 (ICESC)

Adopted and ratified by General Assembly resolution 2200A (XXI) of 16th December 1966 and came into force on 3rd January 1976. ICESCR, which is the outcome of Universal Declaration of Human Rights deals with the right to food more comprehensively than any other treaty. Economic, Social and Cultural rights focus on freedom from want of basic services, they address some of the basic needs of people. For example: freedom from hunger……[24] The Covenant comprises the preamble and 31 Articles divided in five parts. In Article 1clause 2, it provides for the self determination for all people and strongly puts forth that in no case the person may be deprived of his means of subsistence[25]. The state parties to the covenant are required to implement the provisions of the covenant by suitable municipal legislation and other methods. Since, the implementation of the economic rights involves the economic resources and taking the limited economic resources of the developing states into consideration, the covenant gives power to such states to determine the extent of guarantee of such rights to non citizens.

Article 11(1)[26] deals with the right of everyone to an adequate standard of living which includes food, shelter, clothing and to the continuous improvement of living conditions. The covenant enjoins the parties to initiate appropriate steps for the implementation of these rights. Clause 2 of Article 11[27] recognises the fundamental right of everyone to be free from hunger and malnutrition. In order to realize this right fruitfully it urges the state parties to improve methods of production by making use of the technical

---

[24] Human Rights For All International Covenant on Economic, Social and Cultural Rights A Handbook 1st Edition 2015 PWESCR (Programme on Women's Economic, Social and Cultural Rights) New Delhi Page 35

[25] Article 1(2) ……….In no case may a people be deprived of its own means of subsistence.

[26] The States Parties to the present Covenant recognize the right of everyone to an adequate standard of living for himself and his family, including adequate food, clothing and housing, and to the continuous improvement of living conditions. The States Parties will take appropriate steps to ensure the realization of this right, recognizing to this effect the essential importance of international co-operation based on free consent.

[27] The States Parties to the present Covenant, recognizing the fundamental right of everyone to be free from hunger, shall take, individually and through international co-operation, the measures, including specific programmes, which are needed:

a. To improve methods of production, conservation and distribution of food by making full use of technical and scientific knowledge, by disseminating knowledge of the principles of nutrition and by developing or reforming agrarian systems in such a way as to achieve the most efficient development and utilization of natural resources;

b. Taking into account the problems of both food-importing and food-exporting countries, to ensure an equitable distribution of world food supplies in relation to need.

innovations, by adopting scientific reforms in agrarian sector which will result in the proper utilization of the natural resources thus boosting up the production. The Covenant has also tried to make sure that there is equitable distribution of food among all the states in commensurate with their need. Article 12 (2) aims at ensuring better physical health for all the people[28]. The state parties are expected to adhere to the provisions of the covenant strictly and the violations of these rights may be in the form of failure to prevent starvation in all areas and communities in the country (the right to food) and 'failure to prohibit public and private entities from destroying or contaminating food and its source, such as arable land and water' (the right to food)[29].

With this maiden venture of legally binding norm, the Researcher believes that this human-rights idea became overshadowed by the state-centered thinking of the nineteenth century without realizing its full potential and did not receive the necessary attention that it should have, for a long time.

### 6.3.1 The Committee on Economic, Social and Cultural Rights (CESCR)

In order to carry out the provisions of ICESCR the Committee on Economic, Social and Cultural Rights[30] was established under the U N's Economic and Social Council Resolution 1985/17 of 28 May 1985. It is the body in charge of monitoring the implementation of the International Covenant on Economic, Social and Cultural Rights in those states which are party to it. The objective of CESCR was to determine the prevailing situation in the countries concerned with respect to various rights which are economic, social & cultural in nature, to identify the obstacles to its realization, to record its observation and document the same in the form of General Comment. General Comments provide normative interpretation and clarification of the Covenant's provisions[31]. In 1999,

---

[28] Article 12( 1).The States Parties to the present Covenant recognize the right of everyone to the enjoyment of the highest attainable standard of physical and mental health.

[29] HUMAN RIGHTS FOR ALL International Covenant on Economic, Social and Cultural Rights A Handbook 1st Edition 2015 PWESCR (Programme on Women's Economic, Social and Cultural Rights) New Delhi Page 40

[30] The Committee on Economic, Social and Cultural Rights (CESCR) is the body of 18 independent experts that monitors implementation of the International Covenant on Economic, Social and Cultural Rights by its States parties.

[31] HUMAN RIGHTS FOR ALL International Covenant on Economic, Social and Cultural Rights A Handbook 1st Edition 2015 PWESCR (Programme on Women's Economic, Social and Cultural Rights) New Delhi Page 1

the CESCR issued its General Comment No12[32] on the Right to adequate food. This General Comment aims to identify some of the principal issues which the Committee considers to be important in relation to this right. Currently this is the most commanding source of the right to food under the United Nations Human Rights jurisprudence and mirrors the present status of the right under international law. However, after issuing this General Comment 12 dealing with right to food, CESCR has issued six more General Comments dealing with various other human rights dealt under the Covenant, which demands that each of these General Comments cannot be read in seclusion but must be considered and construed in a coherent manner.

#### 6.3.1.1 General Comment No12

The General Comment No 12 of the Committee dealing with 'Right to Adequate Food' provides that the human right to adequate food is of crucial importance for the enjoyment of all rights. As per the tenors of this General Comment 12 the Right to adequate food applies to everyone, thus the reference in Article 11(1) of ICESCR to 'himself and his family' does not connote any restriction upon the applicability of this right to individuals or to female headed households[33]. The committee insists that the right to food should not be interpreted in a restrictive manner and observed that the protection of this right requires that states to take all possible steps to reduce infant mortality and to increase life expectancy, especially in adopting measures to eliminate malnutrition and epidemics. The Committee affirms that the right to adequate food is indivisibly linked to the inherent dignity of the human person and is indispensable for the fulfillment of other human rights enshrined in the International Bill of Human Rights and casts a duty on the state to provide food even in times of natural or other disasters[34].

The Committee also insists upon providing unadulterated & safe food which meets the dietary needs of the people[35]. The committee has gone a step further to define dietary needs[36]. The Committee also advocates for the provision of food which is free from adverse

---

[32] DOC.E /C.12/1995/5 of 12th May 1999 Available at https://www.refworld.org/pdfid/4538838c11.pdf (accessed on 12/3/2017 at 4.32pm)
[33] *ibid* Para 1
[34] *ibid* Para 6
[35] *ibid* Para 8
[36] 'the diet as a whole contains a mix of nutrients for physical and mental growth, development and maintenance and physical activity' Ibid Para 9

substances which meets requirements of food safety and for a range of protective measures to be undertaken by both public and private players to prevent contamination of foodstuffs through adulteration and/or through bad environmental hygiene or inappropriate handling at different stages throughout the food chain and requires that, care must also be taken to identify and avoid or destroy naturally occurring toxins[37]. While discussing the obligations of the state, the Committee has mandated every State to ensure for everyone under its jurisdiction, have access to the minimum essential food which is sufficient, nutritionally adequate and safe, to ensure their freedom from hunger[38].

One of the essences of Human rights is the ideology of non discrimination which is depicted in Article 2(2)[39] and Article 3[40] of ICESCR. This value of non discrimination is imbibed in the instant General Comment[41] and establishes that any discrimination on any ground in access to food is a flagrant violation of the Covenant. The General Comment not only speaks about making the food available during peace times but also casts a duty on the state to make the food available even during the war times and in times of internal disturbances[42].

### 6.3.2 Obligations of the State Parties under the Right to Adequate Food

The Covenant has clearly spelt out the nature of the legal obligations of state parties in Article 2 of the ICESCR and the same has been very well articulated in the Committee's General Comment No 3(1990)[43]. The Committee has imposed three types of obligations at three levels on States parties: the obligations *to respect*, *to protect* and *to fulfil.*

---

[37] *ibid* Para10

[38] *ibid* Para 14

[39] Article 2(2) The States Parties to the present Covenant undertake to guarantee that the rights enunciated in the present Covenant will be exercised without discrimination of any kind as to race, colour, sex, language, religion, political or other opinion, national or social origin, property, birth or other status.

[40] Article 3 The States Parties to the present Covenant undertake to ensure the equal right of men and women to the enjoyment of all economic, social and cultural rights set forth in the present Covenant.

[41] Para 18 Furthermore, any discrimination in access to food, as well as to means and entitlements for its procurement, on the grounds of race, colour, sex, language, age, religion, political or other opinion, national or social origin, property, birth or other status with the purpose or effect of nullifying or impairing the equal enjoyment or exercise of economic, social and cultural rights constitutes a violation of the Covenant.

[42] Ibid Para 19 Violations of the right to food can occur through the direct action of States or other entities insufficiently regulated by States, the prevention of access to humanitarian food aid in internal conflicts or other emergency situations.

[43] The nature of state parties obligations. Doc. E/1991/23 of 1 January 1991.Availablwe at https://www.refworld.org/docid/4538838e10.html(accessed on 12/3/2018 at 6.00pm)

a. **Respect**

Para 15 of the General Comment 12 enunciates that 'the obligation to respect existing access to adequate food requires states parties not to take any measures that result in preventing such access. The obligation to respect requires the state and all its organs, institutions, agents to refrain from indulging in anything that has the effect of violating the integrity of the individual thus depriving him of his freedom to avail of the available material resources that satisfies the personal basic needs.

b. **Protect**

The obligation to protect this right requires the state parties to initiate such measures which ensure that the individuals are not deprived of the access to adequate food. The realization of this right in the true essence is possible only when the state parties endeavour to protect every human being against restrictions from third parties. State parties will have to undertake all necessary and obligatory means to protect the right to adequate to food of the affected and targeted population.

c. **Fulfil**

The duty to fulfil mandates the state parties to undertake the necessary measures to guarantee every one of the opportunities to obtain the satisfaction of those needs which results in fructification of the right to adequate food. More than the obligation to respect and protect, this obligation of fulfilling the right depends on various factors including the availability of required resources.

The Researcher strongly thinks that if there is no commitment of the state to fulfil its obligations, it might as well not undertake any obligation. The recommendations of the Committee urges the State to proactively engage in activities intended to strengthen people's access to & utilization of resources and means to ensure their livelihood, including food security. This obligation to fulfil is not only found in General Comment 12 but all the subsequent General Comments endorse the obligation of the state to ensure that the right to adequate food is considered even in the conduct of public affairs and in any vital decision making course of action. Finally, whenever an individual or group is unable, for reasons beyond their control, to enjoy the right to adequate food, States have the obligation to *fulfil* that right by providing food to the needy. This obligation also applies for persons who are victims of natural or other disasters. The committee has suggested various measures for implementation of this Right at the National level, for monitoring and for Remedies, in case of improper implementation of the right and accountability.

### 6.3.3 The Normative Content of Article 11of ICESCR

As already mentioned, the right to adequate to food can be found in Article 11 of ICESCR which includes two rights, one dealing with the right of everyone to adequate standard of living and the other fundamental right of everyone to be free from hunger. It is the observation of the Researcher that the elementary right under the Covenant is the 'freedom from hunger' as this is a basic right to enjoy all other economic, social, political, civil and cultural rights. The right to be free from hunger connotes the justiciable claim of every human being to food and also casts a corresponding duty on the states parties to employ the necessary action to ease hunger even during the natural calamities and other disasters if any. Article 11 also deals with the right to an adequate standard of living and the same is reflected in Para 6 of the comment[44], which means that this right cannot be interpreted in a narrow and restrictive sense limiting to the provision of a package of some calories, proteins and other essential nutrients. This statement enunciates the meaning of 'Normative content of the right to adequate food'. The Normative content of the right to adequate food involves the following essential factors, namely,

**a. Adequacy**

In General Comment No12, the idea of normative content of the right to adequate food is developed based on the concept of adequacy. Adequate food must be 'sufficient and acceptable within a given culture[45]. Adequacy also involves the dietary needs of an individual and the food should be free from any adverse or extraneous matter which means that the food must be safe for human consumption. This General comment very clearly states that food should be made available not only to present generation but also to future

---

[44] The *right to adequate food* shall therefore not be interpreted in a narrow or restrictive sense which equates it with a minimum package of calories, proteins and other specific nutrients. The *right to adequate food* will have to be realized progressively

[45] General Comment No12,Para 8

generations which means that the General comment dealing with Sustainability[46] should also be dealt with scrupulously for ensuring the Right to food. The Researcher is all praises for the Committee for its concern to see that the food is free from any adverse substances which might creep into the food by any possible way.

**b. Availability**

Para 12 of the General Comment No 12 speaks about the concept of availability of food[47]. But the Researcher is apprehensive about the long term availability of food as agricultural operations in most of the countries are restrained by ecological conditions like monsoon, drought and other natural disasters, coupled with climatic change which may negatively affect the production and distribution of the same to places of need.

**c. Accessibility**

According to Para 13 of the General comment12, accessibility entails both economic and physical accessibility[48]. The food should be made available to those who involve themselves in any economic activity which arms them with economic means to access food. These economic activities may be directly linked to the natural production of food or other resources or means of production. The normative content involving both economic and physical access to food establishes the entitlement to access. The General comment also observes that access to food must be comfortable and convenient even for a common man and the same should not take a toll on his other basic necessities, which means to say that food should be reasonably priced, must not be exorbitant.

General comment also enjoins the states parties to adopt special programmes to vulnerable groups such as landless persons, other impoverished segments of the society and

---

[46] *ibid* Para 7. The notion of ***sustainability*** is intrinsically linked to the notion of adequate food or food *security*, implying food being accessible for both present and future generations.

[47] Availability refers to the possibilities either for feeding oneself directly from productive land or other natural resources, or for well- functioning distribution, processing and market systems that can move food from the site of production to where it is needed in accordance with demand.

[48] Economic accessibility implies that personal or household financial costs associated with the acquisition of food for an adequate diet should be at a level such that the attainment and satisfaction of other basic needs are not threatened or compromised. Economic accessibility applies to any acquisition pattern or entitlement through which people procure their food and is a measure of the extent to which it is satisfactory for the enjoyment of the right to adequate food. Socially vulnerable groups such as landless persons and other particularly impoverished segments of the population may need attention through special programmes.
Physical accessibility implies that adequate food must be accessible to everyone, including physically vulnerable individuals, such as infants and young children, elderly people, the physically disabled, the terminally ill and persons with persistent medical problems, including the mentally ill. Victims of natural disasters, people living in disaster- prone areas and other specially disadvantaged groups may need special attention and sometimes priority consideration with respect to accessibility of food.

victims of natural disasters. Physical accessibility does not involve any credentials but implies that adequate food must be accessible to everyone. Physical accessibility to food must be unqualified. The General Comment 13 the CESR included 'non discrimination' as a new factor to the accessibility which proclaims that the food must be accessible to all including the most marginalized sections of the society without any discrimination on any of the prohibited grounds. Accessibility also encompasses the right to inquire about and give away any information relating to food issues. The information has a distinct meaning for the issue of hidden hunger[49], with regard to information on the nutritional substance of food. Access to information also applies to knowledge and awareness about right to food and related rights among the right holders, i.e. where to claim the right to food when violated[50].

The Researcher has noticed the fact that though the international community has frequently reaffirmed the importance of full respect for the right to adequate food, a disturbing gap between the standards set in article 11 of the ICESCR and the situation prevailing in many parts of the world then, stares straight into our face and the Researcher is convinced that implementation of the Covenant was not up to the mark. The committee itself has admitted that more than 840 million people throughout the world, most of them in developing countries, were chronically hungry and millions of people were suffering from famine as the result of natural disasters[51].

"*Unfortunately, the said covenant cannot come to the rescue of even citizens of signatory states, because several states have not introduced the legislation ratifying the obligations under the covenant. Since the victims can only prosecute their governments under domestic law and not under the covenant. There is no redress available to the citizens aggrieved and the states continue to violate obligations with impunity*"[52]. The acute problems of hunger and malnutrition are not dogging the developing countries alone but

---

[49] Hidden hunger does not refer to the overt and obvious hunger of poor people who are unable to feed themselves, but to the more insidious hunger caused by eating food that is cheap and filling but deficient in essential vitamins and micro nutrients

[50] *The "Breakthrough" of the Right to Food: The Meaning of General Comment No 12 and the Voluntary Guidelines for the Interpretation of the Human Right to food.* Sven Sollner. A von Bogdandy and R. Wolfrum,(eds), Max Planck Yearbook of United Nations Law, (2007)Vol11, p 391-415

[51] Ben Saul, David Kinley, Jacqueline Mowbray *The International Covenant on Economic, Social and Cultural Rights Commentary, Cases and Materials,* Oxford University Press, (1st edn, 2014) available at https://books.google.co.in/books?id=J0jbAgAAQBAJ&pg=PA869&lpg=PA869&dq=more+than+840+million+people+throug (accessed on 1/2/2017 at around 5.00pm)

[52] Dr K V Ravikumar, "Right to food in India-Whether a protection under fundamental rights", Indian Bar review,[July-September, 2012] VolXXXIX No3, p 39-48

even some of the most economically developed countries are also affected. Looking at this deplorable scenario, the Researcher understands that fundamentally, the roots of the problem of hunger and malnutrition are not lack of food altogether but on one hand lack of access to available food due to various reasons like poverty, ignorance and on the other, food is used as a political weapon by the states.

## 6.4 International Covenant on Civil and Political Rights, 1966 (ICCPR)

The ICCPR was adopted by the United Nations General Assembly on 16th December 1966 and entered into force on 23 March 1976. This Covenant consists of a Preamble and 53 Articles divided into 6 parts: optional protocol with 14 Articles adopted with the Covenant in 1966 and second Protocol with 11 Articles adopted in 1989[53]. In the Preamble, the Covenant has reiterated the freedom from want as propounded by the US President Franklin D Roosevelt in his Four Freedoms speech. Article 6 of the Covenant warrants specific guarantee of Right to Food. It provides for the inherent right to life of every human being[54]. Article 2 (2) gives guidelines to the state parties as to the implementation of the provisions of the Covenant domestically. The Covenant makes a distinction between Civil and Political rights. The civil rights cannot be denied while the state in the given situations such as Public emergency, when the life of the nation is at peril, may derogate from the obligations of the Covenant[55]. But, Civil rights such as right to life and right to be free from slavery cannot be denied[56] under any circumstance.

---

[53] James Crawford *Brownlie's Principles of International Law*, Oxford University press (8th edn 2013) p 233

[54] Article 6. 1. Every human being has the inherent right to life. This right shall be protected by law. No one shall be arbitrarily deprived of his life

[55] Article 4

1.In time of public emergency which threatens the life of the nation and the existence of which is officially proclaimed, the States Parties to the present Covenant may take measures derogating from their obligations under the present Covenant to the extent strictly required by the exigencies of the situation, provided that such measures are not inconsistent with their other obligations under international law and do not involve discrimination solely on the ground of race, colour, sex, language, religion or social origin.

2. No derogation from articles 6, 7, 8 (paragraphs I and 2), 11, 15, 16 and 18 may be made under this provision.

3. Any State Party to the present Covenant availing itself of the right of derogation shall immediately inform the other States Parties to the present Covenant, through the intermediary of the Secretary-General of the United Nations, of the provisions from which it has derogated and of the reasons by which it was actuated. A further communication shall be made, through the same intermediary, on the date on which it terminates such derogation.

[56] *Ibid*

### 6.4.1 Human Rights Committee

A thought process to protect and implement the Human rights universally culminated in the setting up of the Human Rights Committee[57] under this Covenant consisting of 18 members, who are nationals of states parties to the Covenant. This was the first of its kind in the history of Human Rights. The State Parties to the present Covenant were mandated to submit reports to this Committee within one year of the entry into force of the present Covenant on the measures they adopted in order to give effect to the rights recognized herein and on the advancement made in the enjoyment of those rights. Subsequently the states were required to send the report as and when the Committee so required indicating the factors and hiccups, if any, encountered in the implementation of the recommendations of the Covenant. By the verbatim of Article 41(1) (a)[58] it is clear that when any state party is not fulfilling its obligations enshrined in the Covenant, the same can be objected to by other states which are abiding by the obligations cast by this Covenant and an opportunity will be afforded to the defaulting state to explain its stand.

If the dead lock cannot be solved within six months then the same can be brought to the notice of the Committee[59]. In the opinion of the Researcher this provision gives an universal jurisdiction in the event of any violation of Human rights which is a welcome step.

---

[57] The Human Rights Committee is the committee, established under article 28, which monitors the implementation of civil and political rights under the ICCPR. The function of the Committee is to examine the reports of the state parties which they are mandated to submit under article 40, on the measures they have adopted to give effect to the rights enshrined in the Covenant and on the break through achieved in the enjoyment of these rights

[58] Article 41 1. A State Party to the present Covenant may at any time declare under this article that it recognizes the competence of the Committee to receive and consider communications to the effect that a State Party claims that another State Party is not fulfilling its obligations under the present Covenant. Communications under this article may be received and considered only if submitted by a State Party which has made a declaration recognizing in regard to itself the competence of the Committee. No communication shall be received by the Committee if it concerns a State Party which has not made such a declaration. Communications received under this article shall be dealt with in accordance with the following procedure:

(a) If a State Party to the present Covenant considers that another State Party is not giving effect to the provisions of the present Covenant, it may, by written communication, bring the matter to the attention of that State Party. Within three months after the receipt of the communication the receiving State shall afford the State which sent the communication an explanation, or any other statement in writing clarifying the matter which should include, to the extent possible and pertinent, reference to domestic procedures and remedies taken, pending, or available in the matter;

[59] Article 41(1) (b) If the matter is not adjusted to the satisfaction of both States Parties concerned within six months after the receipt by the receiving State of the initial communication, either State shall have the right to refer the matter to the Committee, by notice given to the Committee and to the other State;

In its General Comment No 6[60] dealing with Right to life under Article 6 of the instant Covenant, the Human Rights Committee observed that .....the protection of the right to life requires that states take all possible measures to reduce infant mortality and to increase life expectancy, especially in adopting measures to eliminate malnutrition and epidemics[61]. Under the First protocol of 1966 which has been in force from 1976, an individual can make a complaint of human rights violations to the Human Rights Committee[62] but this is possible only when the state has accepted the special jurisdiction of the Committee.

The Human rights Committee performs the vital function of monitoring the enjoyment of the rights set out in the Covenant which is a legally binding international treaty and it is the pre-eminent interpreter of the provisions of the Covenant. Over the years, the Committee's work has resulted in numerous changes in law, policy and practice, both at national level and in the context of individual cases which goes to establish that the Committee's discharge of the monitoring functions entrusted to it under the Covenant has improved the lives of individuals in countries in all parts of the world[63].

The Researcher has noticed that the rights contained in the Covenant are not absolute but suffer from some limitations. The Covenant though guarantees some of the basic rights and freedoms to the people, also confers on the state parties, the privilege of restricting themselves from the application of the provisions in entirety and they are also permitted to derogate from the obligations under certain circumstances. This 'pick and choose' attitude of the state parties can never be in the interest of the community at large.

Though the Committee is empowered to entertain inter-state claims under Article 41of ICESCR whereby a State party may bring to the Committee's attention that another State party is not fulfilling its obligations under the Convention, this procedure was never employed in spite of flagrant violations of the rights and freedoms given under the Covenant by many state parties. The parties to the Covenant were hand in glove to protect each

---

[60] HRI/GEN/1/Rev.9 (Vol. I). Available at https://tbinternet.ohchr.org/_layouts/ treatybodyexternal/ Download .aspx?symbolno=HRI/GEN/1/Rev(accessed on 13/2/2017at 4.30pm)

[61] *ibid.* Para5

[62] Article 1 of the First Protocol A State Party to the Covenant that becomes a Party to the present Protocol recognizes the competence of the Committee to receive and consider communications from individuals subject to its jurisdiction who claim to be victims of a violation by that State Party of any of the rights set forth in the Covenant. No communication shall be received by the Committee if it concerns a State Party to the Covenant which is not a Party to the present Protocol.

[63] http:// www.2/ohchr.org/eng/bodies/hrc (accessed on 10/7/2016 at 11.17am)

other's skin in tune with the saying 'never throw a stone at a glass house if you are living in one'. If this is the implication of the provisions of the covenant, the Researcher is apprehensive if this Covenant made any difference in protecting the right to health, right to life and other basic rights as enshrined in the Covenant.

## 6.5 UN Conference on Food and Agriculture 1943 and the Birth of Food and Agriculture Organisation in 1945

For the first time, a need was felt to establish a separate permanent body for agriculture and food at the international level. US President Franklin D Roosevelt was the force behind this and in continuation of the same, a United Nations Conference on Food and Agriculture was called at Hot Springs, Virginia, from 18th May to 3rd June 1943 in which Representatives of 44 nations participated. This conference was held when the II World war was still going on and security issues were still at stake. The term 'United Nations' in the title of the conference referred to the nations that were working together in the effort to win the war[64].

### 6.5.1 Objectives

As it is evident in the open sentence of the declaration that it had considered the global problem of food and agriculture and declared its commitment to achieve the goal of freedom from want of food, suitable and adequate for the strength and health of all people[65].

### 6.5.2 Outcome of the Conference

It was resolved to establish an Interim Commission with the power of deriving a specific plan for a permanent organisation in the field of agriculture and food. Apart from this primary task, the conference also adopted recommendations for the amelioration of national diets, mal nutrition and the resultant diseases, diets of vulnerable groups like children, women, displaced, dietary standards, cooperation of national nutritional organisations to achieve all these objectives, changes in production in the short term period, adjustment of production in the transition from the short term to long term period,

---

[64] http://www.fao.org/docrep/009/p4228e/p4228e04.htm (accessed on 4/7.2017 at 1.45pm)
[65] *ibid*

agricultural credit, cooperative movements, protecting and conserving water and land resources; special national measures for wider food distribution, intense expansion and earmarking of land for food production, special international measures for wider food distribution, occupational adjustments specially for rural population, accomplishment of an economy of plenty, additions to and improvements in marketing facilities, rising the effectiveness and reducing the cost of marketing and fish & marine products.

These varied recommendations overshadowed and camouflaged much of the important subject matters which ought to have been incorporated for the setting up a permanent body in the form of Food and Agricultural Organisation. This conference popularly known as Hot Springs Conference was a milestone in the history of UN's effort to boost agriculture which was the main vocation of many people in the world. This was the first maiden attempt in the endeavour to feed the growing population. It was decided in this conference that Interim Commission on Food and Agriculture should be set up in Washington by 15 July 1943 to implement all the recommendations of the conference for which the representatives of the governments who took part in the conference could designate a representative. The conference urged the United States to initiate necessary preliminary action for the establishment of the commission. The primary task of the interim commission was to draft a constitution for FAO. This commission was in place till FAO could be formally established in 1945. The first session of the FAO Conference was held at Quebec, Canada from 16 October to 1 November 1945. Its first task was to bring FAO formally into existence, according to the provisions of the Constitution the Interim Commission had drafted.

At the First session of the Conference 'COMMISSION A' was established to outline policies and a program for the FAO. This involved an examination of every field in which FAO may have a stake whether directly or indirectly like every aspect of the production, distribution, and consumption of the products of farms, forests, and fisheries. To accomplish this assignment the commission set up six committees and they were required to submit a report and one of the Committee was on Nutrition and Food management. It has been observed by this committee that a large portion of the world's population is undernourished and there is enormous need for more and better food still opines that much can be done without further delay in the field of nutrition and food management[66]. In Part B of the Report of Commission A to the conference has accepted that it is the primary objective of all the nations who are united in FAO to raise the levels of nutrition of the people throughout the world. It also urged all member nations to not only free the people from hunger, danger of starvation and famine but also ensure that they get the kind of diet essential for good health. In order to achieve this it has solicited the assistance of all member nations.

This was the maiden step at the international level which gave importance to food production and agriculture and it was clear from its objective, the commitment of the UN to fight out food scarcity the world over thus giving importance to agriculture. As on today Article V of FAO's Constitution dealing with the Council of Organisation provides for eight committees which in turn will have to assist the Council and one of the committees is Committee on World Food Security. The main function of this committee is to examine and make necessary recommendations for the effective and efficient implementation of the objectives, principles and guidelines to fulfil the international undertaking on world food security. It is very heartening to state that Binay Ranjan Sen from Assam, India, was elected as the Director General of FAO at the Third Special Session of the FAO Conference on September 18th 1956 and served in this post till the end of 1967. He is known to have undertaken lot of innovations during his tenure and the significant one was his contribution for development of Freedom from Hunger campaign and the holding of the maiden World Food Congress in 1963.

## 6.6 World Food Programme,1961(WFP)[67]

During 1950's there was no proper allocation of food articles in the world, some of the countries had surplus and many more countries had shortage and most of the countries were starving. When this was the reality, UN thought it fit to help those nations who were in need of food by transporting the same from the countries which had surplus. *"The General Assembly, recalling its Resolution 1496(XV)of 27 October 1960 and Economic and Social*

---

[66] http://www.fao.org/docrep/x5584E/x5584e06.htm#vi. report of commission a to the conference (accessed on 4/7/2017 at around 5.30pm)

[67] A/RES/1714(XVI) 1961 Available at http://www.un.org/en/ga/search/view_doc.asp?symbol (accessed on 12/3/2018)

*Council Resolution 832(XXXII) of 2nd August 1960 on the provision of Food surpluses to food deficient peoples through the United Nations System, Having considered the report of the Director-General of the Food and Agricultural Organisation of the United Nations entitled Development through food- A strategy for surplus utilization*[68]*, the report of the Secretary general entitled 'The role of the United Nations and the appropriate specialized agencies in facilitating the best possible use of food surpluses for the economic development of the less developed countries*[69] *and the joint proposal by the united Nations and the Food and Agriculture Organisation of the United Nations regarding procedures and arrangements for multilateral utilization of surplus food*[70].

*'Having reviewed the action taken at the eleventh session of the Conference of FAO on the utilization of food surpluses and specifically, its resolution of 24th November 1961 stating that subject to the concurrence of the General assembly of the United Nations, an initial experimental programme for three years, to be known as the World Food Programme should be undertaken........'.*[71].

In tune with the Resolution of United Nations General Assembly 1714 (XVI) of December 19th 1961 and resolution 1/61 of the 1961 Conference of Food and Agriculture Organization of the United Nations, on 24th November 1961, as a part of the United Nations System established World Food Programme on a 3 year experimental basis[72] . It began its operation in January 1963 and extended by a General Assembly Resolution 2095 (XX) of 20 Dec 1965, on a continuing basis for all time to come as long as there is need for multilateral food aid. The General Assembly and the FAO Conference endorsed resolution XXII of 16th Nov 1976 at the World Food Conference, by which the *United Nations/FAO Intergovernmental Committee (IGC)*, which provided guidance on policy, administration and operations of the Programme, reconstituted as the Committee on Food Aid Policies and Programmes of the UN/FAO WFP[73].

---

[68] Food and Agricultural Organisation of the United Nations, Rome 1961

[69] Official Records of the Economic and Social Council, thirty second session, Annexes, agenda item 8, document E/3509

[70] Official Records of the General Assembly, Sixteenth Session, Annexes, agenda item 28, document A/4907.

[71] 1714(XVI). World Food Programme available at page No 20 , at http://www.un.org/en/ga/search/view _doc. asp? symbol=A/RES/1714(XVI) (accessed on 5/7/2017 at 6.30pm)

[72] A/RES/1714(XVI) 1961 Resolution adopted by the general assembly during its sixteenth session

[73] https://www.uia.org/s/or/en/1100053307 (accessed on 5/7/2017 at 12 noon)

As per the provisions of the General assembly Resolution 48/162, WFP was refurbished in 1995. By the General Assembly Resolution 50/227 (1995), the FAO and the WFP absorbed the functions of the World Food Council, which was established by the FAO initially but discontinued later[74]. As per the General Assembly Resolution 1714 (XVI), in 2008, WFP was metamorphosised from a food aid organisation to food assistance organisation as it felt that the disbursal of food vouchers, coupons and cash transfers are easier than the physical handling of food[75]. At present WFP is one of the Organs of United Nations linked with the General Assembly and United Nations Economic and Social Council. Till date WFP is the world's biggest humanitarian agency fighting hunger throughout the world.

**6.6.1Objectives**

WFP aims to save and protect lives in emergencies, prevent acute hunger, strengthens the ability of the countries to curb hunger, reduce chronic hunger and malnutrition, invest in disaster preparedness and thus mitigate the ensuing destruction, restore livelihoods following any wars or disasters or natural calamities where there is mass destruction, maintain food security coupled with nutrition and help the people, communities & countries thus affected to meet their own needs of food and nutrition.

**6.6.2 Working of WFP**

WFP pursues a vision of the world in which every man, woman and child has access at all times to the food needed for an active and healthy life and works towards that vision with its sister UN agencies in Rome-the FAO and the International Fund for Agricultural Development as well as other government, UN and NGO partners. In emergencies, it gets food to where it is needed, saving the lives of victims of war, civil conflict and natural

---

[74] https://www.unsceb.org/content/wfp (accessed on 5/7/2017 at 12.35pm)

[75] Since the late 2000s, a strategic rethink has seen the World Food Programme (WFP) shift from the concept of food aid to that of food assistance. While food aid is a tried and tested model, proudly woven into WFP history, it sprang from a largely unidirectional, top-down vision: people were hungry; we fed them. Food assistance, by contrast, involves a more complex understanding of people's long-term nutritional needs and of the diverse approaches required to meet them. This conceptual shift has been at the core of WFP's transformation in recent years. While we remain the world's leading humanitarian agency, we have evolved to combine frontline action with the quest for durable solutions. This shift is about recognizing that hunger does not occur in a vacuum. It means we must concentrate time, resources and efforts on the most vulnerable in society. It implies not just emergency interventions, but tailored, multi-year support programmes designed to lift a whole nation's nutritional indicators. We balance the urgency to alleviate hunger here and now with the broader objective of ending hunger once and for all. http://www1.wfp.org/food-assistance accessed on 5/7/2017 at around 12.22pm

disasters. After the cause of an emergency has passed, it used food to help communities rebuild their shattered lives[76]. On an average, each year this Food Programme reaches more than 80 million people with food assistance in 82 countries. 11,367 people work for the organization, most of them in remote areas, directly serving the hungry poor[77]. Its Food-for-Growth programs are directed at vulnerable groups including children, pregnant and nursing women, and the elderly and its Food-For-Work program encourages self-reliance by providing food in return for labour[78]. WFP meets people's food needs through cash-based transfers that allow the people we serve to choose and shop for their own food locally. On a daily basis WFP has 5000 trucks, 20 ships and 70 aeroplanes delivering food and other allied help to the needy. Approximately every year it distributes 12.6 billion rations at the cost of US$9.31/ration[79]. It also administers the International Emergency Food Reserve which is established by the General Assembly with a minimum target of 500,000 tonnes of cereals.

Due to all these efforts, this programme earned the unparalled reputation of being an emergency responder, the one that fixes the problem even in the most convoluted situation. The global community has adopted the 17 Global goals in 2015, for sustainable development to better the lives of common man by 2030[80]. Out of these 17 goals, Goal 2 deals with Zero Hunger and has pledged all its support to end hunger in general, achieve food security, improve the general nutritional levels, and accentuate sustainable agriculture, WFP along with the other agencies of UN is working tirelessly to bring the world closer to Zero hunger. WFP development projects focus on nutrition, especially for mothers and children, addressing malnutrition from the earliest stages through programmes targeting the first 1,000 days from conception to a child's second birthday, and later through school meals[81]. With its humanitarian food assistance programme, it is providing nutritious food to those who are in dire need of it owing to varied factors.

[76] https://www.wfp.org/about/ (accessed on 6/7/2917 at 1.00pm )
[77] *ibid*
[78] World Food Programme (WFP) UN, Karen Mingst, Professor of Political Science, University of Kentucky, Lexington, available at https://www.britannica.com/topic/World-Food-Programme (accessed on 5/7/2017 at 1.07pm)
[79] http://www1.wfp.org/emergency-preparedness-and-response( accessed on 5/7/2017 at 1.18pm)
[80] http://www1.wfp.org/zero-hunger (accessed on 5/7/2017 at 2.25pm)
[81] http://www1.wfp.org/overview (accessed on 5/7/2017 at 2.30pm)

Most of the work undertaken by this programme is in conflict affected countries where people are more prone to undernourishment than those living in other countries devoid of any conflict or vagaries. We have by and large witnessed that WFP is always the First in the scene of emergency or conflict or natural disaster or drought or floods, earthquakes or hurricanes providing food assistance to the victims. It has adopted modern day approach by meeting the food needs of the people through cash based transfers or food coupons which enable them to choose and buy food of their choice as the food habits are local and place specific. Apart from this egalitarian venture, the food programme also involves itself in other programmes to address the prime causes of hunger and encouraging the people to build resilience, so that it does not have to keep the same people and same lives every year. The Researcher is elated to state that The Comptroller and Auditor General of India was the external auditor of WFP till June 2016[82].

### 6.7 First World Food Conference, 1974

Till 1974 many international instruments honestly invested the efforts in seeing the right to food through. All instruments so far enunciated the responsibility of the states to safeguard, respect and protect this right, but the irony was that no efforts were made to spur up the agricultural production though the world was suffering from extensive food crisis mainly in Asia due to drought. All promises and commitments seemed to be superficial and were made in the vacuum in the light of dwindling agricultural production. In view of the serious imbalances and fluctuations in the world food and agriculture economy in 1972 and 1973[83], due to which both developed and developing countries suffered, radically illustrated

[82] http://www.cag.gov.in/content/world-food-programme-wfp (accessed on 3/7/2017 at 2.48pm)

[83] The world food situation in 1973 was more difficult than at any time since the years immediately following the devastation of the second world war. As a result of droughts and other unfavourable weather conditions, poor harvests were unusually widespread in 1972. Cereal stocks have dropped to the lowest level for 20 years. In the neiv situation of worldwide shortage, changes are occurring with extraordinary rapidity. Prices are rocketing, and the world's biggest agricultural exporter has had to introduce export allocations for certain products. World food production in 1972 was slightly smaller than in 1971, when there were about 75 million fewer people to feed. This is the first time since the second world war that world production has actually declined. There have 1101V been two successive years of poor harvests in the developing countries. After a series of encouragingly large harvests (especially in the heavily populated Far East) in each of the four years 1967-70, 1971 brought only a small increase in food production in the developing countries as a whole. In 1972 the Near. East was the only developing region to record a large increase, and with a substantial drop in the Far East (3 percent) no increase occurred in the total food production of the developing countries-page no 8. ..... It would indeed be a blessing in disguise if the precarious world food situation of 1973 could lead to the longer term measures that are required to ensure that such a situation can never occur again. It is intolerable that, on the threshold of the last quarter of the twentieth century, the world should fitzd itself almost entirely dependent on a single season's weather for its basic food supplies. For many years we have been protected

the need for new international approaches[84]. Hence the thought process began to primarily boost the production of food grains. Director general of FAO Addeke Boerma on February 1st 1973 addressed the problem of food crisis at the General Assembly, which resulted in the consensus of the member states that an emergency joint conference of FAO and UNCTAD should be convened to facilitate the formulation of an international co-operation programme to surmount the spiralling shortage of food and to maintain the stability of prices of food products.

Henry Kissinger, the US secretary of State in the Nixon and Ford administrations in a UN General Assembly statement on 24th September, 1973, apprehended the growing threat to the food supply the world over and opined that the General Assembly had to intervene immediately. He urged all states to join hands as he opined that this cannot be done by one state. It was after this that it was decided to organise a World Food Conference in 1974 under the aegis of United Nations to discuss and examine ways and means to maintain sustainable supply of food. The World Food Conference was convened under General Assembly resolution 3180 (XXVIII) of 17th December 1973. The problem of world security against food shortages had become increasingly serious because of important changes in the world cereal situation. Following unfavourable crops in several regions, cereal stocks had been drawn down to levels which gave no assurance of adequate supplies to meet world demand in the event of further crop failures or natural disasters[85].

An endeavour was also made to garner the support of all nations to meet the hunger and malnutrition which may result from natural calamities and natural disasters. Consequently it was agreed by FAO in its resolution 1831(LV) that the proposed conference should aim at resolving the food crisis in the larger context of the world at large to achieve the overall development. This was a timely initiative in view of the widespread depletion of world food stocks.

---

from such a situation by the large surplus cereal stocks accumulated in a few rich countries. These surpluses have now disappeared, and it can hardly be expected that the same countries will deliberately build them up again- Page no 9 The State Of Food And Agriculture 1973 http://www.fao.org/docrep/017/e1900e/e1900e.pdf ( accessed on 20/6/2017 at 12 noon)

[84] http://www.fao.org/docrep/x5590E/x5590e05.htm#Resolution3 (accessed on 21/6/2017 at 12 noon)

[85] http://www.fao.org/docrep/x5590E/x5590e05.htm#world food programme, para117 (accessed on 20/6/2017 at 11.00am)

The term 'Food Security' first originated in the mid 1970s, when for the first time in the World Food Conference 1974 defined food security in terms of food supply- assuring the availability and price stability of food articles both at the national level and international level. A thrust was placed to guarantee the availability of adequate world food supplies of basic foodstuffs at all times to sustain a steady expansion of food consumption and to counterbalance the fluctuations in production and prices if any. The World Food Conference examined the global problem of food scarcity, production and consumption. The conference set as its goal the eradication of hunger, food insecurity and malnutrition within a decade. Many proposals for ensuring food security were put forth by the various states and finally all these proposals were grouped under five main strategies[86]. First, the highest importance was given to ways and means of increasing the production of food in developing countries which would boost the average annual growth rate of food production from the present 2.6percent in the previous 12years to at least 3.6percent in the next twelve years. If this measure does not work then the developing countries as a whole might face scarcity of food products to the tune of 85million tonnes annually in normal years and about 100million tonnes during bad crop years. To increase the food production in developing countries four essential elements were recognized as: agricultural inputs in terms of fertilizers, water, further research in agriculture and technology aimed at the tropical and sub-tropical regions which has housed most of the undernourished people.

The second element was focussed on policies and different programmes for bettering the consumption patterns all over the world which should result in ensuring availability of adequate food mainly to vulnerable groups in developing nations. It was estimated that at least 40% of the approximate 460 million undernourished people in the world are infants. It was strongly opined that at least one quarter of this section should receive supplementary nutrition at a cost of 20dollars which would increase the intake of calories by 600 and 20gram of protein to every infant every day.

The third element was to reinforce world food security through initiatives which included a better early warning and food information system, efficient national and international stock holding policies and better measures for tackling emergency and food

[86] D John Shaw *World Food Security A history since 1945*, PALGRAVE MACMILLAN Houndmills, Basingstoke, Hamshire RG216XS and 175 Fifth Avenue, Newyork10010(1st edn2007), p 143

aid. The fundamental objectives of world food security were identified as ensuring that all countries can

• meet emergencies that occur in an uncertain world without a substantial cutback in supplies of basic foodstuffs to their populations;

• rely on the availability of supplies on commercial or concessional terms when formulating their own development strategies; and

• make agricultural production decisions in the knowledge of reasonable market stability and the continuance of stable trading relationships[87].

The fourth element was specific objectives and initiatives in the sphere of national trade and adjustment which were relevant to food problem including an attempt to stabilize and expand the markets for exports by developing countries. Trade was considered to have dual purpose in ensuring food security. The availability of basic food stuffs for import plays a major role in offsetting the problems of fluctuations in domestic production vice versa if there is excess food stuffs the same may be exported to the needy states.

The fifth element dealt with arrangements for enforcing the recommendations of the conference for which a new body was proposed to be established called 'World Food authority'[88] to implement or coordinate the implementation of the recommendations and decisions of the conference. This body was invested with three functions, it had to muster international financial assistance for development of agriculture in developing countries, had to assist in providing wider system of world food information and food security and aid the observance of International undertaking on World Food Security which was approved by FAO council and the conference.

The main challenge now was the effective implementation of the conference's resolutions. One area where the action of the conference had fallen short of expectations had been the short-term food problem. The most seriously affected countries needed at least seven to eight million tons of additional food grain in the next eight to nine months. Unless that amount was provided quickly, a large number of people would face starvation. The conference had resolved that 'within a decade no child will go to bed hungry, that no family

---

[87] *ibid*, p126
[88] *Ibid*

will fear for its next day's bread, and that no human being's future and capacities will be stunted by malnutrition'[89].

Non-governmental organizations from across the globe also participated in the conference and they also issued their declaration and delivered the same to the Secretary General of the conference[90]. The declaration emphatically stated that every human being has a right to regular supply of food adequate for total development, the need for basic nutrition for infants, people must not starve when there is sufficient food for all, hunger or food should not be used as a political weapon, food must be available as a manifestation of social justice and fundamental human right. It has been recorded that on Saturday 16th November 1974, Dr Swaminathan, Director-General of the Indian Council of Agricultural Research represented India and briefed about our efforts in food security in India[91].

### 6.7.1 Conference Resolutions

Consequently, the conference was a sham and could not arrive at a consensus on the grand strategy and the supposedly overreaching institutional arrangements proposed by the states and the preparatory committee for achieving a much hyped up food security but a reverberating 'Universal Declaration on the Eradication of Hunger and Malnutrition'[92] was adopted, in which the conference solemnly proclaimed that:

*'Every man, woman and child has the inalienable right to be free from hunger and malnutrition in order to develop fully and maintain their physical and mental faculties. Society today already possesses sufficient resources, organizational ability and technology and hence the competence to achieve this objective. Accordingly, the eradication is a common objective of all countries of the international community, especially of the developed countries and others in a position to help'*[93].

The Conference called for four major initiatives for follow up actions[94],

---

[89] *Ibid* p 139
[90] *Ibid*, p 142
[91] WGI/WFC/11 ANNEX VI Page 2, httpssearch.archives.un.orguploadsrunited-nations-archivesf3df3df5fb2c 6517891ce448adb184b33db03c061 page 42 (accessed on 21/6/2017 at 2.00pm)
[92] Adopted on 16 November 1974 by the World Food Conference convened under General Assembly resolution 3180 (XXVIII) of 17 December 1973; and endorsed by General Assembly Resolution 3348 (XXIX) of 17th December 1974
[93] http://www.ohchr.org/EN/ProfessionalInterest/Pages/EradicationOfHungerAndMalnutrition.aspx (accessed on 21/6/2017 at 2.10pm)
[94] https://search.archives.un.orguploadsrunited-nations- archivesf3df3df59fb2c6517891ce448adb 184b33db03c 0 61 (accessed on 21/6/2017 at 2.14pm)

One: The World Food Council, an umbrella-type structure to co-ordinate activities of various international agencies in the agricultural field. If approved by the United Nations General Assembly, the Council would be established at the ministerial or plenipotentiary level to function as an overall co-ordinating mechanism of policies concerning food production, nutrition, food security and food aid as well as the related matters by all the agencies of the United Nations system. Council membership would consist of Member States- of the United Nations or of its specialized agencies, nominated by the Economic and Social Council and elected by the General Assembly. The Council would be serviced within the framework of the Food and Agriculture Organisation at its Rome Headquarters.
Two: An international fund for agricultural development to channel as yet unspecified levels of investment towards the improvement of agriculture in the developing world with contributions to come on a voluntary basis from both traditional assistance - granting nations and those developing countries with ample means, provided that the United Nations Secretary- General determines, in consultation with nations supporting the fund, that sufficient resources are available to give its operations "a reasonable prospect of continuity".
Three: An international undertaking on world food security based on a co-ordinated system of nationally-held cereal reserves, supported by a world-wide food information and food shortage detection service.
Four: A commitment to provide, on a three-year forward plan basis, commodities and financing for- food aid to a minimum level of ten million tons of cereals each year plus certain other food commodities.

The United Nations World Food Conference ended in Rome on the night of 16 November 1974 with the adoption of various resolutions resolving to initiate measures for the establishment of international machinery to increase food production in the world and to channelize more basic foodstuffs to countries that have a shortage of them critically. The key Conference action namely the highlight is, calling for the establishment of a World Food Council as a co-ordinating organ within the United Nations system, received the approval of the Conference at a plenary sitting. The declaration acknowledged that the severe food crisis that was dogging the people at the developing countries was not only saddled with acute economic and social ramifications but also jeopardizes the very basic fundamental values and principles connected with the right to life and human dignity as enshrined in the

*Universal Declaration of Human Rights*[95]. "*the giant 1974 Rome food Conference ambitiously declared its intention to wipe out starvation within a decade, twenty three years later, despite the technological leaps made in food production, the world is still characterized by over 800million chronically under nourished people*"[96]

The Researcher believes that though the United States had an opportunity and the capacity to lead but it did not. Of course the conference was instrumental in usurping widespread interest in generating awareness about the problems of hunger and malnutrition all over the world and focusing the attention of the world at large on the deplorable human incapacity to quench the most basic need. We should accord due credit to this conference for its efforts to publicize the food issue and to institutionalise follow up action. Tim Albert has stated that the 'World food Conference was in many ways a gathering of the guilty'[97]. Global system of food reserves and better deal for agriculture were not secured.

The conference failed to take note of the fact that world food problem was essentially a global political concern of the highest magnitude which could neither be resolved through the efforts of technicians of ministers for agriculture. This conference utterly failed to address the world food security problem including measures to ensure access of the poor to the basic food that they needed. It is very disturbing to note that the Indian delegate who insisted at his press conference that 'nobody, yes nobody, was dying of hunger in his country'[98]. It is very unfortunate that such a mammoth conference miserably failed to address the problem of hunger and failed to institute any permanent mechanism to deal with this basic human right. The bureaucracy will carry on much as before.

### 6.8 Plan of Action on World Food Security-1979

After the World Food Conference in 1974, though there was some improvement in the matters of food at the world arena, in general the food security situation was as rickety as before. The recommendations that were adopted by 75 countries in the world food conference were scarcely implemented sacrificing major portion of it. There was no

---

[95] Adopted by the UN General Assembly in 1948 (UN General Assembly resolution 217 A (III)).

[96] Anuradha Mittal, S Anantha Krishnan, "Politics of Hunger and Right to Food", Economic and Poilitical weekly,[February 1-7, 1997] VolXXXII, No5, p201 -204

[97] '*Failure in Rome*' New Statesman 22 November 1974 httpssearch.archives.un.orguploadsrunited-nations-archivesf3df3df59fb2c6517891ce448adb184b33db03c061 (accessed on 23/6/2017 at 3.45pm) p 70

[98] *ibid*

agreement among the nations at the international level as to the release of national stocks and reserves which had been accumulated so far. Adding fuel to fire was the lack of transport and handling facilities at the time of crisis in most of the exporting and importing countries. The target of 500000 tons of cereals annually for the International Emergency Food Reserve had been partially met and the annual volume of food aid had not yet reached its target of at least 10 million tons from 1975 onwards as agreed at the World Food Conference 1974. Though the developing countries as a whole were able to increase their cereal production, their total stocks had in fact declined slightly since 1977. Out of 39 developing countries which had set the national targets for cereal stocks only 11were able to achieve this target. The resultant effect was most of the developing countries were facing acute shortage of food due to crop failure. Taking stock of the situation and to bail out the situation and to devise a plan of action for bettering the food security, the Committee on World Food Security held its fourth session from 5th April 1979 to 11th April 1979 at FAO headquarters in Rome. This session was presided over by Mr BS Raghavan, from India, the Chairman for the biennium.

In this session the world food security situation and the adequacy of the stocks were assessed on the basis of the Secretariat document CFS 79/7[99]. Upto date information was also provided by the delegates on food production and the supply and stock of food in their respective countries. The committee was convinced that certain aspects of food had improved in many countries owing partly to favourable weather conditions and cereals production world over had exceeded the consumption quantity. There was a significant rise in the investment in agriculture in developing countries which resulted in self sufficiency of food grains. Despite this buoyant increase in food production, food security had not been bettered in many regions like Southern Africa, as a result nutritional levels were precariously low. In the backdrop of this glaring reality, the committee initiated some of the steps for implementing the international undertaking to cure food problem and the lack of nutrition.

### 6.8.1 Steps Taken to Implement the International Undertaking

A. Proposed Plan of Action on World Food Security

[99] CFS 79/7 Assessment of world food security situation and adequacy of stocks. CL 75/10 APPENDIX C, List of Documents. http://www.fao.org/3/a-bn124e.pdf (accessed on 10/7/2017 at 1.04pm)

B. Action taken to adopt national cereal stock policies and targets in accordance with the undertaking.

A. Proposed Plan of Action on World Food Security

This Plan of Action consisted of series of measures which urged all governments to adopt a new grains arrangement with adequate stock, price and food aid policies with special reservation for developing countries as it is essential for the effective world food security system. Once this new grains arrangement was over, this Plan of Action was supposed to be reviewed.

I Adoption of policies governing stock of food grain

a. In conformity with the institutional and respective Constitutional requirements, all governments which have subscribed to the International Undertaking on World Food Security must adopt and implement national cereal stock policies and targets.
b. Governments especially of developed countries should encash the abundant food supply situation to build up the stocks in tune with the national stock objectives by the end of 1979 as foreseen by the Committee on World Food Security at its third session.
c. Governments ought to a take measures to maintain national food stock policies without adversely affecting the structure of domestic production or international trade and also taking into consideration the interests of developing countries which are heavily dependent on food exports.
d. This Plan of Action would later be reviewed and suitable further action would be decided depending on the need and situation.

II Criteria for management and Release of National Stocks held in pursuance of the undertaking Governments should apply the following general criteria to guide the national decisions on the release of stocks maintained to protect the food security.

a. National Stocks held in conformity with the undertaking should be handled in such a way that it contributes to the stability of markets and supplies. Governments were instructed to release these stocks[100] only in the event of crop failure[101], natural or manmade disasters or rise in prices, to

i. Maintain a regular inflow of food supplies both in international and domestic markets at prices which are fair and reasonable to consumers and lucrative to the producers.

ii. To overcome the surfacing of food scarcity

iii. To enable the developing countries to suit their import requirements without negatively impacting their economic development

b. The Director General relying on the Global Information and Early Warning System, in exceptional circumstances may caution the governments about the need for more extra supplies including the requirements of the developing countries that are also the importing countries to meet the emergency consumption requirements.

c. A special session of the Committee may be prorogued by the Director General to facilitate the governments to consider any special action required to meet the sharp and large scale food scarcity.

III. Special measures to assist low-income food deficit countries to meet the current import requirements and emergency needs

i. All donor countries should strive to increase their commitment to the levels as contemplated by the draft Food Aid Convention which was pending before UN Conference

ii. Considering that the developing countries import only a small quantity of cereals through food aid and many of them are confronted with ever increasing gap and balance of payments difficulties, the annual food aid target of at least 10 million tons of cereals should be revaluated by the Committee on Food Aid Policies and Programmes, in the light of FAO Secretariat estimation that by 1985food aid needs would be about 15-16 million tonnes.

---

[100] Specific criteria for management and release of stocks may depend and differ from country to country, depending on the already established rules or guidelines. Page No 11, Document No CL 75/10 http://www.fao.org/3/a-bn124e.pdf (accessed on 11/72017 at 10.00am)

[101] *ibid* An abnormally large decline in the national cereal harvest caused by serious drought, heavy rains, severe floods, pests, plant diseases or other natural hazards which leads to a large scale disruption of the flow of supplies to markets.

iii. International monetary fund should be approached to finance the additional balance of payments for meeting the surge in food import bills of low income and food deficit countries, especial whenever there is domestic shortage of food and rise in the import prices.

iv. Those countries who are in a position to contribute to the international emergency food reserve should do so to achieve the minimum annual target of 500000tons by 1979.

v. The food aid donor countries were required to establish food aid reserves or initiate other measures to maintain the continuity of food aid during the times of shortfalls in supply and high prices.

vi. Apart from providing various assistance to food–exporting developing countries and developed countries, concerned international organisations were also instructed to consider their interests to provide extra assistance to these countries like purchasing food from these countries.

IV. Special arrangements for food security assistance.

i. In order to facilitate the developing countries to effectively take part in this undertaking, the developing countries were required to f give high priority to the framing and implementation of national food security programmes, the developed countries and the other concerned international organisations were enjoined to provide required financial assistance to developing countries to enable them to implement their national food security programme.

ii. The committee on world food security must review the activities of Food and Agriculture Organisation's Food security assistance scheme, review the action initiated to meet the requests of developing countries for assistance to maintain and implement domestic food security.

iii. The committee on world food security in its fifth session was asked to review the necessity to establish a sub-committee to carry out all the functions effectively.

iv. Meeting of interested food aid donors and developing countries should be convened at the country level whenever there is a need for external financing and technical assistance by the developing nations.

V. Collective and self reliance of developing nations

The international community should foster collective self reliance of developing countries and to tackle this effectively, the developing nations were required to establish cooperative arrangements which included the setting up of regional reserves to meet emergency requirements of food. Concerned international financial and technical organisations and the developed countries were urged to extend a helping hand to support the developing countries in this venture of collective self reliance.

B. Action taken to adopt national cereal stock policies and targets in accordance with the undertaking

The Committee reviewed the steps initiated by governments to adopt cereal stock policies in line with the undertaking, based on the document CFS 79/4[102], which contained information on the national stock policies and stock targets in 114 countries. The committee was elated that the number of countries with national grain stock policies had shot to 70 which included almost all the grain importing countries and 45 countries with food security objectives, had adopted explicit grain policies, out of which13 were developed countries and 32 were developing countries. The Committee appreciated the endeavour of the developing nations to institute and execute stock policies and targets in view of the food scarcity due to various reasons.

### 6.8.2 Recommendations of the Committee

The Committee made the following recommendations

i. The governments which had not provided the national data of food policies, food stocks had to do so
ii. The Director general was required to invite the governments of the major food producing as well as importing countries that had not joined the system to participate and provide relevant information which included China and USSR.
iii. To perk up the flow of data particularly from those regions which are acutely vulnerable to crop failure, the governments of developing nations were enjoined

[102] Action taken to adopt national cereal stock policies and targets in accordance with the international undertaking on world food security. CL 75/10 APPENDIX C, List of Documents. http://www.fao.org/3/a-bn124e.pdf (accessed on 11/7/2017 at 10.40am)

to setup and to further strengthen national early warning and crop forecasting systems.

iv. Through multilateral channels and bilateral programme, the developed countries along with the international organisations mainly FAO, had to provide necessary assistance to the developing nations to set up national early warning and crop forecasting systems.

v. The secretariat had to continue to better the methods of storing the food articles, the accuracy of its trade forecasts and also its assessment of the effect of weather on crops.

vi. The secretariat was to continuously assess programme priorities in the light of the general food situation prioritising monitoring the potential food emergency situations.

The committee holistically agreed that it should indulge in periodic review of the functioning of the system to examine if there was any new development and if there was any need for reassessing the programmes priorities.

### 6.8.3 Future Programme of the Committee

In course of the session, number of the suggestions concerning the matters which had to be taken up and discussed at the future session were made, namely, to improve the assessment of world food security and the adequacy of food, it was resolved that long term trends and prospects with regard to production, consumption and trade in basic foods need to be analysed and looked into. It was the need of the hour to analyse the growth of population and food production trends and the nutritional status in specific countries. The delegates had also stressed the need for improving the analysis in the light of the estimates of stock movement in the USSR and China. The Committee may review the steps undertaken by governments and international organisations to implement the Action Plan as a whole. The committee agreed that its Fifth session should be held in March/April 1980 at FAO headquarters.

## 6.9 The World Food Day Established-1981[103]

[103] WA/RES/35/70 1981 Available at http://www.un.org/en/ga/search/view_doc.asp?symbol (accessed on 12/3/2018 at 4.30pm)

On 16th October 1945, at Quebec in Canada 42 countries came to a consensus to constitute the Food and Agriculture Organization (FAO) of the United Nations which was a milestone step towards fighting out the man's perennial struggle against hunger and malnutrition which has been a bane in any civilized society. International society as a whole is fighting tooth and nail against this malady from time immemorial. This was an earnest step in helping themselves and many other nations with a system through which its member nations could handle a similar set of problems which was a matter of grave concern to all people in all the countries of the world. Considering that food is a pre requisite for survival and well being of human being, the General Assembly deemed it fit and appropriate to designate a day in a year as a World Food Day. In its 83rd Plenary meeting dated 5th December 1980, it was unanimously decided by the Conference of the Food and Agriculture Organisation of the United Nations, for the first time, to observe 16th October 1981 as the *'World Food Day'* and every year thereafter on the same day as per the General Assembly Resolution A/RES/35/70[104]. Food and Agriculture Organization celebrates World Food Day each year on 16th October, to mark the day on which the organization was founded in 1945. The General Assembly urged all nations, governments and international organisations to put in more efforts for the efficient and successful implementation of this World Food Day to the achievable maximum extent. The objectives of World Food Day are to[105]

- Encourage attention to agricultural food production and to stimulate national, bilateral, multilateral and non-governmental efforts to this end;
- Encourage economic and technical cooperation among the developing countries;
- Encourage the participation of rural people, particularly women and the least privileged categories, in decisions and activities influencing their living conditions;

---

[104] 35/70. World Food Day,
*The General Assembly,*
*Considering that food is a requisite for human survival and well being and fundamental human necessity,*
*Welcomes the observance of World Food Day, to be held for the first time on 16th October 1981 and annually thereafter, as unanimously decided by the Conference of the Food and Agriculture Organisation of the United Nations at its twentieth session;*
*Urges Governments and national, regional and international organizations to contribute to the effective commemoration of World Food Day to the greatest possible extent.*
*83rd Plenary Meeting, 5th December 1980.*
Available at http://www.un.org/documents/ga/res/35/a35r70e.pdf (accessed on 11/7/2017 at around 2.31pm)
[105] http://www.fao.org/world-food-day/background/en/( accessed on 11/7/2917 at 4.33pm)

- Heighten public awareness of the problem of hunger in the world;
- Promote the transfer of technologies to the developing world; and
- Strengthen international and national solidarity in the struggle against hunger, malnutrition and poverty and draw attention to achievements in food and agricultural development.

In 150 countries across the world, various events and programmes are organised on this day to promote general awareness worldwide and action for those who are deprived of basic necessity and suffer from hunger. These programmes also create awareness about the need to ensure food security and nutritious diets for all though it is a necessity ironically the same has become a luxury to the major portion of the population in the world. These celebrations and programmes make it one of the most celebrated days in the UN calendar. World Food Day is a chance to show our commitment to Sustainable Development Goal (SDG) 2- to achieve Zero Hunger by 2030[106]. It is also a day for us to introspect as to how far we have travelled in our endeavour of reaching and attaining Zero Hunger.

### 6.9.1 Value Underlying the World Food Day

The right to food is a minimal human right. Empowering sustainable food systems and rural development which are two inseparable concepts, two faces of the same coin addresses one of the primary universal challenges, beginning from feeding the world's ever growing populace to protecting the global climate thus solving some of the main causes of migration and displacement. It is high time we all realise that achieving the 17 Sustainable Development Goals cannot materialise unless we end hunger and paucity of food. The Researcher believes that the contributing factors for the failure to achieve these goals are absence of sustainable and resilient, climate compatible with agriculture and food production systems which can ensure the delivery of food for the people of the planet.

### 6.9.2 Achievements Due to the Concerted Efforts of World Food Day and Various Programmes Initiated by FAO

Out of the 129 countries under the ambit of FAO, 72 countries have the distinction of having already achieved the target of halving the proportion of people who were starving by 2015. In the past two decades the chance of an infant dying before the attainment of the age of five has been astonishingly brought down to half. Extreme poverty instances have been

[106] http://www.fao.org/world-food-day/2016/history/en/ (accessed on 11/7/2017 at 4.58pm )

reduced to half since 1990[107]. All these achievements were possible only because of observance of programmes and schemes in lieu of World Food Day by FAO.

The 2018 Global Hunger Index report published by Welt Hunger Wilfe and Concern Worldwide, raises concern about whether the world can achieve Sustainable Development Goal 2, which seeks to end hunger, ensure food security and improved nutrition and promote sustainable agriculture by 2030[108]. In conclusion, the Researcher opines that mere observance of World Food Day may not cure poverty and hunger. In Fact not many of us are even aware of the observance of this day. The observance of this day should be marked by initiating more poverty alleviation programmes and generating more employment opportunities by all the nations in tune with the true spirit of fighting and destroying this problem with its roots. The manifestation of this malady is varied and needs a multifaceted approach and not a mere dedication of the day in a year.

### 6.10 Convention on the Rights of the Child, 1990

We have witnessed that in the history there have been various international treaties and agreements that dealt with the rights of the child. Initially in 1924 the league of nations had adopted the Geneva Declaration of the Rights of the Child. The United Nation's first initiative to take note of the child's Rights was in 1953 by establishing the United Nations International Children's Fund (UNICEF). Two years subsequent to this, United Nations adopted the Universal Declaration of Human Rights 1948, making it imperative that the children's rights need to be protected. "The first UN document specially focused on child's rights was the Declaration on The Rights of The Child, but instead of being a legally binding document it was more like a moral guide of conduct for governments"[109]. The first legally binding document concerning child rights was United Nations Convention on the Rights of the Child 1989, which put an end to a process lasting for almost ten years. This Convention established 4 principles to govern and control the implementation of all the rights enunciated and they were non discrimination, best interest of the child, right to life, survival and

[107] The Committee on Economic, Social and Cultural Rights General Comment No. 12: The Right to Adequate Food

[108] K N Ninan, "Hunger amidst prosperity: of what use is fast growth?", Deccan Herald Dated 29/10/2018 Bengaluru Edition

[109] http://childlineindia.org.in/United-Nations-Convention-on-the-Rights-of-the-Child.htm (accessed on 17/12/2017 at 4.30pm)

development and respect for the views of the child. The implementation of the Convention is undertaken by '*Committee on the Rights of the Child*'.

India has also declared its commitment to protect the rights of the child in its Declaration: "While fully subscribing to the objectives and purposes of the Convention, realising that certain of the rights of child, namely those pertaining to the economic, social and cultural rights can only be progressively implemented in the developing countries, subject to the extent of available resources and within the framework of international co-operation; recognising that the child has to be protected from exploitation of all forms including economic exploitation; noting that for several reasons children of different ages do work in India; having prescribed minimum age for employment in hazardous occupations and in certain other areas; having made regulatory provisions regarding hours and conditions of employment; and being aware that it is not practical immediately to prescribe minimum age for admission to each and every area of employment in India - the Government of India undertakes to take measures to progressively implement the provisions of article 32, particularly paragraph 2 (a), in accordance with its national legislation and relevant international instruments to which it is a State Party"[110].

It is anybody's guess as to the significant role played by nutritious food in the mental, emotional and physical development of a child. Infant malnutrition is a major contributory cause of high incidence of infant mortality and physical and mental handicaps. Mankind owes to the child the best it has to give[111]. States have an obligation to create and maintain adequate measures at the national level, in particular in the fields of education, health and social support for the promotion and protection of the rights of persons in vulnerable sections of their populations[112]. Persons with special needs have been designated as vulnerable section because of the danger to their health if the needs are not met[113] and one such group includes children.

---

[128] https://treaties.un.org/pages/ViewDetails.aspx?src=IND&mtdsg_no=IV-11&chapter=4&lang=en#EndDec (accessed on 17/12/2017 at 4.55pm)

[111] Preamble to UN declaration on the Rights of the Child, General Assembly Resolution of 20th November 1959

[112] Part –II, para 13 of the Vienna Declaration and Programme of Action, 25 June, 1993 adopted by the World conference on Human Rights held at Vienna from 14-25 June 1993

[113] Maxine E McDivit and Sumanti RI, Mudambi, '*Human Nutrition: Principles and Applications in India,* New Delhi : Prentice-Hall, (1969) p 115.

In the convention on the Rights of the Child 1990, two articles address the issue of nutrition. Article 24[114] says that the state parties recognize the Right of the child to the enjoyment of the highest attainable standard of health....(paragraph1) and shall take appropriate measures to combat disease and malnutrition through the provision of adequate nutritious foods, clean drinking water and health care(paragraph 2c). Article 24 says that state parties shall take appropriate measures to ensure that all segments of society, in particular parents and children are informed, have access to education and are supported in the use of basic knowledge of child health and nutrition.... Article 24 paragraph 2a requires that the state parties shall take appropriate measures to diminish infant and child mortality. Further Article 27 provides that state parties shall in case of need, provide material assistance and support programmes, particularly with regard to nutrition, clothing and housing.

This specific instrument on children indicates clearly the right of every child to be adequately nourished as a means to attaining and maintaining health. By nutrition, it means a balanced diet consisting of proper intake of protein, carbohydrates, fats, minerals and vitamins. Convention on the Rights of the Child is the most ratified international human rights treaty which clearly demonstrates the commitment of all nations to advance the rights of the children. This has been an inspiration for the countries to change their laws and practice which has resulted in the improvement of lives of million children.

The child by reason of his physical and mental immaturity needs special safeguards and care. Moreover, the child in all circumstances should be the first to receive protection and relief as provided under the principle of "First Call for Children, and which one of fundamental basis of the United Nations Convention on the Rights of the Child 1989[115]. As the foundation for a strong healthy adult is laid down during childhood, it becomes necessary to attach more importance to the early health care and nutritional needs of the child. Though it is almost been three decades since this convention has come to stay, yet the lives of many children are much to be desired for. Infant mortality especially that of a girl

---

[114] Article 24 (Health and health services): Children have the right to good quality health care – the best health care possible – to safe drinking water, nutritious food, a clean and safe environment, and information to help them stay healthy. Rich countries should help poorer countries achieve this

[115] Dr Harpal Kaur Khehra, "*Legal Regulation of Infant Foods in India*", MDU Law Journal [Vol 9, 2004],issue No 12, p 46-51

child is on the rise. Every 15 seconds a child dies of hunger[116]. 'This milestone instrument must also serve as an urgent remainder that much remains to be done'[117]. There are many children who do not enjoy their rights qualitatively as that of their peers. The Researcher thinks that Convention should become a guiding document for everyone globally.

## 6.11 First International Conference on Nutrition - World Declaration and Plan of Action for Nutrition: 1992

By 1992, it was 50 years since FAO was established. Due to the constant perseverance of FAO through its various schemes and programmes, global food production had been augmented considerably and the offsetting effect was, there was a significant reduction in the number of people suffering from under nutrition. Though these facts are something to be cheered about, it is estimated by Food and Agriculture Organisation that more than 800 million people are still deprived of access to enough food which should meet their daily dietary needs. Rubbing the salt on the wound is the fact that 40 percent of the world's population that is approximately, 2000 million people suffered from deficiencies in one or more micronutrients. At the same time hundreds of millions suffered from chronic diseases caused or exacerbated by excessive or unbalanced dietary intakes or from consumption of unsafe food and water[118].

In this backdrop, taking this crucial situation into consideration and in order to venture out to nip the world wide under nutrition problem mainly in developing countries, Food and Agriculture Organisation jointly with World Health Organisation prorogued the First Global Conference dedicated primarily and solely to address and redress the general universal nutrition problems. International Conference on Nutrition (ICN) was convened at head quarters of Food and Agriculture Organisation, Rome in December 1992. This Conference witnessed a representation of 159 countries along with European community, 15 United Nations Organisations and 144 non-governmental organisations.

---

[116] RuthAlexander' Does achild die of hunger every 10 seconds?'available at http://www.bbc. com/news/ magazine - 22935692 (accessed on 17/12/2017 at 5.11pm)

[117] https://www.unicef.org/crc/index_73549.html (accessed on 17/12/2017 at 5.17pm)

[118] FAO Documentary Repository available at http://www.fao.org/docrep/015/u9260e/u9260e00.pdf (accessed on 12/7/2017 at 1.15pm)

### 6.11.1 Preparatory Work for the Conference

Three years preceding the ICN, vigorous preparatory activities and agendas were undertaken throughout the world. Governments were asked to submit information with respect to food and nutrition situation in their respective countries describing the factors influencing the nutritional status of their people. Experts, policy makers and planners across the countries participated in preparatory meetings at regional and national levels. In one of the preparatory committee meetings that was held at World Health Organisation head quarters in Geneva in August 1992, all the government representatives considered and finalised the draft of *World Declaration and Plan of Action for Nutrition,* which was subsequently unanimously adopted at the International Conference on Nutrition.

### 6.11.2 Objectives of the Conference

- Ensuring continued access by all people to sufficient supplies of safe foods for a nutritionally adequate diet[119]
- To achieve the required nutritional status, it is imperative to make sure that there is a continuous and sufficient supply of safe food which can be easily affordable by a common man especially the not so privileged class and vulnerable groups in the society which guarantees nutritionally adequate diet to everybody. This was a crucial issue of utmost importance to all the people of the world who were afflicted with mal nutrition, constant hunger and diseases caused due to the deficiency of micro nutrients.
- Achieving and maintaining health and nutritional well being of all people[120].
- Good nutritional status depends on many factors including intake of adequate macro µ nutrients coupled with good health, availability of safe drinking water, having access to information and knowledge of proper diets to prevent diseases caused due to under nutrition and diet related non- communicable illnesses. In any development strategy or plans, the focal point must be human development and it is possible only when we achieve nutritional well being.
- Achieving developmental goals that are sustainable, environmentally sound and contribute to improved nutrition and health[121].

---

[119] International Conference on Nutrition 1992, Plan of Action for Nutrition
[120] *ibid*
[121] *ibid*

All governments should adopt sustainable and environmentally sound development policies and programmes which automatically lead to better nutrition and heath for people not only for one generation but for all time to come. The plans and policies must also focus on development in the areas of agriculture, food, family health, family welfare, population and education. The governments should aim its policies in such way that between population and available resources there is a balanced and harmonious relationship.

**6.11.3 Important Policy Guidelines**

- Commitment to promoting nutritional well being

All countries as an integral aspect of their developmental activities, ought to invest in a firm social, economic and political commitment to attain the motto of promoting the nutritional well being of its populace. The states should integrate nutrition objectives in all their programmes and schemes and carry out massive mass awareness programme. The different organs of the state and authorities under different ministries should work in harmony and coordinate with each other to develop nutritional status.

- Environmentally sound and sustainable development

Environmentally sound and sustainable measures should be adopted to ensure access to adequate and safe food supplies, which demands careful and meticulous planning to utilize the natural resources to constantly meet the nutritional needs of ever growing population and also generations to come. In order to achieve this first of all our farmers should be motivated to adopt efficient sustainable farming practices.

- Growth with equity: the need for both economic growth and equitable sharing of benefits by all segments of the population[122]

Development strategy should be oriented to accomplish economic growth with equity, ensure social justice and promote the well being of all focussing mainly on vulnerable section of the society. These strategies must allow equitable access to economic opportunities, adequate, safe and healthy food, clean water and sanitation.

- Priority to the most nutritionally vulnerable groups

Infants, small children, pregnant ladies and nursing women and the elderly and disabled in any poor household are the most nutritionally vulnerable groups, their nutritional well being must be prioritised by the government. Here female members of the society must be

122 *ibid*

accorded special focus as only a healthy mother can give birth to a healthy baby which has a far reaching effect on the society.

- Special focus on Africa[123]

Dastardly deteriorating nutrition status in Africa was alarming and is an image of vulnerability of the African population, which called attention of the international community for tangible and sustained support.

- People's Participation[124]

There cannot be any improvement in the nutritional status of general public unless they also take part in the programmes of the state and awareness should be created by the government to do so. If there is no local involvement in terms of families and households, no programme of the government can reach the poor and marginalized who are the targets of these programmes.

- Focus on women[125]

In most of the countries we can see patriarchal families which give importance to men and relegating the women to the back burner. In patriarchal families, even the nutritional needs of a women are neglected which in turn impacts their general health. Even lactating and pregnant women are not provided with nutritious food when they ought to be.

- Population policies[126]

Through its policies the governmental must keep the growth of population at such a level that it can provide adequate means of livelihood and their basic necessities are taken care of. While deciding the population policies, the government should also take into consideration its wealth of natural resources.

- Allocating adequate resources[127]

For achieving the objective of improved nutritional level, it is mandatory that the government provides much needed adequate financial, technical and other allied support for the effective implementation. Whenever any developing country is not in a position to

---

123 *ibid*
124 *ibid*
125 *ibid*
126 *ibid*
127 *ibid*

allocate the required support, the developed nations are urged to go to the rescue of such nations.

### 6.11.4 Strategy to meet the objectives

The goals set out in this conference was resolved to be achieved through the observance of the following strategies and actions. It was realised in this Conference that protecting and promoting nutritional well being of all can happen only with the involvement of various sectors at different levels of responsibility. The strategies are grouped under eight themes

- Incorporating nutritional objectives, considerations and components into development policies and programmes
- Improving household food security
- Protecting consumers through improved food quality and safety
- Preventing and managing infectious diseases
- Caring for socio-economically deprived and nutritionally vulnerable
- Preventing specific micronutrient deficiency
- Promoting appropriate diets and healthy life styles
- Assessing, analysing and monitoring nutrition situations

### 6.11.5 Implementation of the Recommendations of the Conference

As the ground reality in every country is different, the recommendations of the plan of action adopted in the conference need to be translated in the light of the available resources and must be supported by action at the international level. Taking all these factors into consideration, the national governments should prepare their suitable national plans of action. The responsibility of implementation lies on variety of players or participants beginning from government institutions till the individuals.

1. At National Level

At the national level, the national governments should formulate, adopt the plan of action as mandated in the conference and implement programmes and strategies in achieving the same. Ministries of agriculture, food, health social welfare and planning should prepare concrete proposals for their respective sectors to promote nutritional well being. People or the groups who are at the greater risk, should be supported by allocation of adequate resources, be it by the public or the private sector. Government should aim their policies towards the betterment of nutritional status.

2. At the International level

International agencies be it multilateral, bilateral and nongovernmental are implored to explain the steps that they would undertake in 1993, to contribute to the achievement of the goals and strategies set up in this conference. The governing bodies of FAO, WHO, UNICEF, the World Bank UNDP, UNESCO, WFP must during 1993 devised and finalised the ways and means to give appropriate importance to their nutritional related activities and programmes.FAO and WHO were required to jointly prepare, with the cooperation with member countries, the consolidated reports on the implementation of the International Conference on Nutrition Declaration and plans of action.

The Researcher thinks that International Conference on Nutrition was a milestone in the world nutritional history. So far, all the programmes of UN only focussed on right to food overlooking the nutritional aspects of thus provided food, all this while the national governments and international governments spoke about it in terms of quantity and for the first time they spoke about the quality of food, which ensures the intake of minimum quantity of nutrition in a day. This conference also initiated blanket measures to curb the incidence of diet related non-communicable diseases. Of course the conference made a beginning and toed the line of nutrition to be followed by the member nations by their follow up action in bringing about the nutritional well being of its people.

## 6.12 World Food Summit, 1996 - Rome Declaration on World Food Security and World Food Summit Plan of Action: 1996

In 1974, at the World Food Conference the governments had proclaimed that 'every man, woman and child has the inalienable right to be free from hunger and malnutrition in order to develop their physical and mental faculties'. This Conference had resolved to eradicate hunger, food insecurity and malnutrition within a decade. But due to the inevitable reasons, these goals could not be met owing to the lapses in policy making and funding. Consequently FAO understood the flip side of these consequences that unless progress is made in the direction of fighting out hunger and malnutrition, there could still be around 680 million hungry people in the universe by 2010 and more than 250 million would be in Sub-Saharan Africa alone. In this backdrop, the World Food Summit was called to rein in the continued existence of widespread malnutrition & hunger and accelerate agriculture to meet the needs of the future. The World Food Summit was held between 13th and 17th November 1996 at FAO head quarters Rome. This Summit went on for 5days holding meetings at highest level with representatives from 185 countries and the European Community. This Summit is also called Rome Declaration on World Food Security. This Summit witnessed the congregation of about 10000 participants and the platform was set for deliberations on one of the most significant issue that was haunting the world at large in the millennium, that day- the criticality of eradicating hunger.

**6.12.1 Objective of the Summit**

*'The Rome Declaration calls upon us to reduce by half the number of chronically undernourished people on the Earth by the year 2015..... If each of us gives his or her best I believe that we can meet and even exceed the target we have set for ourselves."* HE Romano Prodi, President of the Council of Ministers of the Italian Republic and Chairman of the World Food Summit[128].

The primary motto of this Summit was to revamp and rejuvenate the universal commitment at the highest political level to curb malnutrition and hunger and to accomplish sustainable food security for all people. This highest vision of the Summit resulted in raising awareness among the decision makers both at the private and public sectors encompassing the media and the society at large. It also lay down the political, conceptual and technical blueprint for an endeavour to exterminate hunger worldwide with the target of bringing down by half the number of undernourished, by not later than 2015[129]. This Rome Declaration set forth seven commitments which laid the basis for attaining sustainable food security for all and the Plan of Action mentions the objectives and modalities of implementation of the seven Commitments[130].

---

[128] http://www.fao.org/docrep/X2051e/X2051e00.htm (accessed on 14/7/2917 at 2.45pm)

[129] http://www.fao.org/WFS/ accessed on 14th July 2017 at 2.25pm

[130] *Convinced that the multifaceted character of food security necessitates concerted national action, and effective international efforts to supplement and reinforce national action, we make the following commitment:*

- *we will ensure an enabling political, social, and economic environment designed to create the best conditions for the eradication of poverty and for durable peace, based on full and equal participation of women and men, which is most conducive to achieving sustainable food security for all;*
- *we will implement policies aimed at eradicating poverty and inequality and improving physical and economic access by all, at all times, to sufficient, nutritionally adequate and safe food and its effective utilization;*

### 6.12.2 World Food Summit Plan of Action

1. The Rome declaration on World Food Security and the World Food Summit Plan of Action laid the foundations for varied paths to a common objective of attaining food security, at the individual, household, national, regional and global levels. As we have understood, food security exists when all people, at all times, have physical and economic access to sufficient, safe and nutritious food to meet their dietary needs and food preferences for an active and healthy life. This calls for a combined action at all levels. Every nation ought to espouse a strategy considering its resources and capacities to accomplish its individual goals and also cooperate regionally and internationally to muster collective solutions to common issues of food security[131].

2. Poverty eradication is crucial to improve the access to food. The undernourished do not have access to means of production and wars, civil strife, natural disasters & environmental degradation have negatively impacted millions of people in the world. Food assistance is not a long term solution to the underlying causes of food inadequacy instead, unbiased access to steady supplies of food should be ensured

3. A stable and peaceful political environment is a primary prerequisite for the realization of sustainable food security and it is upto the government to create such an enabling atmosphere so that common goal of food for all can be reached. This move should be accompanied by the joint efforts of members of society, agriculturists, fishers, foresters and other producers & providers of food[132].

---

- *we will pursue participatory and sustainable food, agriculture, fisheries, forestry and rural development policies and practices in high and low potential areas, which are essential to adequate and reliable food supplies at the household, national, regional and global levels, and combat pests, drought and desertification, considering the multifunctional character of agriculture;*
- *we will strive to ensure that food, agricultural trade and overall trade policies are conducive to fostering food security for all through a fair and market-oriented world trade system;*
- *we will endeavour to prevent and be prepared for natural disasters and man-made emergencies and to meet transitory and emergency food requirements in ways that encourage recovery, rehabilitation, development and a capacity to satisfy future needs;*
- *we will promote optimal allocation and use of public and private investments to foster human resources, sustainable food, agriculture, fisheries and forestry systems, and rural development, in high and low potential areas;*
- *we will implement, monitor, and follow-up this Plan of Action at all levels in cooperation with the international community.*

[131] Para1, 1http://www.fao.org/WFS/ (accessed on 16/7/2017at 1.27pm)

[132] *ibid* Para3

4. In most of the developing countries poverty, hunger and malnutrition are the predominant factors for migration of people from rural areas to urban areas in search of better pastures. Unless this crisis is tackled aptly the overall development of the country can be jeopardized which might in turn disturb the world peace. At the outset the countries must identify the causes and initiate remedial measures to improve the situation. It is imperative for any country to target all its policies towards people and regions which are seriously affected and suffering from hunger and malnutrition[133].

5. The agriculture production must be enhanced through the sustainable management of the available natural resources in order to feed the growing population. Enhanced agricultural production joined with food imports and international trade can reinforce food security. Long term investment in research and conserving & preserving genetic resources is very much needed[134].

6. It was resolved to reduce dangerous seasonal and inter–annual insecurity of food supplies. Progress should be made in minimising the susceptibility to and effect of, fluctuations in the climate& pests. For the transfer of supplies to areas which have deficit and the conservation & sustainable use of biodiversity, climate early warning system should be employed along with many factors like transfer and utilisation of proper agricultural technologies, production, and storage mechanisms. Both natural and manmade disasters can be predicted and prevented by employing suitable measures before hand[135].

7. Unless the national governments and the international community tackle the complicated causes of food insecurity, the number of malnourished and hungry people will shoot up mainly in the developing countries which is not acceptable. This Plan of action envisaged a continuous effort to exterminate hunger in all countries with the immediate aim of reducing the hungry and undernourished people to half of their number no later than 2015 and a midterm review as to scrutinize whether it is realistic to attain this target by 2010 was scheduled[136].

---

[133] World Food Summit Plan of Action Para 4

[134] *ibid* Para 5

[135] *ibid* Para 6

[136] *ibid* Para 7

8. The necessary resources for this venture were supposed to be mustered from domestic private and public sources. The international community has a predominant task of supporting the adoption of suitable national policies, providing technical and financial assistance to help the developing countries and the not so economically good countries in nurturing food security[137].

9. Follow up action to this World Food Summit included measures at the national, intergovernmental and inter agency levels. The international community along with the UN System with all its specialised agencies to contribute immensely to the better implementation of this Plan of action and the FAO Committee on World Food Security (CFS) will have the duty to supervise the implementation of the Plan of action[138].

10. Achieving sustainable world food security is prerequisite for the attainment of social, economic, environmental and human development objectives as settled in the then recent international conferences. The goal of food for all, at all times, can be achieved only if all parties at local, national, international levels make unwavering & indomitable and sustained efforts[139].

11. This Plan of action is in consonance with the objectives and principles enshrined in UN Charter and it makes efforts to combine the results of other UIN Conferences since 1990 on subjects related to food security[140].

12. The implementation of the recommendations in the Plan of action is the sovereign right and responsibility of the respective states through their national policies, programmes, laws and priorities in compliance with human rights and fundamental freedoms, including the right to development and the importance and respect for numerous religious and ethical values, cultural backgrounds and philosophical subscriptions of individuals and their clans have got to contribute to the complete realisation by all of their human rights to conquer the objective of food for all[141].

---

[137] World Food Summit Para 8
[138] *ibid* Para 9
[139] *ibid* Para 10
[140] *Ibid* Para 11
[141] *Ibid* Para 12

"*The heads of state and Government gathered in Rome at the World Food Summit at the invitation of FAO, reaffirmed on November 13, 1996, the right of everyone to have access to safe and nutritious food, consistent with the right to adequate food and fundamental right of everyone to be free from hunger*"[142]. It was proclaimed in this summit that "*Everyone has the right to adequate food and the fundamental freedom from hunger*".

Despite this Summit and its Plan of action to reduce the number of hungry and malnourished by 2015, even today 800 million people in the world are hungry and malnourished. This fact makes the Researcher suspect the conviction and efforts of the international community and the national governments to achieve the objective of ensuring food for all and at all times, which again necessitated for further programmes, Conferences and Conventions.

### 6.13 Millennium Summit of the United Nations 2000[143]

The world was moving towards globalisation, wars were being fought, many people were killed in different wars which questioned the very existence and the efficiency of the UN to curb the war. At this juncture, the world leaders thought it fit to make the world, a better place for living for the people by attaining the overall development of all the countries thus attaining peace everywhere. World leaders resolved to strengthen the rule of law and guarantee the adherence of all nations with the decisions of Inter National Court of Justice to make the resources available to the United Nations for conflict prevention and the pacific means of resolution of disputes and to initiate action against the international drug problem and terrorism.

In the light of this scenario, the world leaders came to a consensus to hold a Millennium Summit in 2000. Accordingly, as per the General Assembly Resolution 55/2: United Nations Millennium Summit 2000 was held for 3days between 6$^{th}$ and 8$^{th}$ of September 2000 at New York, with the purpose of discussing the role of the United Nations at the turn of the 21st century. This was the largest ever gathering of all the world leaders. In the Summit, United Nations Millennium Declaration was adopted and the same was endorsed by 189 countries, who committed their support to a new global partnership to

---

142 Yogendra Kumar Srivastava, "Right to Food: A Human Right", The PRP Journal of Human Rights[June-March 2001],Vol 5 Issue 1, p11-13

143 Available at http://www.un.org/ga/search/view_doc.asp?symbol=a/res/55/2 (accessed on 1/2/2017 at 4.12pm)

reduce extreme poverty and this declaration set out a series of targets to be achieved by 2015. These came to be known as the Millennium Development Goals (MDGs)[144]. The Millennium Development Goals are the world's time bound and quantified targets for addressing extreme poverty in its many dimensions –income poverty, hunger, disease, lack of adequate shelter and exclusion-while promoting gender equality, education and environmental sustainability[145]. These Goals are also the basic human rights like right to health, shelter, education and security.

### 6.13.1 The Millennium Development Goals (MDG)

The Millennium Development Goals are eight goals with measurable targets and clear deadlines for improving the lives of the world's poorest people[146]. These goals range from providing universal primary education to preventing child and maternal mortality and the target achievement date being 2015. The Millennium Development Goal Fund contributed to the attainment of these goals, with the main theme being the eradication of extreme poverty. The Fund initiated an inclusive and comprehensive approach to accomplish the MDGs. The Millennium Declaration emphasises on development as a right, targeting the traditionally marginalised groups like women, ethnic minorities and indigenous groups.

The Millennium Development Goals[147] are to:

(1) Eradicate extreme poverty and hunger;

(2) Achieve universal primary education;

(3) Promote gender equality and empower women;

(4) Reduce child mortality;

(5) Improve maternal health;

(6) Combat HIV/AIDS, malaria and other diseases;

(7) Ensure environmental sustainability; and

(8) Develop a global partnership for development.

As far as the eradication of extreme poverty and hunger is concerned, it was resolved among the member countries that no effort would be spared to free the people from the scourge of abject poverty and to emancipate the people from dehumanising conditions of

---

[144] http://www.unescap.org/resources/ga-resolution-552-united-nations-millennium-declaration (accessed on 17/7/2017 at 3.00pm)

[145] http://www.unmillenniumproject.org/goals/ (accessed on 17/7/2017 at 3.13pm)

[146] http://www.mdgfund.org/node/922 (accessed on 17/7.2017 at 3.22pm)

[147] http://www.un.org/en/events/pastevents/millennium_summit.shtml (accessed on 20/7/2017 at 3.00pm)

severe poverty. The member countries wanted to make the right to development a reality, for everyone by freeing the whole lot of people from the shackles of want. It was agreed to undertake measures to create an environment both at the national and the global levels, which would be conducive to development and to elimination of poverty. Member countries agreed to lend more liberal development assistance to countries which are honestly employing their resources to reduce the poverty. It was further resolved that by 2015, the proportion of the people whose income is less than a dollar/a day and the proportion of people who suffer from hunger would be halved. It was also resolved to reduce the proportion of the people by half, who are unable to reach or to afford safe drinking water. It was solemnly agreed to forge strong partnerships with the private sector and with civil society organisations in pursuit of development and poverty eradication[148]. On a regular basis, General Assembly was to review the progress made in implementation of the provisions of this Declaration and the Secretary General could be required to issue periodic reports for consideration by the General Assembly which could form the basis for further action[149].

The Researcher thinks that the world has made noteworthy improvement in attaining many of the goals of the Declaration. Between 1990 and 2002 overall average income increased by approximately 21%[150]. The number of people in extreme poverty, declined by an estimated 130 million[151]. But the progress was not uniform across the world and with respect to all the goals. There were mammoth disparities across countries and within countries. We could observe that poverty was severe in rural areas compared to urban areas. Sub-Saharan Africa which was the epicenter of crisis, did not witness much of change for good with continuing food insecurity coupled with grueling poverty. Asia witnessed faster progress though hundreds of millions of people were living in extreme poverty. Even the fast growing countries failed to accomplish some of the non-income goals. Some of the other regions had mixed records.

---

[148] 55/2. United Nations Millennium Declaration available at http://www.un.org/millennium/ declaration/ ares 552e. htm (accessed on 20/7/2017 at 4.02pm)

[149] http://www.un.org/en/ga/search/view_doc.asp?symbol=A/RES/55/2 (accessed on 20/7/2017 at 8.54pm)

[150] '*The Millennium Development Achievements and Prospects of meeting the targets in Africa' 2008* Edited by Francis N wonwu, published by Africa Institute of South Africa available at https://books.google.co.in/books?id=tYR2s9xeG4oC&pg=PA20&dq=The+number+of+people+in+extreme+poverty+declined+by (accessed on 20/7/2017 at 9.51pm)

[151] http://www.unmillenniumproject.org/reports/why6.htm (accessed on 20/7/2017 at 9.14pm)

The MDGs were set to expire in 2015 but sadly many countries were not on the path to realise MDGs by 2015. Though the member countries at the summit had attempted to look beyond 2015, there was very little consensus on a global development framework as the role of civil society was crucial including the voices of those who were directly hit by the poverty. Despite these eight MDGs, we still have today 800 million hungry bellies to be fed. Extreme hunger and malnutrition remains a barrier to sustainable development and creates a trap from which people cannot easily escape. Hunger and malnutrition mean less productive individuals, who are more prone to disease and thus often unable to earn more and improve their livelihoods[152]. A profound change of the global food and agriculture system is needed if we are to nourish today's 795 million hungry and the additional 2 billion people expected by 2050[153].

### 6.14 World Food Summit 1996: 5years later[154]

Fidel Castro, the former Prime Minister of Cuba at the World Food Summit in November 1996, criticised the rich countries who spent more than 700$ billion a year on weapons but did very little to satisfy the people's daily need for food. He termed the summit's commitment to reduce the world hunger by half by 2015 'shameful.....if only for their modesty', in that they still accepted hunger in a world of plenty[155]. Even in 2002 there were still 815 million hungry people in the world of plenty, which in the opinion of many leaders including Cuban Foreign Minister, Sr.D.Felipe Perez Roque was a 'crime'[156]. There was not much progress in reducing hunger which was the agenda in the World Food Summit 1996. Many of the leaders felt ashamed that they could not achieve the target that they had set for themselves 6 years back. Even Millennium Development Goals set for curbing hunger had also miserably failed to take off and all the intended agenda only remained in the papers failing to materialise.

---

152 http://www.un.org/sustainable development/ wp-content/uploads/2016/08/2_Why-it-Matters_ZeroHunger _2p. pdf (accessed on 20/7/2017 at 10.01pm)

153 http://www.un.org/sustainabledevelopment/hunger/(accessed on 20/7/2017 at 9.57pm)

154 Available at http://www.fao.org/docrep/MEETING/005/Y7106E/Y7106E00.HTM (accessed on 22/7/2017 at 6.31pm)

155 http://www.gmwatch.org/en/news/archive/2002/2387-world-food-summit-2002-special (accessed on 22/7/2017 at 7.00pm)

156 *ibid*

### 6.14.1 Objective of the Summit

The objective of this summit was to inspect the advancement made in the efforts to eradicate hunger, the vow that was undertaken by 186 nations in the World Food Summit 1996 and devise various means to accelerate the same. The basic purpose of this Summit was to assess the statistical progress towards fighting out hunger and halving it by 2015. This Summit should have been held in November 2001, the fifth anniversary of the World food Summit 1996, but the same was postponed by the Italian Government owing to security reasons.

The Summit was attended by representatives from 179 countries and the members of the European commission. The delegates at the Summit accepted their failure to achieve the target that they had set for themselves in World Food Summit 1996. At the Summit it was consensually agreed to set up an intergovernmental working group to develop voluntary guidelines to accomplish the progressive realisation of the right to food. It was resolved to reverse the general decline in providing the assistance to agriculture and rural development in the national budgets of developing countries and to increase the funding granted by the developed countries and also by the international financing institutions. It was also unanimously agreed to make voluntary contributions to the FAO Trust Fund on Food Safety and Food Security. It was admitted that the vast majority of the hungry and impoverished live in rural areas and the countries should promote the alleviation and emancipation of rural poverty through sustained growth of agricultural production[157].

It confirmed that FAO had a prominent role to play in assisting the countries in the implementation of the provisions of the World Food Summit. Many participants at the Summit expressed their anguish that the advocates of food security and poverty eradication are frustrated with the inability of the political leaders to acknowledge the basic structural problems inherent in globalisation. None of the Food Summit could answer some of the basic questions like why, in a world whose food supplies are increasing and adequate to feed everyone, are one-seventh of these people are denied of their basic right to food. One of the Seven Commitments of World Food Security 1996 was to expand trade in food and agricultural products. The participants deemed this to be a panacea for overcoming food insecurity and thought that more free trade in food and agricultural products would ensure

[157] http://www.fao.org/docrep/MEETING/005/Y7106E/Y7106E09.html (accessed on 27/7/2017 at 3.41pm)

that everybody would be fed and would be relieved of the poverty. But studies conducted since 1996 clearly show that such trade expansion had hurt small farmers in the developing countries and did not reduce the food insecurity in any way[158].

"Apart from Berlusconi, the only common link between the two Summits was the military operation of 16,000 police, carabinieri and soldiers put in place to contain the politicians and exclude the people. Many people from civil society were unable to enter the exclusion zone of half a kilometre around the building, which kept away the 30,000 persons "March for Food Sovereignty: Land and Dignity" organised by Italian social movements with the Civil Society Forum for Food Sovereignty at the start of the Summit. But some intended participants could not even enter the country, because of increased visa problems. FAO became a military zone. And this emphasised the sense of oppression in the Summit. And the one leader who came, Italy's Premier, Silvio Berlusconi, as Chair of the Conference, terminated the final session two hours early so he could watch the football! As the French newspaper Liberation said: FAO was caught playing "off side!"[159]. The final declaration of the Summit "The International Alliance against Hunger" reaffirmed the same recipe but now spiced up little with biotechnology and with the mere words about how the governments should deal with the lack of political will to combat hunger although it did not prescribe any new legally binding measures and also it failed to commit the rich countries to lend more help to the poor countries.

After all the pomp and gaiety, the Summit practically vanished from the news after its extravagant banquet. The summit eventually failed to achieve much as it did not deal with the innermost part of the systematic issue that affected the food insecurity that is unequal power of transnational agricultural business corporations. Social justice demands that we iron out these inequalities. The Researcher genuinely thinks that we should not have any more food Summits because 1974 World Food Conference agreed to abolish hunger in 19years and 1996 World Food Summit swore to halving the malnourished by 2015, but these undertakings and commitments have had little impact on the number of hungry and

[158] http://www.worldhunger.org/articles/global/Endingwhunger/McLaughlin.htm (accessed on 27/7/2017 at 5.07pm)
[159] Patrick Mulvany, ITDG, *Hunger - a gnawing shame* Report from *World Food Summit: five years' later* available at http://www.ukabc.org/wfs5+report.htm (accessed on 27/7/2017 at 5.15pm)

malnourished and the Researcher wonders what the next Food Summit would undertake to do.

"Together with terrorism, hunger is one of the greatest problems the international community is facing," said Italian Prime Minister Silvio Berlusconi, Chairperson of the Summit, at the closing ceremony. FAO Director-General Jacques Diouf, noting the 1996 World Food Summit goal of a reduction by half of the number of hungry people in the world by 2015, said, "*There is a large global consensus on measures to fight hunger. I am still optimistic that the target can be reached by 2015. It is in the interest of all countries to establish a more equitable world*"[160].

The Summit gave no succour to the hungry. Civil Society, including farmers' organisations, rejected this Declaration and noted that it was not a 'lack of political will' but 'too much political will' to establish a global hegemony for trade liberalisation, industrial agriculture, genetic engineering and military dominance that are the main causes of hunger[161]. It is shameful that almost one quarter of the world's population went to bed hungry despite producing enough food to feed everyone in the world. This may be annoying the rich but tragic for the poor. We have witnessed that governments have done little to eradicate hunger and the corporations are continuously extending their control on who gets to eat as the Researcher notices that 'feeding the poor is of little profit or no profit at all' to be precise.

## 6.15 The Right to Food, Report of the Special Rapporteur on the Right to Food, Jean Ziegler, Mission to India, (United Nations Economic and Social Council, 2006)[162]

The Special Rapporteur[163] visited India between 20th August and 2nd September 2005 to study about the realization of the Right to food in India and submitted his report to UNO on 20th March 2006. This visit was for the reason that India is a home for the highest number

---

[160] http://www.fao.org/worldfoodsummit/english/newsroom/news/8580-en.html (accessed on 27/7/2017 at 3.20pm)

[161] http://www.gmwatch.org/en/news/archive/2002/2387-world-food-summit-2002-special (accessed on 27/7/2017 at 5.32pm)

[162] E/CN.4/2006/44/Add.2 dated 20th March 2006

[163] The Special Rapporteur is an independent expert appointed by the Human Rights Council to examine and report back on a country situation or a specific human rights theme. This position is honorary and the expert is not a staff of the United Nations nor paid for his/her work. http://www.ohchr.org/EN/ Issues/ Food/ Pages/FoodIndex.aspx (accessed on 19/3/2017 at 9.00pm)

of undernourished people in the world and also highest levels of child malnutrition prevails here and the irony is despite the economic growth, hunger and food insecurity are increasing like no man's business. This report examines the situation of hunger, food insecurity and malnutrition in India though the public debate on 'hunger amidst plenty' was widely heard everywhere. It then relooks at the legal framework dealing with the right to food and tries to find out if policies and programmes that are in place to meet the country's obligation to respect, protect and fulfil the right to food are adequate[164]. During his visit he travelled through the length and breadth of India, interacting with the agencies, different states and Government of India and was convinced about the commitment of the government which was reflected in wide range of policies and programmes which were implemented to fight hunger and malnutrition. He was overwhelmed by the progress that has been achieved to eradicate drought and the shortage of food which had plagued the history of India[165]and has stated that India has become self sufficient in the production of basic food. It is also documented that though the growth in food grain production has slackened off late, yet it has still remained above the population growth rate, hence India has sufficient food to feed its populace of more than 1 billion people.

Unfortunately, despite the adequate food in prevalence, house-hold food security has been farce, levels of malnutrition, undernourishment and poverty are very high and fact that signs of hunger and food insecurity have spiraled since the second half of the 1990s is baffling. He has appallingly noted that nearly 2 million Indian children die every year as a result of serious malnutrition and preventable diseases and nearly half suffer from moderate or severe malnutrition[166] and the Researcher strongly believes that this must be one of the highest levels of child malnutrition in the world. The Rapporteur has remarked that women and girl children are more malnourished owing to persistent social discrimination against girl children who are fed less than the boys. Schedule caste and tribes who are looked down as lower caste are highly starved and malnourished. The report has taken note that the majority of Indians are still poor, with 25% of them living below the poverty line and 80%

[164] CESCR General Comment No. 12: The Right to Adequate Food (Art. 11)
[165] E/CN.4/2006/44/Add.2 dated 20th March 2006 Page 5 para 6
[166] *ibid* Page 5 para 7

living on less than US $ 2/day which establishes that many Indians cannot even afford adequate food to maintain a healthy life[167].

In urban areas, people who are employed in unorganized sector and vulnerable groups, such as migrants, homeless, slum dwellers, street children, old &infirm and refugees are excluded from the purview of public services and ration cards, which may enable them to feed themselves. Increasing urbanization is closely linked to poverty and food insecurity in rural areas….[168]. The report also deals with the disturbing scenario of the shift from cultivation of food crops like pulses and millets to cash crops which have more export potential thus resulting in the decrease of household food production as well as consumption. But the Rapporteur is appreciative of the fact that India has for a long time maintained the world's largest food –based safety net, i.e. the Public Distribution System which benefits about twenty two percent of the population living below poverty line.

The Rapporteur also recorded the dismal state of affairs that prevailed in drought stricken regions of Rajasthan, during the year 2000, while food grains were getting rotten in government ware houses, people were dying of hunger & starvation and were compelled to sell their children to make both ends meet[169]. He has documented the satire of starvation deaths amidst overflowing state owned granaries. The report also deals in detail with the public interest litigation by the People's Union for Civil liberties against the government in which it was argued that right to food is a part of Article 21 which deals with Right to life and where the Supreme Court ordered for the complete implementation of all the food based schemes throughout India. While dealing with the legal framework for the implementation of Right to Food in India the Rapporteur has noted the obligation of India when it is a party to the ICESCR and ICCPR, the main international instruments defending the Right to food and other international instruments which casts an obligation on the state to ensure the right to food to all its people. He has enjoined the government to respect, protect and fulfill the right to food without any discrimination.

In continuation of this, he has urged the government to initiate positive steps to identify the vulnerable groups and execute need based policies and programmes to ensure their access to adequate food and those who cannot feed themselves for the reasons beyond

---

[167] *ibid* Page 6 para 8
[168] E/CN.4/2006/44/Add.2 dated 20th March 2006 page 7 para 13
[169] *ibid* Page 8 para 16

their control, government is obligated to feed them. In order to give effect to all these recommendations, the Rapporteur has required the government to use maximum of its available resources. The Rapporteur while dealing with the domestic constitutional and legislative frame work to support the right to food, is of the opinion that the social, cultural, civil and political rights are included in the Constitution in the form of Fundamental Rights and Directives principles of state policy and has opined that although the right to food does not find a mention in the Constitution but the same is imperative in Directive principles of state policy which must be used as a tool to interpret the fundamental Rights.

The report has categorically stated that all victims of violations of the right to food ought to have an immediate recourse to effective remedies including access to justice. He is glad that India is one of the countries in the world which provides for the justiciability of economic, social and cultural rights and he has recorded the judicial activism of the Supreme Court in interpreting Article 21 to give effect to Right to Food. He has enumerated all the food based schemes which were the outcome of the ruling of the Supreme Court. The Special Rapporteur was pleased about the work done by the National Human Rights Commission on the right to food in Orissa in 1996 when there starvation deaths were rampant.

The Special Rapporteur has recorded in the report that though many safeguards to guard the right to food is in place, he received a large number of reports of alleged violations of the right to food. Reports of more than 250 starvation and malnutrition deaths in the previous two years were presented to him at the Judicial Colloquium on the Right to food which he attended along with the 70 senior judges across India at Delhi. In the background of all these programmes and policies to wipe out hunger, the key finding of the report states that millions of Indians still suffer from chronic undernourishment & severe micronutrient malnutrition and these programmes are not capable of wiping out the starvation deaths completely. The report emphatically notes the dwindling agricultural wages, rising landlessness and increase in the prices of food have resulted in food insecurity mainly in rural India.

It also identifies the paradox of economic growth which has generated employment in high tech sectors on one hand and on the other the same may not be able to absorb the loss of livelihoods from agriculture on which two thirds of the people are still depending.

With the fast increase in the price of basic food and decline in the agricultural wages, rural India's consumption of basic staple food diminished by 2.14 % with the total calorie consumption coming down by 1.53%[170].

In the Conclusion and suggestions part, the report has positively stated that the Special Rapporteur is buoyant about the commitment of the government of India in ensuring the food security in the country. He has applauded the efforts of the Apex Court in advancing the justiciability of the right to food. Despite all these concerted efforts to ensure food security, he is alarmed by the largest number of chronically malnourished people especially children. In order to strengthen food security further, he urged the government to set up an independent Public Service Commission to monitor the implementation of this right and also to strictly implement the decisions of the Supreme Court in this regard. He has reminded the government of its obligations as set out in General Comment 12 of CESCR. He strongly advocates that everyone must be treated equally before the law and also for setting up of the Human Rights Court and the Special courts for the protection of the human rights. A national early warning system should be established that records starvation deaths to generate emergency response and to improve accountability[171]. He implores for strengthening of Public Distribution System and has opined that food security programmes should also ensure nutritional security.

## 6.16 High Level Task Force on Global Food and Nutritional Security 2008

The brisk and sudden increase in the food prices globally in the early 2008, posed a crucial threat to food and nutritional security throughout the world in turn causing a slew of humanitarian, human rights, socio economic, environmental and security related consequences which posed challenges particularly to food deficit low income countries. Before this sudden soar in food prices, it was estimated that 854 million people throughout the world were undernourished and due to the rise in the food prices, this number shot up to 1 billion by 2008 i.e. to be precise one person out of every six of them was under nourished. Though there was a fall in prices of food by the end of 2008, the same did not happen in most of the developing nations leaving the needy high and dry.

[170] E/CN.4/2006/44/Add.2 dated 20th March 2006 Page 6 Para 8
[171] E/CN.4/2006/44/Add.2 dated 20th March 2006 Page 19 para k

This resulted in a threatening situation, reversing the progress made towards reducing poverty and hunger as enunciated in the Millennium Development Goals. The increase in demand for food worldwide was due to the increase in world population and a drastic and curt drop in the investment in agriculture. The cumulative effect of this crisis was, the underlying structural problems in the food systems of poorer countries were exposed. All these factors led to the establishment of UN System High Level Task Force (HLTF) by the United Nations Secretary General in 2008 with 23 members as a provisional measure to boost up the efforts of UN system and International Financial Institutions in response to the global food security crisis. The mandate of the HLTF was to ensure a coherent system-wide response to both the causes of this crisis and its overwhelming adverse consequences among the world's most vulnerable populations [172]. It was a specialised and a coordination tool with the primary object of promoting a comprehensive and integrated response of the international community to the challenge of achieving the global food and nutritional security which included the creation and implementation of a prioritised plan of action.

By July 2008, the HLTF, in answer to the request for a plan of action, set up the Comprehensive Framework for Action (CFA), was a skeletal framework dealing with the combined position of HLTF members acting like a catalyst for action by providing governments, international & regional organisations and civil society groups with a set of policies, measures and actions. It pursued a twin track approach: It outlined activities related to meeting the immediate needs, like investing in food assistance and food safety nets, as well as activities increasing opportunities for producers, pastoralists and fisher folk to access land, water, inputs and post harvest technologies, focusing on the needs of small holders and enabling them to realise their right to food, sustain an increase in income and ensure adequate nutrition[173]. In December 2008, the Task Force finalised its agenda of work for 2009, lending support for effective action in implementation in member countries, muster adequate funds for both immediate action and long term investment and bettering the accountability of the international system.

[172] http://un-foodsecurity.org/node/135 (accessed on 29/7/2017 at 6.43pm)
[173] http://www.un.org/en/issues/food/taskforce/establishing.shtml (accessed on 29/7/2917 at 7.37pm)

### 6.16.1 The Impact of the High Level Task Force

The outcomes declared by CFA are used as a bench mark for measuring the overall performance of HLTF and its programme of work. It is reported that HLTF member agencies have worked individually, collectively for the implementation of the plan of action. The urgent measures included to afford access to the needy people for the food that they needed, and remarkable encouragement to small holder farmers to boost food production with the help of national authorities. Some of the outstanding achievements are the following.

- During 2008, granted direct support to almost 20% of the world's hungry people
- Shored up the small holder farmers to boost up the food production which extended benefits to nearly 5% of the global 2 billion small holder farmers and their families
- Lent support to more than 15 governments in their financial and tax policy responses to the swell in prices of food by providing supervision and economic resources which prevented the probable second round effect of price rises on inflation.
- Significantly helped to lessen the macro-economic challenges which were the result of volatile food prices, encountered by poor countries.

### 6.16.2 Their Future Programme Included

- Expansion of the social protection system by funding and technical support in more than 60 nations aiming at protecting livelihood with highlighting women and children and aiding the authorities at the national level to design and formulate social protection schemes and safety nets and finance them.
- Continuous impetus to increase the production of food by small holder farmers by offering them increased long term development assistance in at least 35 nations.
- Boosting trade finance and getting bigger negotiations for the attainment of substantial outcomes by the end of 2009 and to work on efficacy and practicality of coordinated stocks of food and a proper regional food reserve system.
- Devise a general reference framework for bio-fuels and analyse their effect on food security, poverty and the environment.

In continuation of the all these programme, by the end of 2008, the HLTF carved a programme of work and set up a Coordination Team to speed up and aid its effective implementation. The HLTF programme of action is organized to achieve the four results, namely. synchronized support for in country action to improve food and nutrition security, mobilization of investment to support immediate necessary actions and longer term regional or local and national plans for food and nutrition security, stimulate the participation and partnership of all the stake holders at local, regional and global levels to promote the joint and sustained assistance to improve food security and lastly to watch the efforts of the international community and tracking the progress of the realisation of CFA results.

The HLTF concentrates its attention on the necessity to set up a coordinated and effectual UN system in every country to amalgamate humanitarian, trade and developmental interests which ensures support for leadership by all the national authorities. The HLTF member as part of their efforts to support action by all the national authorities in respective countries to contribute to the improvement of food and nutrition security, initiated and intensified the inter-agency coordination in 27 countries which later on shot up to 33countries. Even to this day, the harmonization of international support for food and nutritional security by the national authorities in developing countries especially in the least developed countries is still a challenge. In many countries, participation of the civil society and nongovernmental organizations at the national level is dismal and needs immediate attention. Members of the HLTF Coordination Team have recognized factors of good practice and keys to success which can aid the national authorities to work more efficiently with UN teams and other stake holders.

There is a major dependency on external sources for the purpose of monitoring, policy research, information analysis and advice which calls for urgent need for improving both the regional and national capacities. The coordinated efforts by HLTF Members has resulted in a successful development and finance of programmes which have lent support for food and nutrition security in some 60 countries[174] The HLTF worked closely with the officials of the European Commission following the Commission President's commitment to establish a 1 billion euro European Union Food Facility (EUFF) to support urgent action by countries in need in response to consequences of the food security crisis. Coordinated efforts by HLTF entities resulted in a successful collaboration with the Commission to develop and finance programmes of support for food security in some 60 countries. HLTF also contributed to the L' Aquila Initiative on Food Security which was launched in July 2009 at

[174] http://un-foodsecurity.org/node/135 (accessed on 14/8/2017 at 8.26pm)

the G8 L'Aquila Summit at Italy where US $20 Billion were pledged for food security. The joint statement by 26 countries and 14 multilateral organizations by the signatories of the L'Aquila endorsed the HLTF and its CFA as a means to build on the comparative advantage of International Organizations and International Financial Institutions while enhancing their coordination and effectiveness.

But the feedback from those who have made use of this indicates that it needs to indicate options for linked investments in nutrition, social protection and trade and to supply extra analysis on links between food security and access to land and land acquisition, water use and adaptation to climate change which is very crucial. The CFA is a living document but the same will have to be revised time and again to make it applicable to the current scenario. By the time of the L'Aquila summit, and later on by the Pittsburgh G20 meeting in September 2009, it was obvious that several donors were interested in a pooled multilateral financial coordination mechanism. The World Bank, IFAD[175], the ADB[176], FAO, WFP, UNICEF[177] and UNDP being the potential partners, with the World Bank becoming the preferred institution for receiving and handling the funds, the concerted efforts of all these authorities led to the setting up of Global Agriculture and Food Security Programme, which supports government-led proposals, private entities, and technical assistance providers, technical agencies at both the global and the regional levels.

## 6.17 Rome Declaration on World Food Security, 2009

The World Summit on Food Security was held between 16th and 18th of November 2009 at the UN Food and Agriculture Organization (FAO) headquarters in Rome, Italy. The Summit witnessed a congregation of 4,700 delegates from 180 countries, which included Heads of State and Government as well as representatives of governments, UN agencies, intergovernmental and non-governmental organizations.

### 6.17.1 Objectives

This Summit was prorogued to guarantee immediate national, regional and global response to reduce the proportion and the number of hungry people and to halve by 2015, the number of people suffering from malnutrition in the world at large which were actually the essence of the Millennium Development Goal 1 and 1996 World Food Summit Goal.

---

[175] International Fund for agricultural Development
[176] African Development Bank
[177] United Nations Children's Fund

This Summit was held in answer to the ceaseless presence of widespread malnutrition and the mounting anxiety about the agricultural production to meet the future needs for food.

Delegates felt sorry that the participant nations were not on target to tackle the goal adopted at the 1996 world Food Summit to halve the number of under nourished by 2015. Several delegates noted that the number of people affected by hunger had in fact increased by 100million since 2008178. While some nations appreciated the increased attention being given to food security since 2008, rich countries were conspicuous by their absence at the summit which made the sign of the lack of urgency, obvious on their part. Considering the spurt in the number of people suffering from diabetes and heart ailments in developing countries, some delegates stressed on the urgency to ensure the availability of quality food and access to food which is rich in both nutrients and energy.

**6.17.2 Delegates held a series of high level meetings and Four Round Tables.**

*Concept Note for Roundtable 1: Minimizing the negative impact of the food, economic and financial crises on world food security*[179].

This Round Table was held on 16th November focussing their deliberations on lessening the negative impact of food, effect of financial and economic crises on world food security, employment of the reforms for the governance of global food security and to enhance the global food security encompassing the rural development and small farmers. The participants stressed on the role of small holder agriculturists for food security. They observed that the major portion of the small holders' farmers comprises of women and at the outset they highlighted their vulnerability against the effect of economic crisis &agricultural subsidies at the developed countries. They expressed their concern for increasing the investment in local food markets, which might include channelizing the assistance through farmer's organisations and their access to information relating to market and the use of fertilisers. The delegates also emphasised the need to better the early warning systems and to create food safety nets. Urged for better statistical information to reckon the nations in need and ultimately strongly recommended for public and private partnership and national level planning.

---

[178] http://enb.iisd.org/crs/food/wsfs2009/html/ymbvol150num7e.html (accessed on 20/11/2017 at 7.26pm)

[179]WSFS/2009/RT/1 available at http://www.fao.org/fileadmin/templates/wsfs/Summit/Agenda/ Provisional_ List_Docs.pdf (accessed on 15/11/2017 at 8.34pm )

*Concept Note for Roundtable 2: Implementation of the reform of global governance of food security*[180].

This Round Table was held on 17th November 2009 and deliberated on the need to reform the global governance to aid the food security strategies. This Round table met on 17th November and deliberated on reforms that will have to be carried out in global governance to aid the food security strategies. Delegates discussed the reforms that will have to be employed in Committee on World Food Security. They recommended outreach to a greater number of decision making actors and involvement of international agencies181. For the global governance of food security, participants emphasised on the need to bring in new actors mainly civil society and wished that this would bring about a convergence of actions both at the national level and international level and gave a call for stream lining the global governance and coordination among all players and highlighted the role of experts and solicited responses to price volatility of agricultural products especially food products.

Concept Note for Round Table 3: Climate change adaptation and mitigation: challenges for agriculture and food security182

This Round Table was held on November 17 and addressed the association and relationship between agriculture and food security for mitigation and adaptation of policies. The delegates at the summit collectively thought that sustainable forestry and effective agricultural policies can alone reduce poverty and alleviate food security and bring about positive changes in biodiversity and climate change mitigation and adaptation. The participants identified some of the priority areas for immediate action like, resilience of food production systems and ecosystems; conservation and protection of genetic resources and integrated, efficient pest and disease control. Regarding adaptation, they suggested that diversification of food production and water supply can be a cornerstone and called for enhanced cooperation of the most vulnerable countries[183].

They adduced the instances of proactive adaptation planning which included regional scenario building, widening the range of genetic variability, developing new varieties of crops which can adapt to different climatic conditions and build & conserve water reserves

---

[180] *ibid*

[181] http://enb.iisd.org/crs/food/wsfs2009/html/ymbvol150num7e.html (accessed on 16/11/2017 at 1.12pm)

[182] WSFS/2009/RT/3 available at http://www.fao.org/fileadmin/templates/wsfs/Summit/ Agenda/Provisional_ List_Docs.pdf (accessed on 16/11/2017 at 1.23pm)

[183] http://enb.iisd.org/crs/food/wsfs2009/html/ymbvol150num7e.html (accessed on 16/11/2017 at 1.48 pm)

for irrigational purposes& for consumption of livestock. Applauding the contribution of agriculture to mitigation, the delegates stressed on the urgency of meeting the global food needs, called for incentivising agriculture & forest management thus making it sustainable in the changing climate and entreated for research on climate friendly agricultural practices which ensures better rate of food production.

*Concept Note for Roundtable 4: Measures to enhance global food security: rural development, smallholder farmers and trade considerations*[184].

This Round Table was held on 18th November and dealt with the means to increase global food security. Many delegates opined that food security is an integral part of international security. Co-Chair Tina Joemat-Peterson, Minister for Agriculture, Forestry and Fisheries of South Africa, said that agriculture production is the cornerstone of economic development, and suggested partnerships to increase production and ensure food safety[185]. International Federation of Agricultural Producers and Karen Serres, Chair of the Committee of Women Farmers said that the trade policies ought to favour the small farmers in the food value chains. Specifically, other delegates discussed about agriculture being a part of inclusive policy package, integrated rural development, measures to be adopted to increase food production, easy access of small farmers to markets, effect of climate change and the role of women.

### 6.17.3 Declaration of the World Summit on Food Security

The result of the discussions and deliberations among the participants of the Summit boiled down to a Declaration. The declaration called for food sovereignty instead of food security. Food sovereignty involves changing the existing food system to make sure that those who produce food have fair access to and control over land, water, seeds. The declaration has insisted on more support and investment in small holder agriculture as a way to attain food sovereignty. The Declaration consists of an introduction and two sections mentioning the strategic objectives and commitments & actions. The opening statement of the Declaration elucidates the commitment of the heads of state and government and the representatives to initiate urgent action to eradicate hunger from the world. It also urges to reverse the decline in investment and encourage investment in agriculture, rural

[184] WSFS/2009/RT/4 available at http://www.fao.org/fileadmin/templates/wsfs/Summit/Agenda/Provisional_List_Docs.pdf (accessed on 16/11/2017 at 2.13pm)
[185] http://enb.iisd.org/download/pdf/sd/ymbvol150num7e.pdf (accessed on 17/11/17 at 10.24pm)

development and food security in developing countries. The Declaration goads the countries to face the challenges of climate change on food security head on and to mitigate the effect of it on agriculture. In order to attain the objectives of the Declaration, it set forth Five Rome Principles for Sustainable Global Food Security[186].

1. Invest in country-owned plans aimed at channellising resources to well-designed and results-based programmes and partnerships
2. Foster strategic coordination at national, regional and global levels to improve governance, promote better allocation of resources, avoid duplication of efforts and identify response gaps.
3. Strive for a comprehensive twin-track approach to food security that consists of 1) direct action to immediately tackle hunger for the most vulnerable, and 2) medium- and long-term sustainable agricultural, food security, nutrition and rural development programmes to eliminate the root causes of hunger and poverty, including through the progressive realization of the right to adequate food
4. Ensure a strong role for the multilateral system by sustained improvements in efficiency, responsiveness, coordination and effectiveness of multilateral institutions.
5. Ensure sustained and substantial commitment by all partners to investment in agriculture and food security and nutrition, with provision of necessary resources in a timely and reliable fashion, aimed at multi-year plans and programmes.

**Closing of the Summit**

"I am convinced that together we can eradicate hunger from our planet, but we must move from words to actions. Let us do it for a more prosperous, more just, more equitable and more peaceful world. But above all, let us do it quickly because the poor and the hungry cannot wait", the words of FAO Director-General, Jacques Diouf at the end of the World Summit on Food Security well expressed the sense of urgency that prevailed throughout the 3 day summit[187].

The outcome of this Summit, which witnessed various meetings of parliamentarians included the recommendations to adopt legal and legislative frameworks to protect Right to food the world over, by ensuring the access of women to land, credit and markets thus

[186] http://enb.iisd.org/crs/food/wsfs2009/html/ymbvol150num7e.html (accessed on 27/11/ 2017 at 1.42pm)

[187] https://www.un-ngls.org/index.php/un-ngls_news_archives/2009/916-the-world-summit-on-food-security (accessed on 23/11/2017 at 10.13pm)

achieving women empowerment and strive to achieve the Millennium Development Goals by 2015. Jacques Diouf, Director General of FAO, said that important steps have been initiated to achieve the objective of this Summit- world free from hunger and urged all nations to initiate concrete and urgent measures in this direction reminding the delegates that the 'hungry cannot wait', closed the summit.

### 6.17.4 The Key Challenges

The key challenges before this summit was to eradicate hunger from this earth[188] and to ensure adequate production of food to cater to the need of the world population which is expected to grow by 50 percent by 2050 touching 9 Billion and to devise plans to ensure that everyone has access to the food, to establish a better consistent and efficient governance of food security at both international and national levels as well, to ensure that farmers in developing and developed nations can earn their livelihood better compared to those working in secondary and tertiary sectors in other developing nations and developed nations, to equip the developing countries to have a decent chance of sustenance in world market and that agricultural policies do not warp international trade, to muster significant extra public and private sector investment in improving agriculture and rural infrastructure thus ensuring the availability of modern inputs to mitigate food production mainly in the developing nations which are suffering from low income and deficit of food, taking note of the fact that more than 30 countries are presently going through food emergencies, to put more effective mechanisms in place which can respond to the food crises at the earliest, to ensure and equip the countries to adapt to changes in the climate and to mitigate the negative effects of the same.

The cumulative effect of the Summit was a Report of the World Summit on Food Security and a Declaration of the World Summit on Food Security. The Declaration envisaged strategic objectives, commitments & actions and established the Five Rome Principles for Sustainable Global Food Security. The Researcher thinks that like all other Conventions, Summits held and Declarations adopted so far, this summit also ended with all nations agreeing to adopt measures to take concrete and urgent measures to free the world from hunger, but the declaration at this Summit did not define quantified aims or deadlines. The declaration lacked the measurable targets and timeframes for concrete actions with

---

[188] http://www.fao.org/wsfs/world-summit/wsfs-challenges/en/ (accessed on 14/11/2017 at 9.56pm)

respect to short, medium, and long-term approaches to food security. Hence, this Summit though ended with a positive note that 'the hungry cannot wait', failed to achieve much and became another adorning piece at the archives.

### 6.18 RIO+20 - Zero Hunger Challenge, 2012

The Zero Hunger Challenge, an initiative by the UN Secretary-General, invites all countries to work for a future where every individual has access to adequate nutrition and resilient food systems. It sought to frame new policies to achieve global prosperity, bring down poverty and accelerate social equity and environmental protection. The 'Zero Hunger Challenge' is supported by UN FAO, International Fund for Agricultural Development, World Food Programme, UN Children's Fund, the World Bank and Bioversity International[189]. The challenge was not envisioned as a plan but rather as a call to action: eradicating world hunger is a goal that concerns everyone[190].

The UN Secretary General Ban Ki-moon unveiled food security campaign at Rio+20, which consists five points which aimed at a future where everyone's Right to food is guaranteed and protected. For this, Mr Ban Ki- moon, the UN Secretary General called on the civil society and all the leaders of the nation to upsurge the attempts to end hunger emphatically noting the paradox that though the food is available, every day an approximate of 1 billion people still go to bed with empty stomach. He stressed that food security must be a top priority. Acknowledging the challenge, the UK's Deputy Prime Minister, Nick Clegg, vowed that Britain would be allocating 50 million pounds from the International Climate Change Fund to aid about 6million farmers and the money to be used for the adaptation of small holder farmers to the impact of climate change, which can assure the sustainable agricultural output.

Mr Ban said "We can't rest while so many people are hungry in the world while there's enough food for all. Somehow this food is not distributed equally or fairly. Some people are living in prosperity while marginalised people are hungry. We know this has to change"[191].He appreciated the efforts of Brazile's 'Fome Zero Programme' which was launched in 2004 and which bailed millions of people from poverty by backing up the local

---

[189] http://www.un.org/apps/news/story.asp?NewsID=42304#.WhvNbdKWa1s ( accessed on 27/11/2017at 2.08pm)

[190] http://www.fao.org/un-expo/en/the-theme.html (accessed on 27/11/2017 at 2.19pm)

[191] Available at https://www.theguardian.com/global-development/2012/jun/22/ban-ki-moon-zero-hunger-challenge (accessed on 24/11/2017 at 2.00pm)

farmers and community kitchens and by introducing social welfare programmes such as BolsaFamilia[192] scheme.

The campaign has five objectives[193]

1. 100% access to food for all, all year around,
2. Zero stunted children less than 2years
3. All food systems are sustainable
4. 100% increase in small holder productivity and income
5. Zero loss or waste of food.

This challenge deals with the need to address the primary reason for excessive volatility in food prices and tackle the risks involved in high and volatile commodity prices for global food security &nutrition.

Though it is commendable that UN Secretary General initiated this programme, there was no programme outline set but made only the statement in the public to tackle the problem of global hunger. Though this came from the horse's mouth yet no deadline was set for achieving these aims. Intention was good but without a frame work in place to carry out objectives of the campaign.

## 6.19 Food Assistance Convention, 2012

The Food Assistance Convention, an international treaty, was adopted on 25th April 2012 in London. The treaty aims at "addressing the food and nutritional needs of the most vulnerable populations". This Convention was intended to succeed to the Food Aid Convention 1999[194] which was the lone agreement requiring the signatories to grant a minimum quantity of food aid implying the commitment of the participants to address and fight hunger globally. In June 2012, at the 106th session of the Food Aid Committee, members agreed not to further extend the Convention and accordingly it expired at the end of that month. After repeated debates and deliberations on litigious food aid in the Doha Development Round, Food Assistance Convention was drafted. The new Food Assistance Convention came into effect on 1st January 2013. The most important indicator of this change is the name of the treaty itself. The change in name indicated that the new treaty

---

[192] Bolsa Familia Programme, world's largest conditional cash transfer program, has lifted more than 20 million Brazilians out of acute poverty and also promotes education & health care.

[193] http://www.fao.org/un-expo/en/the-theme.html (accessed on 24/11/2017 at around 2.20pm)

[194] FAC 1999

would enable various kind of assistance beyond just food which includes the use of cash and vouchers which not only meets the needs of food but also protects livelihoods during emergencies. These shifts indicate the growing recognition that tied aid is often slower and more expensive, can cause local market distortions and mainly serves donor country interests[195]. The treaty also insisted that food assistance ought to be in grant form and not restricted to food grains grown in the donor country. Local and regionally purchased food aid is cheaper & available faster and can benefit the farmers of developing countries.

This convention was an improvement over FAC 1999 on some of the issues. Primarily its objectives have a far reaching effect. While the FAC 1999 mentioned the specific objective of 'making appropriate levels of food aid available on a predictable basis' [196], the present Convention proposed a wider and more general objective of 'addressing the food and nutritional needs of the most vulnerable population through commitments made by the parties to provide food assistance that improves access to and consumption of adequate, safe and nutritious food[197]'. FAC 1999 only dealt with the supply of food but never mentioned about the nutritive & safety aspect of it.

Secondly, food assistance provided by the Food Assistance Convention 2012could change the lives of vulnerable people only if the worth of commitments raises but not when it decreases during the global price rise of food. But the parody of this commitment could be witnessed during the 2007-08 global food price rise, food aid donations from Food Assistance Convention collapsed to nearly 50year low, compelling the agencies to scurry to prevent reducing the emergency rations and feeding the children at schools. The Convention is oblivious of the measures that it should initiate to enable the donors to withstand another spike in price.

Thirdly, Food Assistance Convention allows upto 20 percent of assistance in the form of loans. This will not be helpful for a recipient of food assistance who is already in distress but instead they become further indebted to meet short term food needs.

---

[195] *'The 2012 Food Assistance Convention: Is a Promise Still a Promise?'* By Jennifer Clapp and C. Stuart Clark available at https://reliefweb.int/report/world/2012-food-assistance-convention-promise-still-promise (accessed on 1/12/2017 at 6.53pm)

[196] Article 1 (a) FAC 1999 available at https://www.foodaidconvention.org/Pdf/convention/iga1995.pdf (accessed on 29/11/2017 at 9.42pm)

[197] Article 1 (a) Food Assistance Convention 2012 available at http://www.foodaidconvention. org/en/Pdf/ FoodAssistance/ FoodAssistance.pdf (accessed on 29/11/2017 at 9.34pm)

The United States will continue to donate food aid largely in kind under the provisions of its Farm Bill that set a minimum 2.5 million tonnes. The European Union agreed to fund a wide array of food-based and non-food-based interventions. So if participants of the treaty are going to do what they would have done anyway, what is the additional value of international treaty commitments and the necessity for another forum for information exchange and consultation, besides all the other places where governments can discuss food security and humanitarian aid effectiveness[198].

The erstwhile FAC1999 had specified the annual minimum commitment for every donor country which was a specified quantity of food. This specification of every participant's contribution was fixed and not ambiguous. The instant Food Assistance Convention no longer specifies the amount of food to be contributed by the participants beforehand but the same will be announced annually by the members of the Convention. This sounded a death knell for the previous predictability of minimum quantity of food aid to be contributed by the members. The Convention allows the donors to specify their contribution to the food aid either in terms of the quantity like tonnes of wheat or its equivalent in other foods or the value of the food and if the donor opts to contribute in terms of value, the currency can also be specified by them.

This privilege to the donors further abrogates and reduces the predictability as the commitment in terms of value may be subject to international food price volatility. Further food assistance will be of crucial importance and most needed when prices are high, but these are the same times when commitments in value will provide less, which in principle defeats the very purpose of the Convention. The Researcher strongly thinks that this Convention is only old wine in new bottle, just a change of forms of words to enable the governments to do what they would have done anyway.

### 6.20 The Second International Conference on Nutrition 2014

More than half of the world's population was adversely affected by malnutrition, ICN2 kept nutrition high on the international and national agendas[199]. ICN2 was the first global inter governmental forum to deal with nutrition problems of 21$^{st}$ century with a goal to better nutritional status of the people through effective national policies and international cooperation.

[198] https://www.odi.org/comment/6656-what-s-use-2012-food-assistance-convention (accessed on 29/11/2017 10.37pm)

[199] http://www.fao.org/about/meetings/icn2/background/en/ (accessed on 1/12/2017 at 10.40pm)

Though, since the first International Conference on Nutrition in 1992, there was a remarkable shift in the global economy, food systems and the nutritional status of the population, yet a need was felt to have a more appropriate measures and policies. The progress in reducing hunger and malnutrition was uneven and alarmingly slow and the challenge was to improve nutrition sustainably by the implementation of consistent and better coordinated actions. It was realised by all the nations that global problems needed global solutions and it is only through inter governmental Conference that commitments of stakeholders to act decisively to tackle malnutrition could be identified.

The Second International Conference on Nutrition 2014[200]was an inclusive high level intergovernmental meeting held in Rome, Italy in November 2014. The Conference focussed the global attention on redressing malnutrition in all its possible forms. Conference witnessed a footfall of nearly 2200 participants including the representatives from more than 170 governments and 150 representatives from civil society and approximately 100 members from business community. As a prelude to the Conference, various pre-conference events, round tables and side events for private sector, civil society and parliamentarians were held to provide a forum for them to investigate deeper into specific issues concerning nutrition. The theme for First Round table was Nutrition in the Post-2015 Development Agenda[201], theme for Second Round table was Improving Policy Coherence for Nutrition[202], theme for third Round table was Governance and Accountability for Nutrition[203] The plenary sessions were held on November19th, 20th and 21st.

India was represented by His Excellency Basant Kumar Gupta Ambassador and Permanent Representative of the Republic of India to FAO[204]. The outcome of this Conference was the adoption of 'the Rome Declaration on Nutrition' and 'the frame work for action', committing countries to frame national policies to eradicate hunger and prevent all forms of malnutrition worldwide thus transforming the food systems to make nutritious food available to all. ICN2 witnessed a culmination of senior national policy makers from agriculture,

---

[200] ICN2

[201] ICN2 2014/RT/1 available at http://www.fao.org/3/a-ml931e.pdf (accessed on 2/12/2017 at around 7.07pm)

[202] Ibid, ICN2 2014/RT/2

[203] Ibid, ICN2 2014/RT/3

[204] Available at http://www.fao.org/3/a-mm272e.pdf ICN2 2014/DJ/3 page 2 (accessed on 2/12/2017 at around 6.57pm)

health and other relevant ministries and agencies, leaders of United Nations agencies, other intergovernmental organisations and civil society, researchers, private sectors and consumers.

**6.20.1 The key objectives of ICN 2[205] were to:**

1. Review progress made since the 1992 ICN including country-level achievements in scaling up nutrition through direct nutrition interventions and nutrition-enhancing policies and programmes;
2. Review relevant policies and institutions on agriculture, fisheries, health, trade, consumption and social protection to improve nutrition;
3. Strengthen institutional policy coherence and coordination to improve nutrition, and mobilize resources needed to improve nutrition;
4. Strengthen international, including inter-governmental cooperation, to enhance nutrition everywhere, especially in developing countries.

**6.20.2 The Scope of the Conference[206]**

- global in perspective, but focused particularly on nutrition challenges in developing countries;
- addressed all forms of malnutrition, recognizing the nutrition transition and its consequences;
- sought to improve nutrition throughout the life cycle, focusing on the poorest and most vulnerable households, and on women, infants and young children in deprived, vulnerable and emergency contexts

It was unanimously agreed that governments have the primary responsibility to initiate appropriate action at the country level after due consultation with the stake holders and affected communities. In order to fulfil the aspirations of this Conference, a set of recommendations were addressed to the government leaders. It was open for the national leaders to consider the appropriateness of the recommendations and make suitable adaptations for improving maternal, infants and young children's nutrition and to reduce the incidence of non communicable disease which are to be achieved by 2025.

Recommendations were made to create an enabling environment for effective action, for sustainable food systems promoting healthy diets, in international trade and investment,

[205] *ibid*

[206] http://www.fao.org/about/meetings/icn2/background/en/ (accessed on 1/12/2017 at 11.40pm)

for nutrition education and information, for social protection, for strong and resilient health systems, to promote, protect and support breast feeding, to address wasting, to address stunting, to address childhood overweight and obesity, to address anemia in women of reproductive age, actions in the health services to improve nutrition, actions on water, sanitation and hygiene, for accountability and actions on food safety and antimicrobial resistance . Recommended actions on food safety and antimicrobial resistance urged the nations to develop, strengthen, review national food safety legislation and regulations to ensure that food producers and suppliers throughout the food chain operate responsibly[207], actively take part in the work of the Codex Alimentarius Commission on nutrition and food safety and adopt the international standards at the national level[208],to exchange food safety information, including managing emergencies[209].

For the first time the nations thought of food safety and accepting the recommendations of Codex Alimentarius Commission while deliberating on nutritional aspect. When 800 million people in the world still have fire in their belly[210], the Researcher thinks that it is a mockery on the part of the nations to think about raising the nutritional level instead of catering to the basic necessity of the affected community or both these aspects of food must have gone hand in hand.

### 6.21 Report of the Special Rapporteur on the Right to Food, 2014

The Special Rapporteur on the Right to food, Mr Olivier De Schutter visited thirteen countries including low income, middle income countries and high income countries and presented the final report on the same titled '*The Transformative Potential of the Right to Food*' on 24th January 2014. In the report, submitted to the Human Rights Council in accordance with its resolution 22/9, the Special Rapporteur on the right to food draws the conclusions from his mandate, establishing the connections between his various contributions. In the report, he has discussed the nature of Right to Food. He has

---

[207] Recommendation 53: Develop, establish, enforce and strengthen, as appropriate, food control systems, including reviewing and modernizing national food safety legislation and regulations to ensure that food producers and suppliers throughout the food chain operate responsibly. Available at http://www.ifrc.org/docs/IDRL/a-mm215e.pdf accessed on 2/12/2017 at 6.34pm

[208] Ibid, Recommendation 54: Actively take part in the work of the Codex Alimentarius Commission on nutrition and food safety, and implement, as appropriate, internationally adopted standards at the national level.

[209] *ibid*, Recommendation 55: Participate in and contribute to international networks to exchange food safety information, including for managing emergencies

[210] www1.wfp.org/zero-hunger (accessed on 2/12/2017 at 7.45pm)

summarised the normative content of the right to food by referring to the requirements of availability, accessibility, adequacy and sustainability, all of which will have to be included into the legal entitlements. He has opined that the same shall be secured through accountability mechanisms.

He has contended that right to food overlaps with the right to work and also right to social security measures. The main focus of his work was on how to reform the food systems to ensure effective realisation of right to food. The beginning of his mandate coincided with the global food price crisis in 2008 and he made it a priority to guarantee that global and national endeavours would be grounded in the right to food. He has observed that most of the measures to espouse the strength of the countries to increase the production of food and meet up a major part of their own need for food only concentrated on supporting small scale farmers but did not incorporate the mechanism for monitoring the progress and accountability and failed to make sure that the food producers and consumers participated in the process of policy making.

He has taken note of the fact that though there is a boost in the food/ agricultural production, neither the number of hungry came down nor nutritional outcomes were satisfactory. It has been stated that even when food intake is adequate, improper diets could result in deficiencies of micronutrients like iodine, vitamin A or of iron. He has deplored the fact that over 165 million children globally are stunted and they are so malnourished that they do not reach their complete physical and cognitive potential. It has been reported that around 2billion people globally lack vitamins and minerals which are crucial for good health. He has painfully noted that not much has been done to ensure adequate nutrition despite the established long term impacts of inadequate nutrition during pregnancy and before the child's second birthday.

He is full of appreciation for Green Revolution which increased the agricultural production which could meet the challenge posed by population growth and dietary transition due to the rising incomes. On the flip side he has noted the over use of chemical fertilizers to boost the production, the ensuing soil erosion owing to repeated cultivation. He has entreated for the important measures to improve the efficiency of food systems by reducing the food losses and food wastage. It is stated that a study conducted in 2011 estimated the wastage or loss of food, meant for human consumption to be 1.3 billion tons which is approximately about one third of the total food produced. It is observed that in low income countries the losses occur due to inadequate storage, packing, processing facilities and poor access of markets to farmers which in turn results in economic losses to the farmers. Ironically, the levels of per capita food wastage are much higher in rich countries than in developing countries. These discrepancies exert pressure on natural resources. The Rapporteur has prescribed several measures to combat these problems. He has urged the nations to ensure the development of all regions so that mass exodus of people from rural to urban could be avoided. He has urged the nations to reduce the poverty which acts like a panacea for most of the food related problems.

He thinks that food systems ought to be reshaped to include small scale food producers who were always disadvantaged due to inequitable food chains and also because agricultural technologies had not considered their specific needs. With this he wanted to balance the food chains. He advocated for contract farming which would ensure benefit to small scale farmers. He dedicated a report to access to land as a component of the right to food. He believes that change can be achieved, though not easy. To democratize food security policies, actions should be launched at three levels, to rebuild the food systems at the local level, multi sectoral strategies must be deployed at the national level and at the international level, greater coordination should be achieved between actions launched at the multilateral, regional and national levels with an intention to create an enabling international environment. At every level, the right to adequate food has an important role to guide the efforts of all actors and to institute suitable accountability mechanisms.

In conclusion, the Rapporteur has opined that the eradication of hunger and malnutrition is an achievable goal subject to the coordination between sectors, across time and across levels of governance. At the local levels the communities should be empowered to enable them to identify the obstacles that they would probably face and the solutions that suit them best. This should be complemented by encouraging policies at the national level. The Rapporteur wants every nation to achieve food sovereignty which is a condition for the full realization of the right to food. The Rapporteur only gave a road map to fight hunger and malnutrition but did not prescribe the mode of action or the policies that will have to be adopted by the nations. The Researcher thinks that we have too many prescriptions for the ailment but the implementation part is slackening. It is like a saying 'water water everywhere, but not a drop to drink'.

### 6.22 Milan Declaration on Enhancing Food Security and Climate Adaptation in Small Island Developing States 2015

The Meeting addressed food security and nutrition from multiple angles: the importance of promoting sustainable approaches to agriculture and fisheries and building resilience to climate change and disasters; the benefits of improving rural livelihoods of smallholders and family farmers. Though small farmers in Small Island Developing States[211] are the most affected victims of climate change, yet their entreats are not properly understood and taken cognisance of in international negotiations. Small farmers in SIDS are vandalised because of the negative impact of climate change. The constantly changing weather and climate coupled with adverse weather upheavals like prolonged drought, hurricanes and cyclones pose newer challenges every day that have adversely affected the productivity, costs and efficiency of farming in these places. Though these are small farm holders yet they are important as they are the foundations of rural communities and their contribution for nutrition security is critical.

The populace at these SIDS, their land holdings and their economy is very small, these places are geographically remote, easily prone to natural disasters which can be devastating. With their meager natural resources and restricted technology and limited accessibility by road or sea, they are away from international markets. Due to all these adverse conditions their economy is choked and not able to cater to the basic necessities of its people. People are undernourished. Poverty and unemployment are the main constraints of access to food in SIDS[212]. Due to the limited opportunities in agriculture and non agricultural sector, the unemployment rate of the youth in most of the SIDS is higher than the average in the world which results in migration of people to urban centers either within the country or outside. In some of the SIDS the wages are so low that they cannot escape from poverty. In many of these SIDS, as opposed to national food production, food imports are major sources of food. Food consumption by the people in SIDS is nutritionally poor.

---

[211] SIDS

[212] State of Food Security and Nutrition in Small Island Developing States (SIDS) available at http://www.fao.org/3/a-i5327e.pdf (accessed on 4/12/2017 at 5.36pm)

Constant instability and vulnerability negates the efforts to advance nutrition and food security in these regions.

In the light of this, the world community deemed it fit to do something about the plethora of problems faced by these SIDS and hence the SIDS Action Platform has been developed to support the follow up to the Third International Conference on Small Island Developing States, SIDS Accelerated Modalities of Action (S.A.M.O.A) Pathway. The international community is committed to providing the required support to SIDS in continuation of this world leaders renewed their commitments at the SIDS conference and pledged an approximate US$ 1.9billion for the implementation of the pathway. Paragraph 63[213] of the SAMOA Pathway specifically mentions about the commitments of the heads of the state to bail the SIDS of their current predicament.

This emergent necessity to develop food security in SIDS was followed by the Milan Declaration[214] on Enhancing Food Security and Climate Adaptation in Small Island Developing States within the frame work of the SAMOA pathway. The Milan Declaration categorically stated that multilateral trading system ought to play a major role in addressing food security. It specifically emphasised the need for the designations of easily prone, vulnerable small economies and the net food importing developing countries. In the light of the easy vulnerability and lack of resilience it mentioned that the

[213] "63. ... we are committed to working together to support the efforts of small island developing States:
a) To promote the further use of sustainable practices relating to agriculture, crops, livestock, forestry, fisheries and aquaculture to improve food and nutrition security while ensuring the sustainable management of the required water resources;
b) To promote open and efficient international and domestic markets to support economic development and optimize food security and nutrition
c) To enhance international cooperation to maintain access to global food markets, particularly during periods of higher volatility in commodity markets;
d) To increase rural income and jobs, with a focus on the empowerment of smallholders and small-scale food producers, especially women;
e) To end malnutrition in all its forms, including by securing year-round access to sufficient, safe, affordable, diverse and nutritious food
f) To enhance the resilience of agriculture and fisheries to the adverse impacts of climate change, ocean acidification and natural disasters;
g) To maintain natural ecological processes that support sustainable food production systems through international technical cooperation."
Available at https://sustainabledevelopment.un.org/samoapathway.html accessed on 4/12/2017 at 9.37pm

[214] We, Ministers and representatives of the Small Island Developing States (SIDS) have met in Milan, Italy, 14-16 October 2015, for a Ministerial Meeting on Enhancing Food Security and Climate Adaptation in SIDS, in the context of the EXPO Milano 2015 on the theme "Feeding the Planet – Energy for Life" available at https://sustainabledevelopment.un.org/content/documents/8537MilanDeclaration.pdf (accessed on5/12/2017 at 6.06pm)

trade policies should not negatively affect the food production. It was realised that to adapt SIDS would need access to more financial resources and the same could come from the traditional sources of development and environment finance and performance based funding like the sale of carbon credits, payments of ecosystem services. A very significant source of finance for small holders is the Adaptation for Smallholder Agriculture Programme which was launched by the International Fund for Agriculture Programme and it focuses on channelling the climate finance to smallholder farmers to enable them to access the information tools and technologies that they require to build resilience to climate change.

To better the economy and to adapt to the challenges of climate change, the other nations urged the governments of SIDS to develop and properly maintain infrastructure, particularly feeder roads, bridges and coastal and flood defences, to disseminate appropriate technical and market research on alternate crops& providing advice to agriculturists, to provide crop insurance for small farmers, to provide a post-disaster reconstruction fund in case of any disaster & the disaster fund should be adequate to recover from the devastations caused. Wherever there is a need the governments of SIDS should not shy away from holding consultations with small farmers and their organisation.

The Milan Convention was of limited application, the target group being the SIDS. The Researcher strongly believes that these states are always at the receiving end and constantly face the nature's fury which makes them more vulnerable and underdeveloped resulting in regional imbalances. Unless the other nations lend them a helping hand in reality not restricted to lip sympathy, there cannot be overall development of the world and the SIDS also should make the optimum utilisation of the privileges extended by the other nations in right earnest and become self sufficient over a period of time as they cannot depend on the other nations for their basic necessities especially food security, for all time to come. Nothing can ensure food security in SIDS but to help them selves and become independent.

### 6.23 United Nations Sustainable Development Summit, Goal 2, 2015

From the time the Millennium Development Goals were adopted there has been a remarkable advancement in the attainment of First Millennium Development Goal that is to 'Eradicate extreme poverty and hunger'. Even at the global level, the poverty target was attained five years ahead of the set schedule which was 2010and the plan was to achieve this by 2015.Though the rapid growth in the economy of some of the big countries chipped distinctly to reach this goal, growth in many parts of the world was not satisfactory and inadequate. In the light of this, the 2030 Agenda for sustainable development highlights the highest commitment of the countries to end poverty in all its manifestations, which includes the eradication of extreme poverty by 2030.

It also stresses on the need to accomplish food security as a matter of priority and to put an end to all forms of malnutrition. From the Millennium Development Goals, the countries had learnt that there can be poverty and hunger eradication only when interconnected factors like livelihoods and employment, food security, nutrition, health, education, access to basic infrastructure and services are addressed comprehensively. Peaceful and just societies are a necessary precondition for success of SDG 1 "Ending poverty in all its forms everywhere"[215]. Since 1990, though the number of people living with less than US$ 1.25 per day has been halved, 836 million people, representing 14 percent of the developing world's population are still extremely poor[216]. Risks associated with climate change and disasters are closely interlinked with poverty eradication efforts, around 42 million human life years are lost in internationally reported disasters each year[217] .

General Assembly's high level plenary meeting 'The United Nation's Summit for the adoption of the post 2015 Development agenda was held between 25th and 27th September 2015 at New York. The United Nations Sustainable Development Summit, held during the 70th Session of the UN General Assembly, will be the culmination of years of

215Interactive Dialogues 1 - Poverty and Hunger.docx available at https://sustainabledevelopment.un.org/content/documents/8141Interactive%20Dialogue%201%20- (accessed on 15/12/2017 at 6.14pm)

216The Millennium Development Goals Report 2015. Available at http://www.un.org/millenniumgoals/2015_MDG_Report/pdf/MDG%20report%202015%20presentation_final.pdf (accessed on 15/12/2017 at 6.30pm)

217 UNISDR (2015). Making Development Sustainable:The Future of Disaster Risk Management. Global Assessment Report on Disaster Risk Reduction. Geneva, Switzerland: United Nations Office for Disaster Risk Reduction (UNISDR) available at https://www.unisdr.org/we/inform/publications/42809 (accessed on 15/12/2017 at 6.20pm)

research and negotiation on the post-2015 development agenda, a bold vision for achieving sustainable development for all[218].

Consented to by the 193 member states of the United Nations, the proposed tentative agenda captioned 'Transforming our world :2030 agenda for Sustainable development' consisted of a Declaration, 17 Sustainable Development Goals (SDGs) and 169 associated targets, a part of it concentrating on the modes of implementation & a framework for review and follow up. This ubiquitous agenda gave a call for action by all countries poor, rich, developing, developed and middle income. It recognised the fact that ending poverty must be coupled with a plan that goes hand in hand to build economic growth and address a range of societal requisites.

Sustainable Development Goal 2 endeavours to put an end to malnutrition and hunger by 2030 in all its manifestation. These Sustainable Goals are also popularly known to be Global goals, are a "Universal-Call" to save the planet earth, wipe out poverty and make sure that everyone leads a peaceful and contented life. These Sustainable development Goals, which are interconnected are the continuation of Millennium Development Goals with some more new disciplines appended to it. 'The SDGs work in the spirit of partnership and pragmatism to make the right choices now to improve life, in a sustainable way, for future generations. They provide clear guidelines and targets for all countries to adopt in accordance with their own priorities and the environmental challenges of the world at large'[219].SDGs, a comprehensive programme combats the core causes of deprivation and poverty. "Poverty eradication is at the heart of the 2030 Agenda, and so is the commitment to leave no-one behind," UNDP Administrator Achim Steiner said[220].

This programme offered a sole chance to place the world and in turn the people on a more sustainable developmental road map the reason why the United Nations Development Programme came into existence for. More than one hundred fifty nations who attended the Summit officially adopted Sustainable Development Goals which was a set of 17 Goals with the three key things to be achieved and they were

[218] http://www.unfpa.org/events/united-nations-sustainable-development-summit-2015 (accessed on 15/12/2017 at 1.03pm)
[219] http://www.undp.org/content/undp/en/home/sustainable-development-goals.html (accessed on 15/12/2017 at 12.59 noon)
[220] *ibid*

firstly, to fight inequality & injustice, secondly to end poverty and third to fix the climate change. These Goals were tailor made to promote the benefit of all people of all countries in the world.

In a statement issued following the consensus reached by Member States on the Summit outcome document on 2 August, UN Secretary-General Ban Ki-moon said: "[the agreement] encompasses a universal, transformative and integrated agenda that heralds an historic turning point for our world. This is the People's Agenda, a plan of action for ending poverty in all its dimensions, irreversibly, everywhere, and leaving no one behind," he said[221]. India was represented by our beloved Prime Minister Mr Narendra Modi in the Summit. In his statement he mentioned "Some progress has been made in preventing distortions in world agricultural markets.

The global agricultural export subsidies were reduced by 94 per cent from 2000 to 2014. In December 2015, members of the World Trade Organization adopted a ministerial decision on eliminating export subsidies for agricultural products and restraining export measures that have a similar effect...........Our attack on poverty includes expanded conventional schemes of development; a huge skill development programme; and a new era of inclusion and empowerment, turning distant dreams into immediate possibilities: new bank accounts for 180 million: direct transfer of benefits: funds to the unbanked: insurance within the reach of all: and pension for everyone's sunset years.......Sustainable development of one sixth of humanity will be of great consequence to the world and our beautiful planet. It will be a world of fewer challenges and greater hope and more confident of its success"[222].

### 6.23.1 Accomplishment of Goal 2 during 2017

Since the year 2000, attempts to end hunger and malnutrition have progressed remarkably. More investments in agriculture, government spending and aid are the need of the hour to increase agricultural productivity which directly strengthens and augments the attainment of SDG 2.

---

[240]*Overview_Sustainable_Development_Summit_Final* available at https://sustainabledevelopment.un.org/content/documents/ 8316overview_Sustainable_Development_Summit_ Final.p( accessed on 15/12/2017 at 8.11pm)

[222]Prime Minister of India Narendra modi at UN in 2016.pdf available at https://sustainabledevelopment.un.org/content/ documents/20265india.pdf (accessed on 15/12/2017 at 9.30pm)

- The ratio of undernourished people in the world decreased from 15% in 2000-2002 to 11% in 2014-2016 that is nearly 793 million people in the world were under nourished which used to be 930 million people during the respective periods.
- In 2016, an estimated 155 million children under 5 years of age were stunted (too short for their age, a result of chronic malnutrition). Globally, the stunting rate fell from 33 per cent in 2000 to 23 per cent in 2016[223]
- Ending hunger demands sustainable food production systems and resilient agricultural practices. One aspect of that effort is maintaining the genetic diversity of plants and animals, which is crucial for agriculture and food production. In 2016, 4.7 million samples of seeds and other plant genetic material for food and agriculture were preserved in 602 gene banks throughout 82 countries and 14 regional and international centres — a 2 per cent increase since 2014. Animal genetic material has been cryoconserved, but only for 15 per cent of national breed populations, according to information obtained from 128 countries. The stored genetic material is sufficient to reconstitute only 7 per cent of national breed populations should they become extinct. As of February 2017, 20 per cent of local breeds were classified as at risk[224].
- The share of sector-allocable aid allocated to agriculture from member countries of the Development Assistance Committee of the Organization for Economic Cooperation and Development (OECD) fell from nearly 20 per cent in the mid-1980s to 7 per cent in the late 1990s, where it remained through 2015. The decline reflects a shift away from aid for financing infrastructure and production towards a greater focus on social sectors[225].
- Some progress has been made in preventing distortions in world agricultural markets. The global agricultural export subsidies were reduced by 94 per cent from

---

[223] Progress Of Goal 2 In 2017 available at https://sustainabledevelopment.un.org/sdg_2 (accessed on 15/12/2017 at 8.30pm)

[224] UNISDR (2015). *Making Development Sustainable:The Future of Disaster Risk Management. Global Assessment Report on Disaster Risk Reduction.* Geneva, Switzerland: United Nations Office for Disaster Risk Reduction (UNISDR) available at https://www.unisdr.org/we/inform/publications/42809 (accessed on 15/12/2017 at 6.20pm)

[225] Report of the Secretary-General, "Progress towards the Sustainable Development Goals", E/2017/66 available at http://www.un.org/ga/search/view_doc.asp?symbol=E/2017/66&Lang=E (accessed on 15/12/2017 at 8.50pm)

2000 to 2014. In December 2015, members of the World Trade Organization adopted a ministerial decision on eliminating export subsidies for agricultural products and restraining export measures that have a similar effect[226].

This Summit which aimed at Sustainable Development Goal to fight out hunger &malnutrition and ensure food security globally is a tall order. The Researcher strongly believes that instead of having a certain plan for long duration, the countries should have opted for short duration plan whose efforts will be palpable at the earliest. To assess the fruits of the long duration plans one must wait for the tenure to elapse and the amendment which are needed to be made may not be possible as the plan may not allow, whereas it is easier to make any modification for the future plans after knowing the worth of short term plans. The Researcher is of the opinion that countries should venture out to make short term plans which have a long term effect.

## 6.24 High-level side event on pathways to Zero Hunger-2016

"Achieving Zero Hunger is our shared commitment. Now is the time to work as partners and build a truly global movement to ensure the Right to Food for all and to build sustainable agriculture and food systems." UN Secretary-General Ban Ki-moon Committee on World Food Security, Rome, 12 October 2015[227].

### 6.24.1 Pathways to Zero Hunger

During the high-level side event to the 71st UN General Assembly[228], discussion about Sustainable Development Goals and accomplishing global Zero Hunger by 2030 were held. "It's been a year since the Sustainable Development Goals (the 'Global Goals') were formally endorsed but there are still 795 million hungry people in the world today. Zero Hunger is one of the Goals at the heart of the 2030 Agenda, and it will be highlighted in a joint side event entitled "Pathways to Zero Hunger" on Thursday 22 September"[229]. "With the 2030 Agenda for Sustainable Development, the United Nations member-states have committed to a comprehensive, integrated and universal transformation. The Agenda is

[226] *Ibid*

[227] http://www.un.org/en/zerohunger/pdfs/ZHC%20-%20Pathways%20to%20Zero%20Hunger.pdf (accessed on 16/12/2017 at 7.36pm)

[228] The theme of the 71st session was "The Sustainable Development Goals: A Universal Push to Transform our World". Available at http://www.un.org/en/ga/71/ (accessed on 16/12/2017 at 6,15pm)

[229] UN General Assembly 2016: Zero Hunger available at https://insight.wfp.org/un-general-assembly- 2016-zero-hunger-94da1e4d14d3 (accessed on 16/12/2017 at 1.20pm)

people-centred and based on human rights and social justice. Achieving the Sustainable Development Goals cannot happen without ending hunger and malnutrition and without having sustainable and resilient, climate-compatible agriculture and food systems that deliver for people and planet. This requires comprehensive efforts to ensure that every man, woman and child enjoy their Right to Adequate Food"[230]. The World Food Programme initiated a week long session and sittings to show the growing support for the attainment of global Zero Hunger, which was the most ambitious target set by world leaders to end hunger by 2030.

"The cost of hunger is not only measured in lost lives and unrealised potential for individuals. It also costs families, communities and countries; it affects their ability to deliver on their social development goals and stunts their economic prosperity. The Sustainable Development Agenda seeks to complete the unfinished business of the MDGs by putting an end to poverty and hunger for all by 2030"[231]. The agenda for Sustainable Development 2030 had set the entire world on the path of wide-ranging, integrated global and complete transformation. All the countries had realised the fact that there could not be the achievement of Sustainable Development Goals unless the world is cured of hunger and malnutrition. Unless there is sustainable, resilient, climate friendly agricultural practice and food systems that covers and caters to both rural and urban areas, attainment of Sustainable Development Goals would be a far cry. Progress in the attainment of 2030 agenda would be pre requisite for achieving a Zero hunger world.

The Zero Hunger Challenge (ZHC) was launched by Ban Ki-Moon, the United Nations Secretary General in 2016 to end hunger and ensure nutritious food for all. It was convened as the second in series[232] of global dialogues on nutrition, food security and sustainable agriculture. This webinar witnessed expert and practitioners perspective about ending rural poverty through sustainable living and decent rural employment. The ZHC has borrowed some elements from the SDGs and the interplay of these elements results in the wiping out hunger and malnutrition and build sustainable food systems. "The Global Movement for Zero Hunger offers a platform to accelerate global action, bringing together

---

[230] '*Transforming our Food Systems to Transform our World*' available at http://www. un. org/ en/ zerohunger/pdfs/ZHC%20-%20Pathways%20to%20Zero%20Hunger.pdf (accessed on 16/12/2017 at 7.53pm)
[231] https://www.acww.org.uk/assets/zhc-brochure---final2.pdf (accessed on 16/12/2017 at 6.46pm)
[232] The first Zero hunger challenge was in 2012

all stakeholders to communicate the importance of inclusive, productive, sustainable and resilient food and agriculture systems to deliver on the promise of the 2030 Agenda. It also serves to amplify collective action to ensure food security and nutrition for all people"[233]. The event reflected the complete makeover in terms of food security, nutrition and sustainable farming that have direct ramifications on the fructification of the 2030 agenda which leaves nobody behind in the world.

The ZHC challenge strives to achieve five outcomes: zero stunted children less than two years, 100% access to adequate food all year round, making all food systems sustainable, a 100% increase in smallholder productivity and income, and zero loss or waste of food[234]. The 'key sessions and the themes of this event'[235] were 'refugees and migrants, pathways to zero hunger, breaking bread, innovations and school meals'. "The Zero Hunger vision is comprised of five elements which taken together, can end hunger, eliminate all forms of malnutrition and build inclusive and sustainable food systems. These five elements are:

1. All food systems are sustainable: from production to consumption
2. An end to rural poverty: double small-scale producer incomes & productivity
3. Adapt all food systems to eliminate loss or waste of food
4. Access to adequate food and healthy diets, for all people, all year round
5. An end to malnutrition in all its forms"[236]

Weeklong discussions and deliberations resulted in the Declaration where all world leaders were made to undertake an obligation that they would strive towards ending hunger, eliminate all forms of malnutrition and build inclusive and sustainable food systems, would support all elements of the Zero hunger challenge, and its vision, mission and principles, particularly to enable all people to realize their right to adequate nutritious food, would the advocate for actions and policies that would achieve zero hunger and contribute to the SDGs and would persuade others also to join and to take the challenge.......[237].

---

[233] Pathways to Zero Hunger – UNGA Side Event available at http://www.un.org/youthenvoy/2016/09/pathways-zero-hunger-challenge-unga71/ (accessed on 16/12/2017 at 5.53pm)
[234] http://www.wfp.org/zero-hunger&sa=U&ei=67_UVLEY5MLLA-_BgOgO&ved=0CCsQtwlw BQ&usg=AFQjCNFMYnL2lkl4rcucuNADg_42-Ep-OQ (accessed on 16/12/2017 at 6.54pm)
[235] UN General Assembly 2016: Zero Hunger available at https://insight.wfp.org/un-general-assembly-2016-zero-hunger-94da1e4d14d3 (accessed on 16/12/2017 at 6.26pm)
[236] https://www.unglobalcompact.org/take-action/action/sdg2 (accessed on 16/12/2017 at 7.07pm)
[237] http://www.un.org/en/zerohunger/pdfs/ZHC-Declaration.pdf (accessed on 16/12/2017 at 7.22pm)

This being the latest and last commitment of the world leaders to end hunger and malnutrition globally, we are still to wait and watch how it takes its shape as it is only about a year and a half old. The Researcher opines that it is too premature to assess the working of any plan when it is half way through. But the effect of implementation is not generally palpable. This instrument also seems to be another arrow towards the target of hitting hunger and malnutrition. The option available for the Researcher as on today is only to wait and watch out for the fruits of its implementation. The Researcher only hopes that at least this instrument makes a difference at the world arena to fight out hunger, malnutrition and ensure food security the world over.

### 6.25 Conclusion

In conclusion, the Researcher strongly opines that holding series of conventions and summits and declarations will not help to salvage the situation. Instead, every state ought to make an effort to combat hunger, ensure there is a good investment in agriculture, improve the produce of agriculture, thus making sure there is enough food in the supply. Due to the industrialization, all states seem to be living in fool's paradise thinking that money can buy everything. But, in reality there is no short cut to agricultural produce but to produce it. Despite plethora of summits, declarations, conferences, schemes and programmes for almost 70 years not much has improved but these meets have only been a bureaucratic meets where only the think tank will meet and come out with impractical solutions.

"A safe global food supply can only be achieved through development of effective food safety management systems, which are essential to the well being of consumers, farmers, processors and manufacturers" [238]. "The true source of world hunger is not scarcity but policy; not inevitability but politics", said Dr Peter Rosset, executive director of Food First. "The real culprits are economies that fail to offer everyone opportunities and societies that place economic efficiency over compassion."[239]. "Tolerance to violations of right to food should be ended"[240]. Despite the international community frequently reaffirming the significance of the right to adequate food, a bothersome/worrisome gap exists between the

[238] Margaret A. Hamburg, M.D., FDA Commissioner
[239] https://www.greenleft.org.au/content/myths-about-world-hunger (accessed on 21/12/2017 at 4.43pm)
[240] Yogendra Kumar Srivastava, "Right to Food- A Legal &theoretical Analysis in International Perspective", The PRP Journal of Human Rights.[October-December 2006] Vol 10 No1, p15-23

ideal standards set by international covenants and the situation prevailing in many parts of the world.

"UNICEF estimates that in the year 2000, about 2,420,000 children in India died before their fifth birthdays. This was the highest total for any country. It was estimated that for the same year about 10,929,000 children died before their fifth birthdays. Thus, more than a fifth of the child mortality worldwide occurs in India alone. The international agencies estimate that about half of these deaths of children under five are associated with malnutrition. Thus we can estimate that more than a million children die in India each year from causes associated with malnutrition. To that number must be added a large but unknown number of adults who succumb for the same reason"[241]."*One extreme view is that the global food crisis has been created by politicians obsessed with external security and new weapons of defence and offence"*[242].The problem is in global political pressures to dilute food security.

The Right to safe food is essentially a second generation human right and is plagued with the same issues of enforceability as all other economic, social and cultural rights[243]. "The right to food and other related economic and cultural rights are not given adequate attention at the national and international levels"[244]. Hunger is caused by decisions made by human beings and can be ended by making different decisions. "To be a part of the answer means to let go of old frame work and grappling with new ideas and approaches. Economic dogma should not remain intact while millions starve amid even greater abundance"[245]. Even under narrow interpretation of Right to food, governments must maintain an environment in which people can feed themselves and be free of hunger. "People have a responsibility to obtain their foods that you cannot blame the state for malnutrition"[246]. People must have adequate wages or have access to land to grow.

---

[241] George Kent University of Hawai'i March 12, 2002 '*The Human Right To Food In India*' available at http://www. earthwindow.com/grc2/foodrights/HumanRightToFoodinIndia.pdf (accessed on 27/12/2017 at 3.41pm)

[242] Kamala Prasad, "Politics of Food" Mainstream, [July5, 2008] Vol XLVI no 29, p 9-15

[243] Ms. Uma Sud, '*The Right to Food as an International Human Right*', Social Action, January-December [Vol 54 No4 2004]. P 296-310

[244] FAO, Right to Food in Theory and Practice,(Rome: FAO. United Nations, 1998)

[245] Atul Vishwanathan and ketan Makhija, '*Arrest Hunger The Right to Food*', Lawyers Collective [Vo19, No2, February , 2004], p13-17

[246] FAO legal officer Margret Vidar. http://www.fao.org/worldfoodsummit/english/newsroom/ focus/focus6 htm (accessed on 26/12/2017 at 6.09pm)

The Researcher has taken the stock of all the Declarations and Conventions and the various programmes supported and conducted by the United Nations with the help of member nations. The United Nations should be commended for its painstaking efforts for its attempt to see that basic necessities are ensured for every individual globally. This commitment of the state has been misinterpreted, misconceived and misunderstood by the administrators to their advantage and to the detriment of the society in general. No international instrument ordains the state to provide food to everybody at their doorstep but urges them to create enabling atmosphere where their Right to Food is taken care of and protected. Let us take the example of Indian context, where National Food Security Act 2013 has been passed. As per the tenor of this Act the state is obligated to supply food grains at the subsidised price to the target group who are below the poverty line. The quantum of beneficiaries is 75%of the rural population and 25%of the urban population. This shows that there are more poverty stricken families in rural areas than urban areas.

Our country became independent about 70 years back and statistics show that the proportion of the population below poverty and malnourished is only increasing year by year. Then the Researcher wonders what the successive governments have done to the country, to push the people from poverty to penury. Looking at the funds spent on poverty alleviation programmes and schemes, the improvement on this front is not satisfactory. The Government in Karnataka State has gone one step ahead of other states in terms of providing ready to eat food at Indira Canteens where the consumers are presumably below poverty line but no mechanism to ensure the same.

The Researcher thinks that this acute problem can be overcome by killing the greed and incorporating selflessness that eludes the dominant few presently. Economic efficiency should not be placed over compassion. A change in the state's present policies like world trade practices & more comprehensible communication between national and international law is the need of the hour. The Researcher is of the opinion that instead of state feeding the young robust people should undertake to feed only the old, infirm, destitute, pregnant and small children irrespective of their economic status. Instead of trying to achieve the superficial satisfaction of feeding the targeted group, the state should introspect, states should initiate permanent measures to create more capital infrastructure in the country.

The fulfilment of the right to food should be closely associated with appropriate economic and social policies and specifically in the efforts towards eradication of poverty. It is very important to reflect the right to food in all poverty related programmes and policies and be used as a guiding principle in the preparation of all poverty reduction strategy. Specific efforts should be made include the right to food approach in all the ongoing national reform processes. Such process of change provides a platform for reaching out to various groups of stakeholders through national consultations, thus influencing policy formulation. The implementation of the right to adequate food at the national and international levels may be guided by General Comment 12[247]. The basic problem is not the scarcity of food but the absence of effective entitlements by everyone to food or means of production. The right to food should not be viewed only in the context of state obligations but also consider the responsibilities of different actors, participants and the concrete implementation of these various responsibilities need to be facilitated and enforced by respective national governments.

But what hurts the Researcher is that there is no comprehensive approach to solve this problem. At the outset, the UN should urge all its member nations to devise ways and means to alleviate poverty which is a scourge in most of the developing countries and more so in under developed countries. "A government's primary responsibility is not to interfere with individual's efforts to provide for themselves, but to seek to ensure an enabling environment for such efforts. However, within every state, there will always be some persons who need direct assistance and in this context, the efficacy of existing social safety nets and social legislation should also be reviewed, taking into account the role of local authorities"[248]. The Researcher submits that 'food, obtained from mother earth is nobody's monopoly and what is more is, it tastes better and sweeter when shared and give a man a fish and you feed him a day, but teach a man to fish and you feed him for life'.

---

[247] The Right to Food, *Report on the Third Expert Consultation on the Right to Food,* Commission on Human Rights Fifty-seventh session, Chairman –Rapporteur Mr Asbjorn Eide, E/CN.4/2001/148, 30th March 2001 Economic and Social Council

[248] Ms Uma Sud, "The right to food an international human right", Social Action,[July-September, 2004] Vol 54, p 296-310

# CHAPTER –VII

# CONCLUSION AND SUGGESTIONS

## 7.1 Conclusion

The preamble of the Constitution of the World Health Organisation, projects a vision of the ideal status of health as the eternal and universal goal. It establishes the indivisibility and interdependence of rights as they are related to health of the people. It recognises the enjoyment of the highest attainable standard of health as a fundamental right of every human being. Despite the tremendous increase in agricultural production, reaching adequate standards of food security and food safety at the household level is still a goal to be achieved[1]. The Right to safe food is essentially a second generation human right and is plagued with the same issues of enforceability as all other economic, social and cultural rights[2]. Right to safe food is multi dimensional, hungry bellies on one hand and unsafe food that the market has been swarmed with, on the other. Of course, we have legislations to tackle both these aspects but ironically the authorities under these Acts are not coordinating with each other. There has to be a harmonious working of these two sets of authorities.

Food adulteration is common in India. Even milk, consumed primarily by children, is not spared. What's particularly worrying the Researcher, is the kind of substances employed to adulterate, including toxic chemicals. This shows that the trade off between the risk of getting caught and the 'reward' of huge profits is skewed heavily in favour of the adulterator. The government must focus on raising the risks to the adulterator. One way of doing this is by hiking the penalty, including making it analogous to attempt to murder in extreme cases. It's equally important to regularly check foodstuff for adulteration and ensure speedy trials. In India, food industry is of different sizes such as the organised sector, small scale and unorganised sectors. The requisites of standards in each sector are different. The small scale sector and domestic market in itself is quite large.

The food chain in India is infested with various stake holders ranging from a small farmer to street vendors to retailers to the big industrialists. The protocol for

[1] Kanchan Dwivedi, Research Scholar, Singhania University, Rajasthan, "Human Right to adequate food", International Journal of Innovative research and Development [July 2012], Vol 1, Issue no 04, p196-204

[2] Ms. Uma Sud, '*The Right to Food as an International Human Right*', Social Action, [January-December 2004] Vol 54 No4, p 296-310

standardisation of food articles should keep in mind the actual users of these standards, environment, the culture and the present infrastructure of the country which seems to have been overlooked by the legislators. Instead of setting high standards and eliminating the small time FBOs, the state should provide logistic support to them and see that they will also be competent to maintain the standards. Eliminating them altogether will not mitigate the problem.

The interest of the society can never be allowed to perish. In *Hobbs* v *Corporation of Winchester*[3], the court aptly held that it was no defence for a butcher to say that he could not have discovered the disease unless he had an analyst on the premises. Roscoe Pound, the chief protagonist of the sociological jurisprudence has logically explained the central theme of the statutory crimes in the following words: *"the good sense of court has introduced a doctrine of acting at one's peril with respect to statutory crimes which expresses the need of society. Such statutes are not meant to punish the vicious will but to pressure on the thoughtless and inefficient to do their whole duty in the interest of public health or safety or morals*[4]*"*.

It was held by *Lord Wright in McLeod* v. *Buchanan,*[5] that the intention to commit a breach of the statute need not be shown but the breach in fact is enough. It is understood that one is expected to get for what he pays. The interest of innocent purchaser is paramount because he pays the money and in turn honestly desires that he be given unadulterated food articles. On the same lines, Kerala High Court in *State of Kerala* v *P.K Chenu*[6] accorded a positive gesture to the clarion call of the gullible and innocent purchaser and observed "*when on analysis, it is found that an article of food does not conform to the standard prescribed, the sale of such article is an offence under the Act which does not provide for exemption of marginal or border line variations of the standard from the operation of the Act. To condone such variation on the ground that they are negligible will amount to altering the standard itself prescribed by the statute*".

---

[3] (1910) 2KB 471

[4] '*The Spirit of Common Law*' Roscoe Pound with a new introduction by Neil Hamilton & Mathias Alfred Jaren Transaction Publishers New Brunswick(USA)and London (UK) available at https://books .google. co .in/ books? d=g7El0rDEkrkC&pg=PA52&lpg=PA52&dq=the+good+sense+of+court+has+introduced (accessed on 11/3/2018 at 10.45pm)

[5] (1940) 2 ALL E.R. 179, 186 (H.L.)

[6] 1975 CrLJ 411(Ker) at 413

In *Jagdish Prasad* v *State of West Bengal*[7], the Supreme Court has reiterated that once standard having been fixed, any person who deals with articles of food which does not conform to it contravenes the provisions of the Act and is liable to be punished. But, the moment liability is attached to an innocent vendor who is ignorant of the adulteration, for any marginal deviation from the prescribed standards, there is every chance of clamorous screaming by the vendors.

To fully realize the benefits of the FSSA, FSSAI needs to maintain total transparency in rule framing. Involvement of industry and other stakeholders and transparent public consultation, directly or through representative bodies, during the preparation, evaluation and revision of food law is essential for setting of sound scientific standards. The Researcher is appalled that though the object of this Act is to maintain the standards for the food and make it safe for the human consumption, yet the same has remained a distant mirage. The death and injury to human beings due to contaminated or adulterated food is not a stray case, questioning the very purpose of the Act. The struggle of a common man to ensure safe food has become a constant litany in vain. The business of making food appear appealing and attractive often spoils the quality of what we eat.

To make nation healthy, every citizen must be able to buy food that is free from contamination. Food security cannot be guaranteed merely by provision of a certain quantity of grain to each family but by ensuring that every grain that is distributed is wholesome and nourishing and not noxious. The ideology of food safety is a composite one beyond merely making grain available physically. There is no point in having a legislation which has failed to give reliefs as contemplated by it. The authorities established under this Act have turned out to be only white elephants and eating up a major portion of our economic resources without doing much.

Serving safe food is not an option, but an obligation. "Each and every member of the food industry, from farm to fork, must create a culture where food safety and nutrition is paramount"[8]. Control of quality of food stuffs by preventing their adulteration is a gigantic problem all over the world. It is much more so in an economically developing and under developed states owing to their backwardness in

---

[7] (1973) 2SCJ 397:AIR 1972 SC 2044

[8] WilliamBill Marler, An accomplished attorney and national expert in food safety, he has become an accomplished attorney and national expert in food safety, he has become the most prominent food borne illness lawyer in America and a major force in the food policy in the Us and around the world. Marler Clark, The Food safety Law Firm has represented thousands of individuals in claims against food companies whose contaminated products have caused life altering injury and even death.

technology in means of production and preservation and packing of food articles. Standards play an important role in the daily life of a common man. The standards in some form or the other add value to the life thus making it easier, safer and more comfortable for everyone. Standards are developed in response to needs of life[9].

India being a developing country has to improve upon its infrastructure and technology to keep pace with the international standards to protect the consumers and to compete in trade with developed countries. In advanced countries like USA, USSR, Japan, France or Germany, it is possible to exercise rigid controls over the quality of food articles in view of the high order of general awareness in the people and a sense of realisation of their rights, privileges and responsibilities in the task of cooperation with the law enforcing agency[10]. One of the great weaknesses of this protective legislation is that in our country, the complacent attitude of the people, may be due to illiteracy, ignorance and dejection has led to a state of mass sluggishness with the result that the entire burden of controlling the quality of food stuffs rests with the state and enforcement machinery. Though the erstwhile PFA and the current FSSA have provided for the punishment of life imprisonment, yet there is no instance of the court awarding the life imprisonment to any accused for his guilt. Not even those prosecuted for deadly articles of food taking a heavy toll on lives of human beings nor manufacturers of poisonous liquor or spurious drugs or any other fatally adulterated edibles or medicines have so far been sentenced to life imprisonment[11]. It is difficult to assume that the law has fully succeeded in attaining its objectives.

Policy makers are absolutely confused as to whether they should eradicate hunger first or to make sure that only safe food is available in the market. Hunger is caused by decisions made by human beings and can be ended by making different decisions. Despite the efforts at the international level as well as at the national level to curb hunger, recent report by Welthungerhlife and Concern worldwide, India has been ranked at the 103rd position among 119 countries on the Global Hunger Index[12]. For the

---

[9] Dr Ashok R Patil, '*Products Safety Standards & consumer Protection*', *Dissertation series-5* KILPAR Bangalore

[10] C Gopala Krishnamurthy, '*Enforcement of Prevention of Food Adulteration Act' in Ramesh V Bhat &B S Narasinga Rao's(ed) book ' National strategy for Food Quality Control National Industries Institute of Nutrition*', Hyderabad, 1985 p97

[11] LM Singh "*Strict Liability under Prevention of Food Adulteration Act 1954:Legislative Consciousness and Judicial Activism*", Supreme Court Journal, 1987 Volume 3 (September-December)

[12] '*India has serious levels of hunger :report*' Deccan Herald daily news paper Bangalore Edition dated 16/10/2018

existence, growth and development of the society, the interest of the members should be protected.

It is, therefore in the interest of the existence and growth of a society that consumer protection is extended as a right of the members of such society[13]. In case, it is not protected, the result may be chaos in the society which may result in disintegration and degeneration of the society. In such a scenario, the pre-contractualist situation as the natural law thinkers advocate, the state shall lose its existence. The nation made to consume sub-standard and adulterated food-stuffs will naturally tend to grow sub-standard and weak. If we undertake a reality check, most grocery stores and kitchens in hotels, restaurants, catering houses & bakeries, especially those in rural areas will fail abysmally if subjected to hygiene and food safety tests. This is in terms of meals/drinks served and cleanliness of the place where the food is served, utensils used for cooking, water that goes into preparation and servers.

As Margaret A. Hamburg[14], has opined that 'A safe global food supply can only be achieved through development of effective food safety management systems, which are essential to the well being of consumers, farmers, processors and manufacturers'. "The true source of world hunger is not scarcity but policy; not inevitability but politics", said Dr Peter Rosset, executive director of Food First[15]. The real culprits are economies that fail to offer everyone opportunities and societies that place economic efficiency over compassion. But what hurts the Researcher is that there is no comprehensive approach to solving this problem. At the outset, the UN should urge all its member nations to devise ways and means to alleviate poverty which is a scourge in most of the developing countries and more so, in under developed countries.

Millions of people in the country continue to go hungry and malnourished. They are deprived of their right to food with disastrous and devastating consequences including the death. The conditions of poor people, landless peasants and the unemployed urban dwellers who are all the marginalised section of society is steadily deteriorating. At the same time, the efforts should be made at the international level to ensure that only safe food is available to the people. The UN should constitute a body

---

[13] Dr Subhash C Sharma, "Consumer protection and the Prevention of Food Adulteration Act", Central India Law Quarterly, Annual Index [1995]Vol VIII, p 187-194

[14] Margaret A. Hamburg, M.D., became the 21st Commissioner of Food and Drugs on May 18, 2009, only the second woman to serve in the position. Among her initiatives during her tenure at FDA, Dr. Hamburg oversaw the modernization of a food safety system to reduce food borne illness.

[15] The Institute for Food and Development Policy, better known as Food First, working since 1975 to end the injustices that cause hunger through research, education and action.

which should mandate all the states to report the statistics about the people suffered due to unsafe food, adulterated food stuffs and any untoward incidents relating to food safety, measures undertaken to curb this menace and how effective they are. As there is a Special Rapporteur for Right to food who visits different countries to oversee the working of food programmes, similar arrangements should be instituted to ensure that people are supplied with safe food and the available food conforms with the standards stipulated by the Codex Alimentarius Commission.

## 7.2 Testing of Hypothesis

### Hypothesis 1

1. This Research lies on the premise that though, Right to qualitative and quantitative food in India is the upshot of innumerable international instruments and constant judicial interpretation of the constitution, yet, the human right to food can be effectively ensured only when a parliamentarian explicitly recognizes it under Part III of the Indian Constitution.

**The answer to the above hypothesis is in the affirmative**. The right to food is the outcome of constant efforts of United Nations beginning from the times of Universal Declaration of Human Rights,1948. Followed by which several attempts were made in the form of declarations, conventions, summits and conferences to give it the status of basic human right. As this is not enough even judiciary in series of cases starting from *Kishen Patnaik* v. *State of Orissa*[16], till today many cases have been decided to give it a status of the right. Though initially, the Constitution of India did not deal with the basic right of right to food, the apex court has reiterated in many judgements that Article 21 Right to life and personal liberty includes not only bare minimum necessity but entails the life with dignity. Directive principles of state policy have ordained the state to provide its people with safe food. For a person to lead the life with dignity, food is essential. Today the Constitution of India is abound with judgements from the Apex court giving a legal sanction to Right to food. But unless an amendment is carried out to our constitution, incorporating clause (B) to Art 21, right to qualitative and quantitative food will still be a partially accomplished dream. Hence, Right to qualitative and quantitative

---

[16] 1989 Supp (1) SCC 258

food can be effectively ensured only when a parliamentarian explicitly recognizes it under Part III of the Indian Constitution as a fundamental right.

**Therefore the Hypothesis is proved Right.**

**Hypothesis 2**

2.The Research is based on the assertion that though FSSA has been carefully crafted in such a way that it could subsume Prevention of Food adulteration Act, 1954 and eight other different legislations, due to the blatant lacunae that they all suffered from, yet, the provisions of FSSA are inadequate to provide panacea in India, for the bane of food adulteration which is a curse for the humankind.

**The answer to the above hypothesis is partially affirmative**. The erstwhile PFA Act,1954 never attempted to nip the bane of food adulteration in the bud but it was the duty of the food inspectors to take stock of the situation only if came to their notice. PFA Act tried to curb only the adulteration and other dimensions of making the food safe for human consumption did not find a mention in the Act. FSSA was passed after carefully incorporating the basic features of Prevention of Food Adulteration Act and other eight legislations. The objective of this Act was to fix the scientific standards for food, make the food safe for human consumption apart from ensuring the availability of only safe food in the market. The Act is self-explanatory. There is a paradigm shift under FSSA, where the Act casts a duty on all the stakeholders in the field i.e. all food business operators which includes manufacturers, traders, dealers, wholesalers and retailers to abstain/refrain from dealing and selling unsafe/adulterated food. Not only selling, even if they have unsafe food in their place of business meant for sale for human consumption, they can be prosecuted under the Act. With the introduction of the supply chain concept under the FSSA, the focus will not be on mere inspection but on each person in the chain- sourcing, manufacturing, storing, distributing- assessed by Food Safety Officers. The introduction of preventive measures at all stages of the food production and distribution chain, rather than only inspection and rejection at the final stage, makes better economic sense.

The Act has gone one step further to doubly ensure that unsafe food is not consumed by the general public by introducing Food Recall Procedure, where, if the food in the market is found to be unsafe /obnoxious to the health of the consumer, the same will have to be withdrawn from the market. FSSA regulates food hygiene and safety laws in the country in order to systematically and scientifically develop the food industry. The Act aims to make the food safe for consumption from farm to fork, casting the responsibility on all the players in the field. The motto of the Act is 'prevention is better than cure'. Despite, all these marathon efforts, we still witness cases of people suffering from consumption of unsafe foods. We can still see the presence of unsafe food in the market. In the light of the above adduced facts, the Researcher strongly thinks that provisions of the Act are inadequate and hence, this Hypothesis is only partially fulfilled and there is still a lot much to be desired for.

**In the light of the above facts, the Hypothesis is proved Partially Right.**

**Hypothesis 3**

3. In the light of the hostile heterogeneous nature of the food business in India and the perennial challenges encountered in the implementation of the Food Safety and Standards Act, stray cases of people suffering from food adulteration questions the commitment of Food Safety and Standards Authority of India in ensuring safe food to the people.

**The answer to the above hypothesis is partially affirmative.** FSSA has laid emphasis on consolidating the laws related to food. The Act has cast on FSSAI, the responsibility of stipulating science based standards for food articles as per international norms-Codex Standards, and to closely monitor their manufacture, storage, distribution, sale and import. The administrative control of the FSSA has been assigned to the FSSAI thereby establishing a single reference point for all matters and eradicating any possibility of multiplicity of orders or any coordination problems that would have been caused with other authorities. But, FSSAI has been extending the deadline given for registration and licensing of the food business taking the heterogeneity of the food business into consideration. We have food being cooked at home and sold to consumers elsewhere, food caterers, hotels, small eateries, bakeries, street vendors, dabbawallahs in Mumbai in particular, temples serving food to the pilgrims, star hotels and food being distributed under akshaya patra yojana, mid day meals schemes in the schools and anganwadi centres.

When we have variety of food business operators, uniform standards becomes impossible and unreasonable to apply and maintain. Due to this feature, FSSAI has not been able to curb the large scale adulteration of food and articles of food. FSSAI does not even have the statistics as to how many small time hawkers, who only need to register and how many food business operators whose income is more than 5lakh & need to obtain license are operating in the market. FSSAI has been a failure with respect to compelling the FBO's to adhere to its orders and notifications, for instance FSSAI has utterly failed in compelling all the FBOs to display the helpline number in their premises. This is the least that FSSAI could have done, and if this is not done, the Researcher thinks that the future of food safety is bleak and FSSAI turns out to be another white elephant and rehabilitation centre for the bureaucrats.

**Hence, the Hypothesis is proved Partially Right.**

### 7.3 Findings of the study

1. The major challenge for the implementation of the FSSA is the population of our country which is 127,42,39,769, and is 17.25 per cent of the global population[17] and accounts for a third of the world's poor as remarked by the World Bank[18] earlier this year. Though, the proportion of people living below poverty line (BPL)[19] in India has come down from 37.2 per cent in 2004-05 to 21.9 per cent in 2011-12- a decline of 15.3 percent as reported by the planning Commission, yet poverty and illiteracy have become major primary challenges in India for maintaining food safety and standards, in general. It is often said that the poor will consume 'anything' to mitigate their hunger. This may or may not be true. On one hand, survival depends mainly on access to a minimum quantity of food, on the other hand, consumption of food which does not meet minimum safety standards can also jeopardize the very survival. When one third of the people are living below poverty line, access to food in itself is a major challenge and the concept of 'food safety' is a far cry. The cascading effect of poverty is illiteracy because of which not many people are aware of FSSA. Though FSSA was enacted in 2006, a decade ago, till today consumer awareness about this legislation is very poor among the literates and nil among the illiterates. In India, people came to know about the existence of this Act only after the Maggi noodles incident which took place in June 2015. Successive governments have failed take the necessary steps to ensure that food quality and safety considerations form an integral part of the food security system.

2. Cascading effect of population explosion is poverty and illiteracy. Illiteracy is a stumbling block for the development of a country. It is a shame that even after India's 70years of independence, we still find 22.5% of the population living below poverty

[17] Jansankhya Sthirata Kosh or National Population Stabilisation Fund (NPSF) 2017

[18] planningcommission.nic.in/reports/genrep/rep_hasim1701.pdf (accessed on 21/12/2017 at 8.00am)

[19] An income of less than $1.25 per day per head of purchasing power

line. Researcher submits that, unfortunately, most of the government projects/schemes are short sighted with no permanent solution to the dogging problem in the near future.

3. Developing countries, may not be able to afford expensive food imports, it is important for them to place renewed emphasis on self sufficiency to ensure food security which consequently results in the availability of safe and standard food. prioritising agriculture is the panacea for this. We should pursue and participate in sustainable food, agriculture, forestry and rural development policies in high and low potential areas so that there is balance in productivity and development of all regions.

In India, there is a wide opinion that the agriculture crisis is a product of shifting goalpost of economic reforms initiated in 1991 away from structural problems and the investment needs of agriculture. Procurement price by the state from farmers is a disincentive to the farmers, who increasingly perceive farming to be an unremunerative vocation. This is an unhealthy indication for the country's economy and needs a serious rethink. Neglect of farmers and farming sector and the current food inflation has been further exacerbated by a misplaced zeal in detaching liberalisation of food trade from investment needs to expand food security to incorporate the same in right to food.

4. The FSSA has exempted plants prior to harvesting, live animals and any animal feed from the purview of the definition of food. "Several large scale outbreaks of diseases in the human population and among farm animals in India have been traced to food contamination. Scientific studies have established how contamination is passed through animal feeds to the human food chain. There has been a worldwide increase in public awareness of the potential health hazards posed by environmental pollution and sub-standard animal feed entering the human food chain"[20].

5. The adoption of the international level standards[21] without preparing the domestic food industry will pose challenges for effective implementation of the Act. Harmonisation of domestic scenario with international regulations, keeping national interests in mind is the need of the hour. The small and medium scale industries may not keep track of the regulatory changes which make it difficult for them to identify the procedural and compliance changes introduced by the Act. First the government must upgrade the local set up before trying to achieve the higher targets as set out by the provisions of the FSSA. The local system should be upgraded in a phased manner before

---

[20] Priya Deshingkar, "Political Economy of Animal Feed and Food Contamination Debate in India", Political and Economical Weekly, [April 6, 2002] Vol XXXVII No 14, p1437-1446

[21] Codex Standards

trying to take a leap. The gap between the existing system and the standards set by the FSSA will have to be cemented first. The government must realize that there is no point in setting higher targets and not able to reach them.

6. Street Food hawkers in India are a major challenge in the implementation of FSSA, who are generally unaware of food regulations and have no formal training in food handling. They also lack support services such as good-quality water supply, sanitary facility and waste disposal systems, which hinder their ability to provide safe food. A survey made by the Health Ministry in 16 cities has found that 90 % of the street food is unsafe for consumption. An FSSAI study conducted in Kolkota found that the street food is cheap and gives nutritional value in terms of calories but hygienic standards are very unsatisfactory. The quality of water used, the manner of cooking, display and handling of food and waste disposal are all problem areas. The FSSAI is planning to make an intensive study of street food in Lucknow and Varanasi after the pilot study in Kolkota[22].

7. Although standards are specified for water to be used as an input in the processing/preparation of food, the FSSA does not specify standards for potable water, which is usually provided by local authorities. Thus, the food providers have to shoulder the responsibility of ensuring that clean water is used, even when tap water may not meet the required safety standards. This is a tall order for small food enterprises and street food vendors. Costs also rise if each vendor invests in water purification systems. If such facilities were provided to food vendors by the state authorities as it happens in Malaysia and Singapore, India might be more successful in ensuring that this sector also maintains acceptable standards of hygiene and cleanliness.

In a survey conducted by the Maharashtra government health department in 2006 it was shown that up to 20-25% of household food expenditure is incurred outside the home and some sections of the population depend entirely on street foods. This is one of the consequences of rapid urbanization, with millions of people having no access to kitchen or other cooking facilities. There are millions of single workers without families and a large floating population who move in and out of the city for various purposes. These people largely depend on street foods for their daily sustenance from places of work, hospitals, railway stations and bus terminals. The food at these shanties are generally prepared and sold under unhygienic conditions, with limited access to safe

[22] "Ensure Food Safety" Deccan Herald Daily News Paper dated 18/9/13 Bangalore Edition

water, sanitary services and garbage disposal facilities. Hence street foods pose a high risk of food poisoning due to microbial contamination, as well as improper use of food additives, adulteration and environmental contamination.

8. Shortage of testing laboratories and equipment has hampered the implementation of the Act and Government of India may also find it difficult to identify, recruit and then continually train people for the accredited laboratories. All food testing will have to be carried out in accredited food laboratory. 'Presently all the public sector food laboratories are not accredited. FSSAI commissioned a gap analysis study for upgradation of 50 food laboratories under the central and state government. The study indicated that there is an urgent need to upgrade the infrastructure, strengthen staffing and training inputs and put in place more reliable laboratory management and operational procedure'[23] Where Rules are being observed, it appears to be driven by international trade norms/requirements/parameters than concern for the health of the local people.

9. Our Indian society can be divided into two tiers. One, where a handful of exporters of articles of food are leaving no stone unturned for abiding by the international stipulations/guidelines to maintain and further expand their market niche. The other tier comprises of millions of small scale milk, poultry and meat traders/producers, who are oblivious of the concept of food safety, are passing on contaminated products to millions of gullible consumers. In the second tier, the producers, traders and consumers are not even aware that there is a problem which could impact their health and rural livestock.

10. Hazarad Analysis and Critical Control Point need to be accompanied by an increase in improved access to scientific research findings which would create the much needed general awareness that is currently lacking in India. The shocking paradox is that people are more affected & disturbed by the reports of synthetic milk than by the reports about the presence of Aflatoxins[24] and DDT in natural milk, although the latter may carry a bigger health risk. This indicates lack of scientific knowledge & understanding among the general public.

---

23 Sheeba Pillai, " Right to Safe Food: Laws and Remedies", The Banaras Law journal, [2012] , Vol 41, p 119-135

24 Any of a class of toxic compounds produced by certain moulds found in food, which can cause liver damage and cancer.

11. The FSSA also mentions certain terms like 'Good manufacturing practices', 'Food Safety Management System'[25] good hygienic practices' and 'Hazard Analysis and Critical Control Point' without according any details to them because of which confusions galore. It is not clear from the Act what these terms imply and whether the CODEX definition of such terms is to be followed. The expression 'sub-standard[26]' is not well articulated and states that even if the food article does not meet the standards set by this Act, the food is still not considered unsafe. If this is the interpretation, then there is no point in subjecting the food to test to see if they will meet the requirement under the law.

12. The FSSA authorises FSSAI to implement the Act with the help of Scientific Commissioners of Food. But there is an ambiguity as to who will control it-whether it is the Ministry of Food Processing Industries or the Ministry of Health and Family Welfare.

13. No specific standards in relation to use of potable water has been provided under the Act.

14. An important Codex Alimentarius Commission guidelines for food processing companies is to follow a food quality management system called Hazard Analysis and Critical Control Points (HACCP). Most of the countries have adopted this Codex Alimentarius Commission's guidelines, but unfortunately in India FSSAI has not been successful in ensuring that all food processing units follow this guide line.

15. The Tiffin suppliers who are popularly known as Dabbawalas in Mumbai[27], how will the same food safety laws be applied to them. The Act will restrict the manufacture

---

25 FSSA Section 3 (s) 'Food safety management system' means the adoption of 'Good Manufacturing Practices', 'Good Hygienic Practices' and Hazard Analysis and Critical Control Point' and such other practices as may be specified by regulation, for the food business

26 *Ibid* Section 3(1)(zx) 'Sub-standard', an article of food shall be deemed to be sub-standard if it does not meet the specified standards but not so as to render the article of food unsafe.

27 A dabbawala is a person in Mumbai, India, whose job is carrying and delivering freshly-made food from home in lunch boxes to office workers. They are formally known as MTBSA (Mumbai Tiffin Box Suppliers Association), but most people refer to them as the dabbawalas. The dabbawalas originated when India was under British rule. Since many British people who came to India did not like the local food, a service was set up to bring lunch to their offices straight from their home. The 100-odd dabbas (or lunch boxes) of those days were carried around in horse-drawn trams and delivered in the Fort area, which housed important offices. Today, businessmen in modern Mumbai use this service and have become the main customers of the dabbawalas. In fact, the 5,000-strong workforce (there are a handful of women) is so well-known that Prince Charles paid them a visit during his recent trip to India. Several academic institutions regularly invite the dabbawalas' representatives for discussion, and to complement and enhance their academic content. At times, businesses find it useful to illustrate the application of how such a system uses Six Sigma principles to improve its operations. White Paper Prepared By Mba Students At The University Of North Carolina's Kenan-Flagler Business School. Nishesh Patel and Naveen Vedula http://mumbaidabbawala.in/( accessed on 12/4/2016)

and supply of economically priced food that serve millions of poor consumers and office-goers every day. The Act does not differentiate between the food products being manufactured by the agribusiness companies and the food products being sold by street hawkers and dhabas[28] that dot the national highways. "Both organised sector and unorganised sector are required to follow the same law with regard to specifications on ingredients, traceability and food recall procedures which is very difficult for unorganised sector entities like street vendors and small hawkers"[29]. While the general prescriptions and regulations spelled out by the FSSA, meet the requirements of the agribusiness companies, the same cannot be blindly applied to the small-time hawkers that provide cheaper food to the working class in the urban centres. These food vendors are the lifeline of economically disadvantaged people in urban areas who have no access to kitchen.

16. It is a terrifying challenge to tackle the huge variety of food safety- related issues in the context of the country's sheer size, diversity and complexity of food markets. For example, the central government has created some islands of excellence in the dairy sector by launching nationwide annual food safety and hygiene audits of dairy plants. The National Productivity Council is a partner in those efforts. But unless they are transformed into a mass movement inculcating food safety concerns and consciousness among all sectors and the general public, a significant nationwide impact will not be felt.

17. The Indian food industry is dominated by microenterprises and Home-based units who prepare food products like condiments, traditional and ethnic food. It is thus imperative for policy prescriptions to address these sectors too, before an impact on the overall food safety scenario is felt.

18. The business culture and consumer participation are two key factors determining the success or failure of food safety campaigns. The business culture is simply the attitude of entrepreneurs towards all stakeholders in the food supply chain. Ideally, that culture should be characterized by the ability to welcome and adjust to change, efforts for excellence and putting consumer's best interest at the top of the business agenda. An ideal policy environment should inspire the food industry, especially the Small and Medium Enterprises Sector, to adopt the best possible food safety assurance practices

---

[28] Dhaba is the name given to roadside restaurants in India and Pakistan. They are situated on highways and generally serve local cuisine. www.oxforddictionaries.com/definition/english/dhaba

[29] Ravulapti Madhavi, "Is Food Safety Lurking in The Food Safety and Standards Act 2006?", Supreme Court Journal, [8th May to 12th June, 2008]Vol 4, p 17-20

not only to gain a competitive edge but also to fulfil its social responsibility. This means that a thorough understanding of country-specific business cultures is essential before launching large-scale food safety campaigns. In large & diverse countries like ours, an understanding of region/province-specific business cultures is required.

19. While the professional manner in which the constitution of various committees is spelled out is commendable, what is intriguing the Researcher is the lack of adequate peoples' representation in FSSAI. Only one member each from a consumer organization and agribusiness companies are in the Authority, which is a gross inadequate representation. Overstuffed by bureaucrats and also headed by a senior bureaucrat, there is no space in the authority for the informal food sector.

20. Some economists have apprehended that the objectives of the Act appears to be directed at eliminating the competition that informal food sector including the dhabas and tiffin carriers pose to the agribusiness companies. As long as food is being sold at such cheap prices by the dhabas and the tiffin carriers, the agribusiness companies will find it difficult to gain a strong foothold in the Indian market. The food offered by the dhabas and the hawkers has generally been found to be more hygienic and fresh than what is offered in the top posh hotels, where many of the dishes are usually cooked from frozen foods loaded with preservatives. These posh hotels offer cuisines of all vegetables and fruits at all times throughout the year irrespective of the prevailing season, which automatically implies that cuisines of unseasonal fruits and vegetables are made out of frozen fruits and vegetables. Dhabas and street food vendors always buy what is seasonally available in the market and make the cuisine out of it. They do not have and cannot afford facilities to store unseasonal fruits and vegetables and they cannot buy the frozen vegetables as they are very expensive which will add to the cost of the food. When one eats at these dhabas or street food hawkers, he is sure that he is consuming food prepared out of fresh fruits and vegetables and is not loading oneself with unnecessary preservatives and additives.

21. Unlike in the European countries, in India, hotels, restaurants and supermarkets are not the only sources of food but home cooked food is the major source of people's daily need and therefore regulations on industrialized and pre processed food alone won't help much on food safety front in India.

22. Another serious lapse is the labelling of food products specifying whether they are genetically modified not, though there is a Regulation to state whether a particular food item is genetically Modified or not, as of now not all FBO's are strictly following it.

This lenience is also in tune with the commercial interests of the multinational companies. They have been in the forefront of a global campaign that does not allow a choice between a normal product and a genetically modified food to the consumers - Which in reality establishes the stoic attitude of the government towards genetically modified crops being developed and consumed by the people. But it has been proved that these genetically modified crops can impact food safety and health of the consumers negatively in the long run. On 26/7/2018, a study by an advocacy group claimed that genetically modified processed foods, including infant foods were being illegally sold in the market across the country[30]. According to this report, the Centre for Science and Environment (CSE) tested 65 imported and domestically produced processed food samples, which included a mix of oils, packaged foods, infant foods and protein supplements, for presence of genetically modified foods. "Overall 32% of the food product samples tested were GM positive, while most of the genetically modified foods that were tested did not disclose GM on their labels, few also made false claims of being GM free, 46% of imported food products tested positive which were made of used soya, corn and rapeseed and were imported from Canada, the Netherlands, Thailand, the UAE and the US and also 17% of the samples manufactured in India also tested positive", CSE Deputy Director General Chandra Bhushan said. Despite the Act, the provisions regarding labelling is not strictly adhered to by the people and the same is not being enforced by the authorities. The provision regarding imported foods are also not enforced by the Act. We have a beautiful idealistic legislation but sans the committed authorities.

23. The time limit for prosecution also has been fixed as the trial to start within a year from the date of commission of an offence. As of now more than one lakh cases relating to food standard offences are pending in various courts across the country[31].

24. Deadlines have been extended many times for licensing and registration of food business operators, frustrating the very purpose of the Act. But, Section 63 of FSSA clearly envisages punishment of imprisonment for a term which may extend to 6 months and also fine which may go upto Rs 5 lakhs, in case any FBO carries on food business without obtaining license. Despite this provision not many EBOs have obtained license but carrying on their business in full glare.

---

[30] 'GM foods illegally sold across India, says CSE', Deccan Herald, Bangalore Edition Dated 27/7/2018

[31] Aarti Dhar, "Food Safety Act Takes Effect from Today", The Hindu, 5th August 2011, New Delhi edn availablehttps://www.thehindu.com/todays-paper/food-safety-act-takes-effect-from-today/article2323850.ece (accessed on 29/7/2018 at 7.31pm)

25. Section 66 of FFSA which deals with offences by companies, exonerates the person from liability if he proves that the offence was committed without his knowledge or that he exercised all due diligence to prevent the commission of the offence. This provision reflects the double standards adopted by the legislators. The Researcher is puzzled that, if they can make an individual responsible for any offence under the Act, why the same cannot be done with the company's representative. The Researcher is convinced that, this provision will allow the company to go scot free and the very idea of casting the Absolute liability on the offender is defeated.

26. All of us have been reading in the print media and listening to & watching on the electronic media about the artificial ripening of fruits by using the Carbide gas. Though there is a specific provision i.e. Regulation 2.3.5 prohibiting the use of Carbide gas in ripening of fruits in The Food Safety and Standards (Prohibition and Restrictions on Sales) Regulations, 2011, still the fruits are ripened using the same method. Why cannot the authorities take a suo moto action and save the public health. When the reports of rampant use of Carbide gas in ripening the fruits is aired on mass medias day in and day out, how can the authorities turn a blind eye and sit tight on this menace. Another Regulation 2.3.6 bans the sale of fruits and vegetables which are coated with waxes, mineral oil and colours. It is public knowledge that imported fruits especially the apples are coated with wax so that it has long shelf life and does not lose moisture and looks fresh. What are the authorities doing when a common man knows about the ill effects of consuming fruits and vegetables which are coated with wax.

27. The other challenge is the kind of materials used for storage of food. Often food articles are packed in plastic or polythene wrappers which may not be of food grade. The beverages and water packing bottles have no controls which can result in seepage of toxins into liquids and the food packed. Regulation 2.3.14(2), The Food Safety and Standards (Prohibition and Restrictions on Sales) Regulations, 2011, expressly prohibits the use of plastic articles in commercial establishments, for the purpose of sale or serving of food. Plastic articles of any form may be the plastic articles used in catering and cutlery, unless the plastic material conforms to the food grade. Even today, food establishments ranging from street hawkers to the posh hotels have the fancy of serving the food in plastic plates and packing the food parcels in plastic sheet or plastic containers. For instance, Indira Canteens in Karnataka, is a state owned venture, feeds the people with subsidised food. At these canteens, hot food is served in plastic plates though FSSAI's Regulations clearly demonstrate that plastic should not be used either in

packing or serving the food, as the hot food, when it comes in contact with plastic, plastic will react & release toxins and the food will turn carcinogenic and will not be fit for human consumption. The 'charity should begin at home first'. First, the state owned ventures/enterprises should give impetus to the guidelines stipulated by FSSAI.

## 7.3 Suggestions

In the light of the findings and hypotheses being partially proved, the Researcher strongly advocates that there is a scope and need for improvement of the working of the authorities under FSSA. The goal should be establish a forum for advocacy of right to safe food in the public context. The approach for this should be 'right-cum-duties'. The people need to be sensitised about the fact that right to safe food is their fundamental right and they ought to take complete advantage of the governmental support available to them. The Researcher thinks that as the 'access to adequate food is fundamental for the right to adequate food, it is equally important that accessed food must be adequate in terms of safety and quality'.

Agriculture is the most important economic activity. At the outset, agriculture should be given impetus and the government ought to fix scientific prices for the agricultural produce. This move will lure the farmers back to their original occupation which they have abandoned now, due to the apathy of the government policies. When even a match box comes with a price which is predetermined before it hits the market, how can we allow the fruits of the sweat and blood of our farmers to be lying in the market and for which the price will be fixed by the middle men. In case a particular agricultural crop is produced in large quantity and as a result the prices are slashed, the government instead of compelling the farmers to throw it onto the streets and suffer loss, should fix a reasonable support price, thus allowing the farmers to survive the losses. The government should concentrate more on adding the capital assets to the country's infrastructure in terms of roads, canals, railways, irrigation projects, dams and other facilities which can boost the agrarian sector, thus increasing the agricultural production. We should not turn a blind eye to the lawful demands of this sector, which ultimately ensures food security and food safety in turn can be achieved, which is basic need of the society.

Only in the healthy and conducive scenario of free flow of food, we can dream of food safety and not in scarcity. 'Kautilya was of the view that cultivable land is better than mines because mines fill only the treasury while agricultural production fills both treasury and store houses and states that the King should understand the intricacies of agriculture'[32]. In India, we have witnessed many times the vegetables and perishables being strewn onto the streets and roads by farmers because the offer price will be below the transportation cost. The Researcher wonders why collection and processing mechanism for vegetables and perishable goods cannot be setup and developed at rural levels. Only if there is enough supply of agricultural produce in the market, we can contemplate safe food and not when we have hungry bellies to feed as it is a common scenario that poor will consume anything. As an indirect approach to this, poverty alleviation programmes should be initiated profusely to improve the living conditions.

The following are the few Suggestions, which if adopted can add to the effectiveness and efficiency of the working of the FSSA. The basic requirements for the effective implementation of the Act is

- ❖ In limine, people should be made aware of the importance of the regular intake of micro nutrients, calories, vitamins and minerals and they should be encouraged to consume safe balanced diet. People should also be sensitised about the value of food and not to waste the same.
- ❖ Right to food must be made a fundamental right in the respective Constitutions of the states and the same may be included even in Indian Constitution. The Researcher believes that food is the primary need compared to education, hence Indian Constitution must be amended and the Right to food ought to be brought under Article 21-A and the Right to education should be rearranged under Article 21-B.
- ❖ Tolerance to violations of the right to food should be ended.
- ❖ State must create more employment opportunities so that everybody can have economic access to food.
- ❖ Infrastructure in the form of roads, railways, irrigation facilities/project, hydel power projects are the need of the hour. All these infrastructure ensures better agriculture production and the transportation of the same from the place of production to other places. Undisrupted power supply needs to be ensured and sanitation needs to be taken care of.
- ❖ Policies aimed at eradicating poverty and inequality and improving physical and economic access by all to sufficient, nutritionally adequate and safe food and its

[32] Summary on Kautilya's Arthashastra: Its Contemporary Relevance Published by Indian Merchants' Chamber (2004) https://www.esamskriti.com/essays/docfile/11_359.pdf (accessed on 30/1/2018 at 2.06pm)

effective utilisation must be undertaken by the state. Poverty alleviation programmes should be embarked upon on a war footing.

❖ Education plays great role in educating the people about the overall well being of the society, hence, states should make sure that everybody is educated. Awakened citizens are one of the national powers and they form national assets.

❖ Our country is lacking in terms of infrastructure to tackle the devastating effects of any natural calamities thus hampering the food security of the affected. We should strive to prevent and be prepared to face natural calamities, natural disasters and manmade emergencies and also to meet transitory and emergency food requirements in ways that encourage recovery and rehabilitation.

❖ An international supervisory mechanism should be set up in FAO or the United Nations Organisation, making it responsible for monitoring the implementation of the Conventions and Declarations. The states and the concerned authorities should be made to report to it, in case they encounter any difficulty in the implementation of the right to food.

❖ The major stumbling block is the changing scenario of food standards and their rigid requirements do not reach the rural India. Only few farmers are aware of the procedure. Though agricultural extension services are provided all across the rural India, neither they provide any information regarding the existing/prevailing national and international standards nor do they help the farmers in imparting techniques about the changing cultivation practices to meet the standards. We should strive to ensure the food and agricultural trade and overall trade policies are conducive to fostering food security for all through a fair and market oriented world trade system.

❖ At the outset, the Act should have integrated agriculture and food industries. Indiscreet, rampant and unscientific use of chemical fertilizers, insecticides & pesticides in agriculture and injecting the dairy animals with hormones to increase their yield would naturally find their way into the end product which is meant for human consumption. The Act excludes plants prior to harvesting, standing crops and animal feed from its purview. Any harmful input[33] that could affect the safety standards of end food products is not effectively covered by the Act. An effort to achieve food safety must be a comprehensive venture. Food safety can be achieved only by taking measures in reforming agricultural practice & rearing of live stock for human consumption.

[33] Pesticides in vegetables or antibiotics in animal feed

Integrated farm to table concept should be adhered to. There is a need for greater involvement of rural sector in food safety issues as most of the food articles for food processing industry are sourced from rural areas.

❖ The human right to food cannot be guaranteed without a significant degree of socialization of food production and distribution both at the global level and the national level.

❖ The state should provide the people with local seeds and land for landless families who are willing to cultivate in rural areas, as the local breed will be immune to vagaries that they may encounter in that local area.

❖ It is difficult for food processing industries to take the onus for ensuring that such standards are within the acceptable levels in processed food when raw material itself fails to meet these standards. The Act should have also banned the use of some of the insecticides and pesticides so that the same does not find its way into the raw material for food processing industries. Recently, the government banned the use of Endo sulfan, a pesticide[34], after lot of agitation by the public. The constant exposure of people to Endosulfan has resulted in people getting crippled and facing health hazards like severe neurological and congenital deformities.

❖ It is also important that the FSSA takes into consideration not only food safety but also sets up criteria for monitoring the nutritive and human health aspects of the food products being sold at the market. Food safety has often been generally misconstrued as nutritionally fit. If the junk food sold through the fast food joints is good for human consumption i.e., even if it conforms to safety standards as per law, there is no reason why obesity has emerged as the biggest killer worldwide. In the United States, where junk food is an obsession, resulting obesity has now emerged as the biggest threat to nation's health.

❖ At all stages of the food production and distribution chain, preventive measures should be introduced in-limine rather than inspection and rejection at the final stage, which results in wastage of food and resources, makes better economic sense. Food hazards[35] and quality loss may occur at variety of points in the food chain. A well structured, preventive approach which monitors production and process ought to be the

---

[34] Endosulfan is a harmful insecticide and can cause several health hazards in human beings. http://www.indiaenvironmentportal.org.in/files/Effect_of_endosulfan.pdf

[35] FSSA Section 3(u) 'Food hazard' means a biological, chemical or physical agent in or condition of food with the potential to cause an adverse health effect

preferred method for improving food safety and quality. Potential food hazards could be curtailed along the food chain through the application of good practices, like,

- Using clean and hygiene water in the preparation of food
- First, the authorities must supply clean water
- Discourage the manual handling of preparation of food and automating the same
- Raw material which goes into food preparation must be clean and safe
- The use of anything made of plastic must be discouraged in any stage of food production and distribution
- The place where the food is cooked should be kept clean and open for public viewing
- Facilities for washing utensils used for cooking/processing must be mechanised
- Persons who are engaged in food preparation must not be suffering from any contagious diseases
- Wearing of safety gear and clean apron must be made compulsory for those involved in preparation of food
- The premises where the food is cooked/manufactured/processed, at the entrance itself, hot air blower and air sucker of moderate intensity should be fixed & the same should be linked to an automated sensor, so that when any person enters the premises, he would not be taking the dust and foreign particles inside the premises, which may affect the quality of food/articles of food
- Where ever fresh vegetables and fruits are used in the preparation, at the outset, adequate & proper facilities for storing them and facilities for thorough washing should be installed
- Use of frozen food articles in the preparation should be discouraged, as, as it is, it is loaded with preservatives and the food business operator in order to give long shelf life adds some more additives which ultimately results in the presence of both food additives and preservatives. Though they may be within the permissible limit, yet it is natural that it may be injurious to human life in the long run.
- The state should provide training to interested FBOs, especially those who are in the business of producing the food in large mass about the nuances of food handling, producing, marketing and storing to prevent food borne diseases
- Effective mechanism for waste disposal is the grave need of the hour

- Better sanitary facilities should be provided
- Only non toxic grease and lubricants should be used in the machineries
- The authorities must carry out the periodic check of the machineries and give a certificate to that effect. Old and worn out machineries should not be allowed to be used as traces of metals, which they are made up of, can find entry into the end product.
- More than all this, food business operators should be sensitised about their role in the building of society and about their responsibility towards the society which plays a pivotal role in making the food safe. If this is done, then there is no scope for any other measure, the rest will follow suit.

❖ To set up fully equipped laboratories and providing trained manpower to operate them. All testing has to be conducted only in accredited laboratories as at present not all the public sector food laboratories are accredited. FSSAI commissioned a gap analysis study for up gradation of 50 food laboratories under the Central and State government. The study indicated that there is an urgent need to upgrade the infrastructure, strengthen staffing and training inputs and put in place more reliable laboratory management and operational procedure. The sub-group addressed that a network of efficient laboratories is the backbone of any credible food safety initiative. We need many more state of the art testing laboratories mainly to analyse/examine the imported food articles. Failing which we will not be able to use the SPS[36] and TBT[37] clauses to protect ourselves against the harmful effects of contamination/adulteration in imported articles of food.

❖ An effective strategy for laying emphasis on introduction of food safety preventive measures at different levels of food sector must be devised. The task of reaching the masses throughout the country could be achieved by identifying the centre from where such message could be spread. The ideal place for such activity could be the market place, where producers and buyers interact through sellers of

[36] The Agreement on the Application of Sanitary and Phyto sanitary Measures came into force on 1st January 1995with the establishment of the World Trade Organization. An agreement on how governments can apply food safety and animal & plant health measures (sanitary and phytosanitary or SPS measures) sets out the basic rules in the WTO. They aim to ensure that a country's consumers are being supplied with food that is safe to eat — by acceptable standards — while also ensuring that strict health and safety regulations are not being used as an excuse to shield domestic producers from competition. Available at https://www.wto.org/english/tratop_e/sps_e/sps_issues_e.htm (accessed on 7/8/2018 at 2.50pm)

[37] The WTO Agreement on Technical Barriers to Trade (the "TBT Agreement") came into force on 1st January 1995with the establishment of the World Trade Organization. It aims to ensure that regulations, standards, testing and certification procedures do not create unnecessary obstacles to trade.

food commodities. Stakeholders should be involved for shifting emphasis from end product quality control to preventive measures throughout the food chain and concepts like traceability need to be introduced. Key personnel at the grass root level for implementing the food safety preventive measures should be identified.

❖ Food safety should find a place in the curriculum at the school level. It is always easier to teach young minds and make them inculcate it in their life style than making an effort to teach the adults whose 'habits die hard'. Whatever is taught to a child will become the habits of his life for all time to come. All children should be apprised of the necessity of food safety and the need to make the food safe for consumption as these children only become entrepreneurs, food business operators, importers, processors, food handlers, if not anything consumers, when they grow big.

❖ Food safety should be popularised through mass media. Any hotel or restaurant that we visit, we can witness that the hot food is being served in a plate made of melamine which is a hard and thermosetting plastic material. The crockery made of melamine is extremely durable and unbreakable which is a reason for the commercial establishments to use them widely. But, like any other plastic, dishes made of melamine can potentially pose health risks by leaching chemicals into the food. Though the FSSAI in its Regulations has prohibited the use of plastic at any stage of food production or distribution. Mass awareness programme should be held to disseminate the danger of eating hot food from a plate made up of melamine. Everyone should be made aware of the necessity of food safety and the necessity to ensure that the available food is safe for consumption and right to safe food is an extension of Right to Life as contemplated in Article 21 of our Constitution.

❖ It is heartening to note that food testing science and technology is continuously evolving each day. The advanced and hitech instrumentation and techniques to detect the minute levels of adulterants and undesirable substances in food articles demands sophistication right from the beginning like say, sampling and handling of sampling. With the advancement of technology, globally the legislations dealing with food have become more stringent than ever before and demanding the food industry to cope up with the thus set high standards as far as the quality of food is concerned. But lack of skilled human resource and adequate infrastructure circumvent this. The authorities need to look into this and initiate suitable action in this direction.

❖ Food poisoning is absolutely preventable if only we have to follow *the five key principles of food hygiene as stipulated by world Health Organisation*[38] and they are,

- Prevent contaminating food with pathogens spreading from people, pets, and pests.
- Separate raw and cooked foods to prevent contaminating the cooked foods.
- Cook foods for the appropriate length of time and at the appropriate temperature to kill pathogens
- Store food at the proper temperature.
- Use safe water and safe raw materials.

❖ The FSSAI and its allied committees namely, scientific committees, scientific panels, the central advisory committee require very efficient and highly skilled staff which are hard to come by.

❖ Absolute liability should be attached for the offences which are committed by the companies under the Act. The representative of the company should not be allowed to go free just because he proves that the offence has taken place without his knowledge or he has taken all precautions to prevent the offence.

❖ FSSAI should take measures on war footing to initiate the FBOs to register themselves or to obtain licence as the case may be. FSSAI must also require the FBOs to display their Registration certificate or the License in a prominent place of business, so that a common man will also know whether the FBO has registered himself or has obtained the licence or not. FSSAI should also have a dedicated Helpline only for addressing the grievances of the consumers of food which they can call in the event of suffering after consuming food at a particular place, if the FBO is flouting the Rules and Regulations of FSSA or anything which the consumer wishes to bring it to the knowledge of the authorities. This will help in speedy action than allowing the law to take its own recourse. As on today a single helpline bearing No 1800112100 is available for consumers grievances, food import clearance system issues, food registration and licensing system issues, queries on website and FSSAI initiative' sites. This helpline contact number should be mandated to be displayed in a prominent manner in all premises where food is the concern of the business.

❖ Adding fuel to fire is the changing scenario of food standards and their typical rigid requirements do not reach rural India where major portion of Indian population

[38] Available at http://www.who.int/foodsafety/publications/consumer/manual_keys.pdf (accessed on 11/7/2018 at 5.11pm)

resides. Farmers are unaware of the unhealthy inputs that should be avoided in the farming practice in the form of insecticide, pesticides and fertilisers, as it is common knowledge that whatever goes in has to manifest in the final product, if the raw material for the food processing industries are adulterated, it is imminent danger that the end product will also be laced with adulterants. "Adulteration and chronic effects of low level exposure to pesticide residues and toxins can be dangerous to human health in the long run. Carcinigenic effects of DDT, lead and aflatoxins may become apparent only in the long run"[39].Organic farming should be popularised and the government should provide the required economic assistance for the same.

❖ Food safety records are invariably poor when consumers take back seat & are ill informed, unorganized and not vocal. Mass food safety movements are, to paraphrase the definition of democracy, for consumers, by consumers, and of consumers, must be undertaken. Their success can only be ensured with active consumer involvement. Most food campaigns fail because they are excessively controlled by government functionaries with little or no involvement of consumers or consumers' organizations. The organization and empowerment of consumers coupled with timely redress of grievances form an integral part of any meaningful food safety move. While consumers have every right to expect uncompromising food safety standards at competitive prices, they must also be willing to pay extra money. The most influential and widely quoted statement on consumer rights was from US President John F. Kennedy in 1968, who highlighted consumers' "right to safety, right to information, right to choose, and right to be heard." Consumer International defined eight basic consumer rights: satisfaction of basic needs; information; choose; safety; representation; redress; consumer education; and a healthy environment. Thus consumers and consumers' rights organizations need to be in the forefront in exercising their right to food safety and involved in policy formulation. This is possible only when they are alert, well organized, and present everywhere instead of only in large cities.

❖ The government must allow representation of consumer organizations in regulatory bodies and on consultative committees so that their views are heard and reflected in policies. The real challenge lies in accepting and honouring the rights of

[39] Sathish Y Deodhar, "WTO pacts and Food Quality Issues", Economic and Political Weekly,[July 28-August3, 2001], Vol XXXVI No 30, p 2813-2816

consumers and educating them on those rights. Alert, organized consumers are essential for creating a food safety chain reaction and turning it into a mass movement. Isolated legislative efforts and export centric initiatives may succeed at best only in creating some islands of excellence, leaving the majority of the population untouched.

❖ Though agricultural extension services are established across the country, neither they provide information about the prevailing national and international standards nor do they assist the farmers by imparting technical knowledge about changing the cultivation practices and patterns to meet these standards[40]. In order to set the standards for food, the availability of true and updated data not only on consumer related indicator but also on ingredient related indicators is crucial. Farmers should be encouraged to desist from the rampant use of chemical fertilizers and the government should incentivise the farmers to employ organic farming by providing logistic support.

❖ Causes for hunger are lack of access to adequate food which varies from region to region, due to which different solutions will be needed in each case. In one of the studies conducted by the Federation of Indian Chambers of Commerce and Industry, the interviewed respondents strongly opined that since reliable data for the consumption of food is not available, the FSSAI should be mandated to embark on a comprehensive monitoring of data relating to the levels of food additives, contaminant levels and health survey and information &data regarding the hazards in the food industry and their source must also be collected and acted upon at once[41].

❖ Campaigns should be conducted by both the state and central government creating mass awareness among the people about the need for food safety which ultimately results in healthy citizenry.Good interaction with the food industry and with stake holders while ensuring compliance as well as understanding the limitations that may exist.

❖ In Europe, US and other developed countries, only food grade lubricants and greases are mandatory to be used in plant and machinery which manufactures, processes and packs food, drinks, water and dairy products. Industrial lubricants are manufactured using petroleum base oils and additives which generally have toxic

---

[40] Sheeba Pillai, '*Right to Safe Food: Laws And Remedies*' Vol. 41, No. 2, Ban.L.J. (2012) 119-135

[41] FICCI, '*Study on Implementation of Food Safety Standards Act- An Industry Perspective*', May 2007 available at foodsafetynews.filters.wordpress.com (accessed on 9/2/2017 at 8.00pm)

substance which might dangerously contaminate food items causing severe health hazards. In US, to avoid this toxic contamination only food grade lubricant like NSF42 approved in US, is mandatory by Government statuary bodies. In India no such regulations are mandated by Food Ministry and FSSAI, hence most plants of food process industry are using toxic industrial grade lubricants. Many smart food processing plants procure small pack of food grade lubricant to show during inspection and audits but 99% plants are using non food grade lubricants which are very toxic. The industrial lubricants are not safe to use in food processing plants because they are formulated using many toxic material having toxic ingredient while certified food grade lubricants are formulated using non toxic materials and also approved by third party like NSF stating that these products are safe for food process industry.

❖ Indian food industry ought to adopt hazarad analysis critical control point as a strategic food quality management system which should emphasise hygiene and prevention of contamination in the production process. The government should provide subsidies for the initial fixed costs incurred for the implementation of HACCP and the food business operators can bear the recurring costs. HACCP measures which are appropriate to each sector need to be recognized after a vigilant study and must be implemented.

❖ The adoption of preventive methods at every stage of food production and food distribution chain rather than inspection and dismissal/rejection at the final-stage makes-better –economic- sense. It is common knowledge that food hazards and loss in the quality occurs at various stages in food chain. A-well=structured-preventive-approach that could control the process and the production ought to be the preferred mode to bring positive changes in an effort to make the food safe.

❖ It is a misconception that urbanites can contribute a great deal for food safety but, in reality, there is a need for greater involvement of rural sector in food safety issues. If the farmers are sensitised about the need to make the food safe from the production stage, half of our battle against unsafe food is won. More food related jobs need to be created in rural areas.

---

[42] National Sanitation Foundation is an independent, non-profit organization that certifies food service equipment and ensures it is designed and constructed in a way that promotes food safety. NSF is internationally recognized.

❖ The Indian food industry suffers from lack of trained manpower to handle the post harvest quality management practices and food processing activities. The Researcher submits that there is an urgent need for trained manpower to cater to these needs. Setting up of farm schools on the lines of existing industrial technical institutes should be prioritised so that essentials of hygiene, food handling practices and processing can be taught as syllabi in the certificate courses. Such training must be made mandatory for those who are working in food processing sector.

❖ Many of the imported food products contain instructions on the label in their language and not in English which causes inconvenience to us to know the ingredients and if they contain any ingredient which is prohibited by FSSA, or if the additives are in order with the standard prescribed by FSSA. Hence the FSSAI should mandate that all imported food products to contain the information about the food article in English. Stricter vigilance of imported foods must be emphasised by al the enforcement agencies. Rules regarding effective labelling must be implemented.

❖ "More often than not, we have no representation in the CAC meeting when the standards for various food products are fixed. Owing to lack of participation, standards which may be unfavourable for the developing countries get fixed"43. For effective participation in the CAC meeting, in addition to the civil servant, India should be represented by an eminent team of food scientists, legal experts and economists. The Ministry of Commerce and Ministry of agriculture should also be involved in CAC matters.

❖ Food safety programmes must be undertaken at all stages of food production, food processing and distribution chain so that food from farm to finally landing on the table is made safe i.e. from farm to fork.

❖ Immediate task before the Government of India is to organise and conduct workshops and symposiums which can help in evolving a better food policy and to create awareness among the people to join hands in this noble venture. All the concerned authorities like officers from the concerned ministries and quasi-government bodies dealing with different aspects of food, processors, food producers, trade and industry representatives and consumer groups academicians and researchers need to be implicated. An exuberant task force required to be constituted including

[43] Sathish Y Deodhar, "WTO pacts and Food Quality Issues", Economic and Political Weekly,[July 28-August 3, 2001], Vol XXXVI No 30, p2813-2816

interministerial representatives to draw up a food safety policy document. That document need to be placed before-the public for comments and only then the policy can be finalised. In continuation, appropriate programmes, policies need to be developed and implemented reinforcing the existing infrastructure at the centre as well as state levels. Uniform implementation of these measures throughout the country may ensure quality, safe and hygienic food for domestic consumption as well as for export.

Printed in the USA
CPSIA information can be obtained
at www.ICGtesting.com
LVHW050443020324
773349LV00011B/198